Soil Science Working for a Living

David Dent · Yuriy Dmytruk
Editors

Soil Science Working for a Living

Applications of Soil Science to Present-Day Problems

Springer

Editors
David Dent
Chestnut Tree Farm, Forncett End
Norfolk
UK

Yuriy Dmytruk
Soil Science Department
Chernivtsi National University
Chernivtsi
Ukraine

ISBN 978-3-319-45416-0 ISBN 978-3-319-45417-7 (eBook)
DOI 10.1007/978-3-319-45417-7

Library of Congress Control Number: 2016949115

Printed on acid-free paper

This Springer imprint is published by Springer Nature
The registered company is Springer International Publishing AG
The registered company address is: Gewerbestrasse 11, 6330 Cham, Switzerland

Editorial Introduction

This selection of papers from a symposium at Chernivtsi, in Ukraine, deals with gritty issues that society faces every day: food and water security; environmental services provided, almost accidentally, by farmers—and taken for granted by urban dwellers; the capability of the land to provide our needs today and for the foreseeable future; and pollution of soil, air, and water. The contributions are arranged in three broad communities of practice in which soil scientists work to solve these problems.

It is not all the same out there! Assessment of land capability spells out this *cri de coer*; what is more, soil survey and land evaluation show how every patch of land is different and how it will respond to management. But soil survey and land evaluation depend on knowledge of soil processes and the relationships of soils with the wider landscape—so we deal with these issues in Part I: *Soil Development: Properties and Qualities.*

Dokuchaev's insight on the relationships between soils and their landscape, nowadays expressed as the Factors of Soil Formation, is the very foundation of soil survey so we begin with a timely reassessment of Dokuchaev's concept by Sergiy Kanivets. Reading the landscape, or a soil profile, depends on picking out clues from a plethora of detail; understanding the processes at work in the past as well as the present; and deducing their continuing effects on the performance of the soil under our management. This understanding comes from an accumulation of perceptive research using many and various methods and techniques. Volodymyr Nikorych and his Polish colleagues highlight pedological features that, in a sense, are the memory of the soil. In *Redoximorphic Features in Albeluvisols from Southwestern Ukraine*, they draw upon multiple scales of observation from the field, to the microscope, to the electron microscope, seeking to understand the process of development of the iron–manganese mottles common in alternately wet and dry soils—and what they can tell us about the soil water regime. In another micromorphological examination, *Fractal Properties of Coarse/Fine-Related Distribution in Forest Soils on Colluvium*, Volodymyr Yakovenko applies the mathematics of fractals to establish lithological homogeneity, or lithological breaks, in soil profiles—a crucial step in interpreting how they have developed.

Part II: *Assessment of Resources and Risks* includes applications of the soil scientist's toolkit from the global to the most detailed scale and from the most recent to the oldest identified features related to land use. *The Last Steppes: New Perspectives on an Old Challenge*, by David Dent and Zhanguo Bai from ISRIC—World Soil Information, interprets global and regional land degradation using a long time series of coarse-scale satellite imagery. They argue that, within a generation, without a radical change of policy and management, the Chernozem—the best arable soil in the world—will be no more. Seeking to provide an operational system for diagnosis and monitoring of eroding Chernozem, Tatiana Byndych, from the Sokolovskyi Institute in Kharkiv, explores the potential use of detailed multispectral data from the Ukrainian *Sich-2* satellite. And seeking to identify the first steps in man's conquest, Yuri Dmytruk teams up with Vadim Stepanchuk of the Institute of Archaeology, applying microelement analysis to confirm the identification of hearths in the archaeological sequence at Medzhybozh, making this the oldest proven use of fire in Ukraine—400,000 years ago.

Back to the present, we find that our fundamental information on the soil pattern is broadscale, dated, and incomplete. There seems no immediate likelihood of a nationwide resurvey, but two contributions from Yuriy Fedkovych Chernivtsi National University demonstrate how we can update and improve heritage soil maps using modern techniques and technology. Vasyl Cherlinka *Using Geostatistics, DEM, and Remote Sensing to Clarify Soil Cover Maps of Ukraine* and Olga Stouzhouk and Yuriy Dmytruk *Making Better Soil Maps Using Models of Tangential Curvature* make use of large-scale digital terrain models and apply landform analysis of varying degrees of sophistication, according to the job in hand. None of these developments does away for the need for the reality check of fieldwork, but they certainly enable us to make efficient use of our time in the field.

For long-term management of soil resources and the design of erosion control measures, it helps to compare actual rates of soil erosion with a defined, tolerable value. In *Determination of Soil Loss Tolerance for Chernozem of Right-Bank Ukraine*, Sergiy Chornyy and Nataliya Poliashenko from Mykolayiv National Agrarian University draw upon a wealth of regional data on the relations between individual soil qualities and soil productivity to develop a modified productivity index (MPI). According to the change in MPI and its rate of decline as a result of soil erosion, calculations for *Ordinary* and *Southern chernozem* indicate a soil loss tolerance of 5–7 t/ha per year but, also, significant differences in outcomes for already eroded and not-eroded soils.

Finally, in this section, we note the dearth of systematic, publicly available information on the environmental impact of the agro-industrial complex that dominates the country's economy. In a short article *Assessment of Problems of Soil Contamination using Environmental Indicators*, Yevhen Varlamov and Oksana Palaguta from the Ukrainian Scientific and Research Institute on Ecological Problems, in Kharkiv, propose a system for assessing the condition of the natural environment. This is based on ecological indicators that are already in use in particular regions of the country, within the framework of environmental performance indicators (EPIs) used to assess the state of the natural environment in

Eastern Europe, Caucasus, and Central Asia. The actual measurement of such EPIs is by no means straightforward, but in *Mathematical Tools to Assess Soil Contamination by Deposition of Technogenic Emissions*, Olexandr Popov and Andrij Yatsyshyn from the Institute of Environmental Geochemistry provide an elegant example of the application of mathematics to arrive at a realistic environmental impact assessment.

Soil fertility, Degradation and Improvement focusses on the constraints on agriculture across Eastern Europe and ways to mitigate them in sustainable farming systems. But, first, Anatoly Khristenko points out that there is no unambiguous interpretation of any of the terms commonly used to characterize the provision of nutrients by the soil. This is not pedantry—Confucius made the point in plain words:

> It is important to use language correctly. If language is used incorrectly then what is said is not what is meant. If what is said is not what is meant then what should be done remains undone.

If we are not clear about such a fundamental issue, how can we expect decision makers to know what we mean and take the appropriate action?

Drought always stalks the steppe. Irrigation counters the uncertainty of water supply, but it needs skill to avoid rising groundwater, slaking, and salinity, and maintain well-functioning biodiversity. In *Criteria and Parameters for Estimating Directions of Irrigated Soil Evolution*, Sviatoslav Baliuk and colleagues from the Sokolovskyi Institute distinguish three main pathways of soil evolution under irrigation, depending on initial soil conditions, the quality of irrigation water, and the farming system. Degradation processes that can occur under adverse irrigation conditions are characterized, and a system of criteria and parameters is proposed for assessing the situation, the extent, and nature of degradation of irrigated land, so as to avoid pitfalls.

It is increasingly obvious that farming in Ukraine, and not just in Ukraine, is mining soil organic matter and nutrients. Sustainability requires compliance with the fundamental laws of agriculture—in particular, sound crop rotation and return of nutrients to balance their removal by the crops. In *Sustainability of soil fertility in the southern steppe of Ukraine Depending on Fertilizers and Irrigation*, Valentyna Gamajunova reports long-term field experiments on *Kastnozem* and *Chernozem* maintained by the Mykolayiv National Agrarian University. They demonstrate that combined use of adequate organic and mineral fertilizers is the most effective way to stabilize crop yields and soil fertility and, also, soil structure—which enhances infiltration of rainfall so that rain use efficiency is increased by 20–30 %, in very dry years by 30–40 %. Seeking to understand how fertilizers maintain nutrient status in the long term, Yevheniia Hladkikh investigates the after-effect (over 25 years) of potassium fertilizers on *Typical chernozem*. The modified soil is characterized by elevated contents of mobile potassium and transformation of exchangeable and non-exchangeable forms. Systematic monitoring of available potassium allows accurate recommendations on fertilizer application, especially for demanding crops that respond to potassium fertilizer.

Liming, too, has been sadly neglected but Yuriy Tsapko and his colleagues report an innovative and cost-effective treatment devised at the Sokolovskyi Institute. Compared with conventional liming, applying alternative sources of lime in localized bands along with organic manure reduces leaching from coarse-textured *Sod podzolic* soil: lime by almost six times and soluble organics and nitrate by 1.8 and 2.9 times, respectively. In *Podzolized chernozem*, the same technology increases the population of earthworms and microorganisms, thereby activating self-renewing and regulating processes.

The quite different soils in the Carpathian region present different problems and opportunities that get a lot of attention from the Institute of Biology, Chemistry, and Bioresources at Yuriy Fedkovych Chernivtsi National University. Specific factors of soil formation in the region drive a prevalence of iron and aluminum phosphates and a low incidence of calcium phosphate. Given that the present phosphate status is unsatisfactory for farming, Tatiana Tsvyk argues that optimization can be achieved only through fundamental change of regime. Changes in the space-profile of phosphate fractions in *Albeluvisols* are examined under different drainage conditions and after application of lime and phosphate. It is confirmed that tile drainage not only improves soil aeration but, also, significantly influences the fractional composition of phosphates.

Thirty years ago, the explosion of reactor no. 4 at Chernobyl Nuclear Power Plant blew away the credibility of a system of authority whose claims included safe mastery of technology. We are still dealing with its legacy. Our final part on *Soil Contamination, Monitoring and Remediation* deals with a range of pressing issues from criminally negligent storage of persistent organic pollutants and out-of-date agrochemicals to long-term soil contamination by heavy industry. The litany of contaminated sites documented by Valerii Kovach and Georgii Lysychenko in *Toxic Soil Contamination and Its Mitigation in Ukraine*; Lysychenko, again, with Iryna Kuraieva and her colleagues in *Heavy Metals in Soils Under the Heel of Heavy Industry*; and Lyubov Maslovska and her colleagues on *Dangerous Mercury Contamination Around the Former Radikal Chemicals Factory in Kyiv* makes grim reading.

If we are to avoid such pitfalls in future, we need forethought and a good understanding of the risks associated with industrial developments. Forethought includes establishing the situation in advance of industrial developments. In this vein, we include three studies from the Institute of Environmental Geochemistry, in Kyiv. Two investigate the projected development of the Bilanovo iron ore deposit in Poltava Region which will require a quarry 600–700 m deep that will destroy natural habitat and intercept the water table—so we should assess its likely environmental impact. In *Comparison of Groundwater and Surface Water Quality in the Area of the Bilanovo Iron Deposit*, Oleksa Tyshchenko measures the chemical and bacteriological characteristics of groundwaters and surface waters and identifies various indicators that exceeded the maximum allowable concentration. Yevhen Krasnov and colleagues measure gamma radiation and radon flux density in soils overlying the ore deposit; their maps identify areas of increased radiological risk which, though not critical for the most part, need to be considered and monitored.

And in *Estimation of Soil Radiation in the Country Around the Dibrova Uranium–Thorium–Rare Earth Deposit*, Yuliia Yuskiv and her team present a sequence of radiological measurements and maps of the distribution of the main nuclides, identifying areas of radiation risk. Apart from locally high radon concentration, most parameters are not critical, but the measured levels should be considered as a baseline for regular radiological monitoring.

To put things right, we need better understanding of the mechanisms of contamination and natural remediation. Oksana Vysotenko and her colleagues from the Institute of Environmental Geochemistry report on *Lead and Zinc Speciation in Soils and Their Transfer in Vegetation.* Although these metals behave quite differently, the degree of pollution significantly affects the rate of immobilization of heavy metals and their vertical migration. Tatyana Bastrygina and her colleagues examine the behaviour of heavy metals in soil around the copper and bronze artifacts at archaeological sites that provide analogues of the modern contamination of the environment. And in an elegant pot experiment backed up by field trials: *Environmental Assessment of Soil Based on Fractional-Group Composition of Heavy Metals*, Salgara Mandzhieva's team from universities in Rostov and Moscow Lomonosov and the Sokolovskyi Institute demonstrate the increase in the environmental hazard when soils are contaminated with heavy metals, a decrease when ameliorants are applied, and the roles of both strongly and loosely bound metal fractions in the mobility of heavy metals in soils.

Finally, new developments in the remediation of oil-contaminated soils are highlighted by Elena Maklyuk and colleagues from the Sokolovskyi Institute and Man Oil Company of Switzerland, using chemical oxidation to create optimum conditions for subsequent bioremediation, and by Viktoriia Shkapenko and her colleagues from the Institute of Environmental Geochemistry, exploring the natural humification of oil hydrocarbons, and aiding and abetting the process by specially formulated clay–biodecomposer inoculum.

Contents

Part I
Soil Development: Properties and Qualities

Chapter 1
The Factors and Conditions of Soil Formation: A Critical Analysis of Equivalence

Sergiy Kanivets

Abstract New knowledge of soil patterns depending on geography, ecology and agricultural activity has renewed debate about Dokuchaev's factors of soil formation: parent material, vegetation, climate, position in the landscape, age of the soil, subsoil water and agricultural activities. Considering the southern flank of left-bank Ukrainian Polissya, the dominant factor is the parent material: *Sod-podzolic* soils occur on sandur, *Grey forest soils* and *Chernozem* on loess. In the Forest-Steppe, the dominant factor is vegetation: *Chernozem* under grassland, *Grey forest soils* under oak forests. In the brown forest soil zone and high-mountain meadows of the Carpathians, the climate determines *Burozems*. In the rolling lands of the transition from forest-steppe to steppe, the determining factor is topographic position: *Typical chernozem* on the northern slopes, *Ordinary chernozem* on the southern slopes. In the transition from central to southern dry steppe in Prichernomorya, the dominant factor is redistribution of surface runoff: *Chernozem* in bottomlands adjacent to *Dark chestnut soils* on the heights. And in areas of naturally acid soils, the determining factor can be agriculture. Thus, soil formation is influenced by all the natural factors and conditions but, in any particular landscape, only one or two factors amongst them are determinant and dominant.

Keywords Factors of soil formation · Equivalence · Soil pattern · Soil landscape · Dokuchaev

Introduction

In our educational literature, reference is often made to Dokuchaev's thinking about the equal influence of the various factors of soil formation in determining kinds of soil and their classification—the idea is presented to students, and some editors of

---

S. Kanivets (✉)
Sokolovskyi Institute of Soil Science and Agrochemistry Research,
4 Tchaikovsky St, Kharkiv, Ukraine
e-mail: serg.kanivets@gmail.com

D. Dent and Y. Dmytruk (eds.), *Soil Science Working for a Living*,
DOI 10.1007/978-3-319-45417-7_1

scientific publications abide by it. Geographical research on soils of transitional zones leads me to a different conclusion but, first, we should read Dokuchaev's own words. In *Contribution to the theory of natural zones* (Dokuchaev 1899, *Collected works vol. 6*, p 406), he states "… soil formation agents are in substance equivalent values and take equal part in normal soil formation." In this formulation, *equal* does not mean the same as equivalent, and *equivalent* is hedged with the caveat "in substance".

Dokuchaev's concept of equivalence follows from his perception of the role of the factors and conditions of soil formation in the phenomenon of zonality; and *Contribution to the theory of natural zones* is a collection of examples of the dominant agent in the formation of a particular type of soil, or of deviations from normal latitudinal zone boundaries in specific soil landscapes. For instance, in respect of vertical zonality, altitude is considered to determine the humus content in *Chernozem,* and we read in the same work that, in high mountains, aspect determines xeromorphic conditions and, even, the formation of a different type of soil.

However, in his second *Lecture on soil science* Dokuchaev (1949) writes: "… soil is the function of parent material, climate and organisms—and they are the main soil-formers" and attempts to single out the dominant factor of soil formation are "idle … guesses leading nowhere" (*Collected works vol. 3*, p 347). This contradicts the argument presented in *Contribution to the theory of natural zones* and, yet, in the formation of zonal subtypes of chernozem (Typical, Ordinary and Southern) climate is, surely, acting as the determining agent. So we have to interpret *equivalence* of the factors of soil formation in a particular area as equal participation, not equal power of influence; the determinant factor is climate in one zone, vegetation in another, and so on.

Ever since Dokuchaev, the idea of equivalence of factors and conditions of soil formation has provoked brisk discussion. Glinka (1923) paid special attention to climate and vegetation and, in separate cases, to soil parent material; Zakharov (1927) and Rode (1947) emphasized the different role of particular factors and conditions; an overview of the ideas of outstanding pedologists appears in the well-known textbook edited by Kovda and Rozanov (1988), and, beyond the Russian school, the Swiss-American Hans Jenny (1941, 1980) examined the issue in depth and concluded that, under specific conditions, any factor may play the determinant role.

A Personal Viewpoint

I acknowledge that all known factors and conditions of soil formation are essential, exist and interact. However, analysis of the landscape of several areas (landscape components involve all the factors of soil formation) and observation of their soil patterns lead to the conclusion that, in the overwhelming majority of cases, there are only one or two determining factors. With this weight of evidence, we can apply the fundamental concept to a range of applied tasks and open the prospect of further productive research.

We may rely on physiographic zonality (indeed, Dokuchaev considered zonality simultaneously with the role and importance of "soil formation agents"). Thus, the highest rank—geographic zones—are generally distinguished by latitude. The key factor is the amount of solar radiation received but the global belts may be modified by the local influence of mountain ranges that deflect the movement of air masses and ocean currents, and by stable anticyclones and depression tracks, etc. The trajectories of air masses of different moisture and temperature bring their influence to bear in certain regions: for instance there are dry subtropics with brown steppe soils and moist subtropics with subtropical brown forest soils; in the temperate zone in Ukraine, there is forest-steppe and, at the same latitude, a brown forest soil region —each with their respective soils, and in the tropics there are hot deserts and biologically highly productive wet tropics.

Zonality is commonly recognised in the formation of soil types and subtypes. However, in determining the zonal soil type, we should not make the mistake of drawing zonal and regional boundaries to avoid distortions of the soil cover in limited areas. Aberrant soils, from the zonal point of view, may be formed under the influence of external (azonal) agents—so that soils representative of adjacent zones may be found within any normal zone. In such cases, the determinant factors may be the soil parent material, local climate, slope and aspect. The soils of the East European Plain exhibit a distinct zonal sequence as well as geo-botanical zonality but in Western Europe, including the Ukrainian Carpathians, zonality is blurred by the influence of warm moist air from the Atlantic. So the territory of Ukraine encompasses the province of mixed forests (Polissya), forest-steppe and steppe zones. We also observe vertical zonality in the Carpathians and Crimean mountains.

Polissya and the Forest-Steppe

Examples may be drawn from the diverse transitional zone stretching from Polissya to the forest-steppe zone of left-bank Ukraine. Anomalies are widely encountered on the right bank of the Desna river-valley: *Typical sod-podzolic*[1] (thoroughly characterized by Kanivets 2010) and *cryptopodzolic*[2] soils were formed here under pine-forest on coarse-textured glacial outwash and podzolisation penetrates deep into the forest-steppe along pine-forest terraces on coarse, quartzose parent material. But on inliers of loess under the same climatic conditions, other types of soils were formed: *Grey forest soils*,[3] *Podzolized* and *Leached chernozem*,[4] sometimes *Dark grey podzolized* soils.[5] Under virgin steppe, all chernozem have thick humus

[1] *Umbric Albeluvisols* (IUSS WRB 2006 and in WRB equivalents below)

[2] *Cambic Arenosols*

[3] *Luvic Phaeozems*

[4] *Luvic Chernozem*

[5] *Albic Phaeozems*

horizons with granular structure, over a mixed layer distinguished by crotovinas, on subsoil of calcareous loess or loessic loam—but there are anomalies. Hulyk (2011) found that in the deciduous forest zone of Western Podillya, marshland on strongly calcareous soils has produced areas with alkaline steppic chernozem. Thus, the dominant factor in the formation of soil types of the southern transitional zone from Polissya to the forest-steppe zone is soil parent material.

Moving on to the Forest-Steppe, it is beyond doubt that vegetation is the dominant soil-forming factor in the loess that blankets the uplands: *Grey forest soils* under oak forests, *Typical chernozem*[6] under grassland. There are aberrations such as Chernozem found under dry oak forests or *Podzolized chernozem* and *Dark grey soils* under deciduous forests, but these can be easily explained (there are relevant well-known papers but this is not the place to dwell upon them).

Carpathian Brown Forest Soil Region

Here, *Brown forest soils*,[7] are found on colluvium derived from both acid and basic rocks and spreading from warm, lowland zones of oak, hornbeam and beech forests, through cool temperate zones with beech, spruce and fir forests. Above the tree line, brown soils are also encountered under subalpine and alpine meadows (Kanivets 2012a). Within the mountain region, the humus content of the topsoil increases with elevation, in the same way as in mountain chernozem noted by Dokuchaev, and we sometimes find intense sod pedogenesis—*sod brown soils*[8] (Kanivets 2012b) on base-saturated colluvium. In the Carpathians, the dominance of brown soil at all elevations and under different types of vegetation is determined by a humid temperate climate and free-draining colluvial parent material. In the foothills and the adjacent plains, impeded drainage causes gleying and brown soil formation with an eluvial–illuvial differentiated profile.

Under different hydrothermal conditions in the dry Crimean mountains, dark-coloured soddy soils occur in the mountain meadow belt but, in places with enough moisture, brown soils also occur.

The Steppe

Across the transition zone between Forest-Steppe and Steppe, Topolnyi (2009), like Dokuchaev, observed aspect to be crucial in the formation of soil subtypes: on the more humid northern slopes, *Typical and Podzolised chernozem*[9]; and on the

[6] *Haplic Chernozem*

[7] *Dystric* and *Haplic Cambisols,* respectively

[8] *Eutric Cambisols*

[9] *Haplic* and *Luvic Chernozem*

warmer southern slopes, *Ordinary chernozem*. In the zone of transition from the dry to the middle Steppe of the North-West Black Sea region, Moroz (2010) postulates redistribution of surface run-off according to topography to account for the alternation of *Southern chernozem*[10] in the bottomlands, adjacent to *Dark chestnut soils*[11] on the heights.

Agrogenic Soils

Studying this issue, I should like to add some researchers' suggestions that agriculture be recognized as one of the determining factors in soil genesis—and to establish a new order of *Agrogenic soils*, first suggested by Andrushchenko et al. (1958) for acid brown soils of the Carpathian region that have acquired a high saturation of calcium and magnesium under cultivation.

Conclusions

- Dokuchaev considered that each and every agent of soil formation has the possibility to exert a dominant role in the formation of the soil types or other taxa in certain regions. It is necessary to understand the *equivalence* of factors of soil formation in this way, not excluding the indispensability of agents and their availability.
- Outstanding researchers paying special attention to a single factor only, or to several, have drawn upon limited geographical observations. But if we take account of a wide range of new observations on geography and soil genesis, together with the analysis of Dokuchaev's facts and opinions, we may draw our own conclusions regarding the formula of equivalence of soil-forming factors and conditions.
- All the natural factors and conditions of soil formation occur and interact. But taxonomic units, including soil types and subtypes, within a particular landscape are formed under the special influence of certain determinant, dominant agents. In this respect, the factors of soil formation are equal, but for a specific landscape they are not equivalent.

---

[10] *Calcic Chernozem*

[11] *Haplic Kastanozem*

References

Andrushchenko GO, Hrinchenko OM, Hrin GS, Krupskyi MK (1958) Technique of large-scale studies of soil in collective and state farms of the Ukrainian SSR. Derzhsilhospvydav, Kharkov (Russian)

Dokuchaev VV (1899) Contribution to the theory of natural zones. Collected works, vol. 6. Academy of Sciences of the USSR, Moscow-Leningrad (1951) (Russian)

Dokuchaev VV (1949) Lectures on soil science. Collected works, vol 3. OGIZ, Moscow (Russian)

Glinka KD (1923) The soils of Russia and adjacent countries. Gosizdat, Moscow-Petrograd (Russian)

Hulyk SV (2011) Retrospective analysis of meadow-steppe landscapes of Western Podillia, their current state and the direction of their development. Thesis, Candidate of Geographic Sciences (Physical Geography, Geophysics and Geochemistry of Landscapes), Lviv (Ukrainian)

IUSS Working Group WRB (2006) World reference base for soil resources 2006. World Soil Resources Rept 103, FAO, Rome

Jenny H (1941) Factors of soil formation—a system of quantitative pedology. McGraw Hill, New York

Jenny H (1980) The soil resource. Origin and behavior. Springer, New York

Kanivets SV (2010) The sod-podzolic soils of Polissya. Representative profile: peculiarities of structure and formation. Agrochemistry and Soil Science Collected Papers 73. NSC ISSAR, Kharkiv (Ukrainian)

Kanivets VI (2012a) Brown soils (brown forest soils) in Ukraine and in the world: classification. In: Processes of soil development in the Brown soil-Forest Zone and classification of Brown soil. State Institute of Economics and Management (CSIEM), Chernihiv (Ukrainian), pp 224–241

Kanivets VI (2012b) Some regularities of formation and prevalence of soddy process in mountainous brown soil of the Trans Carpathian region. In: Processes of soil development in the brown soil-forest zone and classification of brown soil. CSIEM, Chernihiv (Ukrainian), pp 61–72

Kanivets SV (2013) Chernozem of Polissian Opillia. Maodan, Kharkiv (Ukrainian)

Kovda VA, Rozanov BG (eds) (1988) Soil science. Part I soil and soil formation. High School Press, Moscow (Russian)

Moroz HB (2010) Geographico-genetic characteristics of soils in the dry to medium-dry steppe transition belt in North-Western Prichernomorya. Thesis, Candidate of Geographical Sciences (Biogeography and Geography of Soils), Lviv (Ukrainian)

Topolnyi SF (2009) Soil research in Forest-Steppe to Steppe transition area between the Bug and Dnipro rivers: Thesis, Candidate of Biological Sciences (Soil Science), Kharkiv (Ukrainian)

Chapter 2
Redoximorphic Features in Albeluvisols from South-Western Ukraine

Volodymyr Nikorych, Wojciech Szymański and Michał Skiba

Abstract Polarizing microscopy and scanning electron microscopy were used to determine the distribution, size, shape, and chemical composition of Fe–Mn nodules in *Albeluvisols (Retisols)* from south-western Ukraine and the relationship of these pedological features with redox processes. Intrusive, impregnative, and depletion features were observed. The highest content of hard nodules is present in the upper part of the solum, indicating that cyclic redox processes occur most often above the illuvial horizon. The nodules are enriched in Fe and Mn (4–14 and 7–40 times, respectively). The chemical composition of the surface of the nodule in contact with microorganisms indicates their crucial role in the accumulation of Fe and Mn and nodule formation.

Keywords Fe–Mn nodules · Fe–Mn concretions · Redox · Albeluvisols · Micromorphology · Scanning electron microscopy

Introduction

Concentrations of iron and manganese compounds in soils may occur as pedofeatures such as nodules and coatings of various shapes and sizes. Fe–Mn nodules and concretions are solid and easily observed and are important in soil diagnostics

V. Nikorych (✉)
Department of Soil Science, Yuri Fedkovich Chernivtsi National University,
L. Ukrainki 25, Chernivtsi 58000, Ukraine
e-mail: v.nikorych@chnu.edu.ua

W. Szymański
Department of Pedology and Soil Geography, Institute of Geography and Spatial Management, Jagiellonian University, Gronostajowa 7, 30-387 Kraków, Poland

M. Skiba
Department of Mineralogy, Petrology and Geochemistry, Institute of Geological Sciences, Jagiellonian University, Oleandry 2a, 30-063 Kraków, Poland

D. Dent and Y. Dmytruk (eds.), *Soil Science Working for a Living*,
DOI 10.1007/978-3-319-45417-7_2

—especially in identifying redox conditions (Zhang and Karathanasis 1997; Vepraskas 2001; Gasparatos et al. 2005). They accumulate elements of variable valency and control the distribution and mobility of cations in the soil (Suarez and Langmuir 1976). In a sense, they are the memory of the soil and throw light on its genesis (Zaidelman and Nikiforova 2010).

Many and various Fe–Mn pedofeatures are described in the literature: glaebules, septaria, ferricretes (Brewer 1964), stains, ortsands, ortsteins, concretions, and nodules (Zaidelman and Nikiforova 2001, 2010; Lindbo et al. 2010; Szendrei et al. 2012; Szymański and Skiba 2013; Nikorych et al. 2014). The different terms refer to the variety of form, composition, genesis and, sometimes, national traditions in the absence of a common classification of these features. For instance, Zaidelman and Nikiforova (2001) distinguish classes (concretional and non-concretional), types (based on chemical composition), families (based on form), and species (based on morphology), but there are several other informative and logical systems.

Most Fe–Mn redoximorphic pedofeatures are discrete, hard bodies characterized by a specific colour and often called ortstein, concretions, and nodules. Various terms are used synonymously, but recent studies with polarizing and scanning electron microscopes reveal two distinct types: (1) showing a relatively uniform distribution of Fe (oxyhydr)oxides and Mn oxides and (2) showing internal structure with concentric bands or layers of different content of Fe (oxyhydr)oxides and Mn oxides. Hickey et al. (2008) distinguish nodules and concretions, arguing that the different forms indicate different genesis; we might also consider differences in parent material, moisture regime, climatic conditions, as well as the quantity and kind of soil organisms.

Fe (oxyhydr)oxides and Mn oxides adsorb various elements including trace metals (Ti, Cu, Co, Cr, Ni, Pb, Zn), phosphorus and carbon, and play a role in many soil chemical reactions (Schwertmann and Fanning 1976; Latrille et al. 2001; Palumbo et al. 2001; Tan et al. 2006; Huang et al. 2008; Cornu et al. 2009; Sipos et al. 2011; Szymański and Skiba 2013).

Here, we determine the distribution, size, shape, and chemical composition of Fe–Mn pedofeatures and consider their relationship with redox processes in Albeluvisols from the south-western part of Ukraine. To understand their formation and evolution, we calculate the enrichment factor (EF) as the ratio of the concentration of the chemical elements in the studied feature to their concentration in the soil matrix. Many studies indicate that EF depends on the large specific surface and pH-dependent surface charge of Fe (oxyhydr)oxides and Mn oxides (Schwertmann and Fanning 1976; Dawson et al. 1985; Tan et al. 2006; Zaidelman and Nikiforova 2010; Szymański and Skiba 2013).

Materials and Methods

Study Area

The study was carried out in the Precarpathians of south-western Ukraine (Fig. 2.1)—an undulating landscape with broad hills rising to 600 m above sea level. The bedrock is flysch, covered by up to 25 m of non-calcareous loess. Mean annual temperature is 6–8 °C; mean annual precipitation is 650–800 mm. Most of the area is forested with oak (*Quercus* spp.), beech (*Fagus sylvatica* L.), fir (*Abies alba* Mill.), hornbeam (*Carpinus betulus* L.), lime (*Tilia cordata* Mill.), larch (*Larix decidua* Mill.), and willow (*Salix* spp.), but quite large areas are cultivated. The soils are mostly Luvisols and Albeluvisols (IUSS Working Group WRB 2006) or Retisols (IUSS Working Group WRB 2014); the lack of carbonates in the parent material and the humid climate encourage wash-down of clay from the upper to the middle part of the soil profile (Quénard et al. 2011).

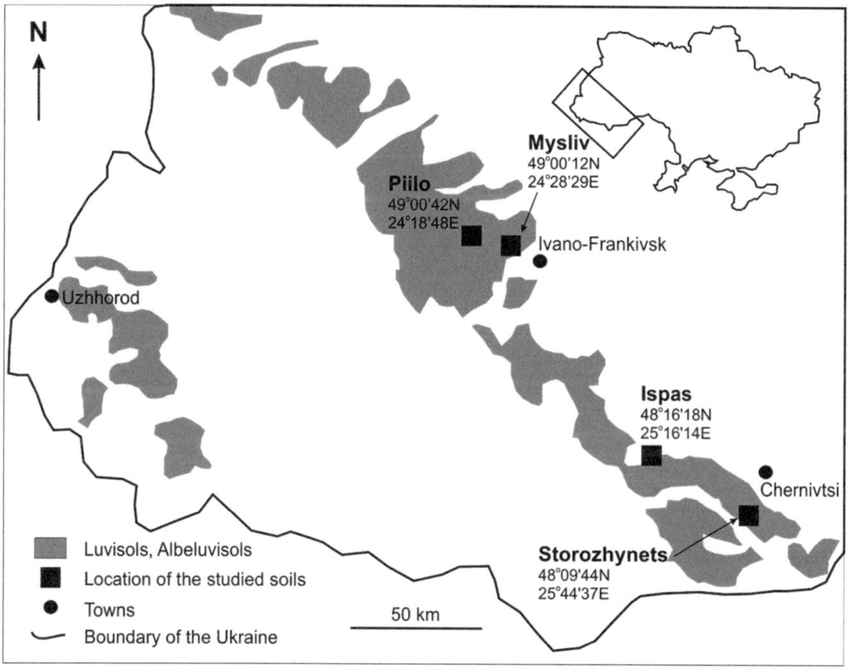

Fig. 2.1 Occurrence of Luvisols and Albeluvisols in SW Ukraine and location of soil profiles

Field and Laboratory Methods

From 20 described soil profiles, four representatives were selected for detailed analysis (Fig. 2.1; Tables 2.1 and 2.2), all classified as Stagnic Fragic Albeluvisols (Siltic). The Piilo profile is under deciduous forest, the Ispas and Mysliv profiles are under mixed forest, and the Storozhynets profile is under coniferous forest. Detailed profile descriptions are given by Nikorych et al. (2014).

From each soil horizon, undisturbed samples were collected for micromorphological study and bulk samples for physical and chemical analysis. The latter were air-dried and sieved through a 1-mm sieve; laboratory analyses were conducted on the fine earth fraction (<1 mm).

Fe–Mn nodules were separated by boiling 40 g of fine earth for 30 min in 600 ml distilled water containing 1.5 g Na_2CO_3 as a dispersant. After boiling, the samples were wet-sieved through 0.5, 0.25, and 0.1-mm sieves and oven-dried at 105 °C for 24 h. The separated fractions were weighed. Coarse Fe–Mn nodules (1–0.5 mm) were hand-picked under a magnifying glass. These were weighed again. The medium (0.5–0.25 mm) and fine (0.25–0.1 mm) fractions were treated with 10% HCl for 24 h to remove Fe–Mn nodules, washed three times with distilled water to remove dissolved Fe (oxyhydr)oxides and Mn oxides, then oven-dried at 105 °C for 24 h, and weighed. Soil organic carbon content was determined by rapid dichromate oxidation (Nelson and Sommers 1996); pH was measured potentiometrically in distilled water using a 1:2.5 soil/water ratio; the concentration of Fe and Al in ammonium oxalate extracts prepared according to Tamm's procedure was determined using a UV–Vis spectrometer. Particle size distribution was determined by sieving and hydrometer methods. Bulk density and total porosity were measured on undisturbed cores.

A Nikon Eclipse E600POL polarizing microscope was used for micromorphological studies; thin sections were described following Stoops (2003). Image analyses of micrographs employed MultiscanBase v.18.03 software using the colour of the studied pedofeatures. Separated coarse Fe–Mn nodules (and aggregates for comparison) were observed and analysed under a Hitachi S-4700 field emission scanning electron microscope (FESEM) equipped with a Vantage Noran energy dispersive X-ray spectroscopy (EDS) system. Details of sample preparation for FESEM–EDS studies can be found in Szymański and Skiba (2013). The relative chemical composition of the nodules and aggregates (point and area analyses in not less than 10 repeats) was determined with the application of an acceleration voltage of 20 kV, an emission current of 10 μA, and a 100 s count time.

Table 2.1 Profile descriptions

Horizon	Depth (cm)	Munsell colour (moist)	Structure	Consistence	Roots	Clay coatings
Storozhynets						
O	0–5	n.a.	n.a.	n.a.	n.a.	n.a.
AEg	5–22	10YR 5/3	Subangular blocky	Soft	Few	Absent
Eg	22–33	10YR 5/4	Subangular blocky	Slightly hard	Few	Few
Btx	33–60	10YR 5/6; 10YR 6/2	Prismatic	Hard	None	Many
2Btg	60–100	10YR 5/6; 10YR 6/2	Prismatic	Hard	None	Many
2BC	>100	10YR 5/6; 10YR 8/1	Massive	Very hard	None	Common
Ispas						
A	0–6	n.a.	n.a.	n.a.	n.a.	n.a.
AEg	6–15	10YR 5/3	Subangular blocky	Slightly hard	Many	Absent
Eg	15–32	10YR 5/4	Subangular blocky	Slightly hard	Few	Few
Btx1	32–52	10YR 5/6; 10YR 6/2	Prismatic	Very hard	Few	Common
Btx2	52–110	10YR 5/6; 10YR 6/2	Prismatic	Very hard	None	Many
Btg	110–140	10YR 5/4	Prismatic	Very hard	None	Common
Mysliv						
O	0–2	n.a.	n.a.	n.a.	n.a.	n.a.
A	2–14	10YR 4/2	Subangular blocky	Soft	Many	Few
Eg	14–30	10YR 5/4	Massive	Slightly hard	Common	Absent
Btx1	30–49	10YR 5/4; 10YR 6/2	Blocky	Very hard	Few	Common
Btx2	49–57	10YR 5/3; 10YR 6/2	Prismatic	Very hard	Few	Many
Btx3	57–120	10YR 5/3; 10YR 6/2	Prismatic	Very hard	None	Common
Piilo						
O	0–6	n.a.	n.a.	n.a.	n.a.	n.a.
A	6–16	10YR 4/2	Subangular blocky	Slightly hard	Many	Absent
AE	16–31	10YR 5/2	Blocky	Slightly hard	Common	Absent

(continued)

Table 2.1 (continued)

Horizon	Depth (cm)	Munsell colour (moist)	Structure	Consistence	Roots	Clay coatings
Eg	31–43	10YR 5/2	Blocky	Slightly hard	Few	Few
Btx1	43–72	10YR 5/3; 10YR 6/3	Prismatic	Very hard	Few	Common
Btx2	72–110	10YR 4/3; 10YR 6/3	Prismatic	Very hard	None	Many
BC	110–123	10YR 4/3; 10YR 6/3	Massive	Very hard	None	Common

Results and Discussion

Distribution of Redoximorphic Pedofeatures

Table 2.3 presents the size distribution of sieved Fe–Mn nodules. The greatest content of the nodules is generally in the upper part of the profile (A, AEg, and Eg horizons). However, the Storozhynets profile is partially formed from the flysch residue and, in this profile, the greatest content of nodules (16.41 g kg^{-1}) is present in the 2Btg horizon, coinciding with the highest bulk density—which is responsible for periodic stagnation of water and frequent redox processes. A second maximum of nodules in the AEg horizon (12.98 g kg^{-1}) indicates the influence of the fragipan on the infiltration of water (Table 2.2).

In the other profiles, the concentration of Fe–Mn nodules in the upper solum suggests that the most frequent redox cycles occur above illuvial horizons. On the one hand, this is connected with the low permeability of the fragipan which causes a perched water table; on the other hand, the upper layers show the highest content of organic matter and are the most active biologically. The distribution of nodules also suggests that eluvial horizons have a higher content of hard nodules resistant to dissolution, which agrees with the data for Alfisols and Albeluvisols presented by Lindbo et al. (2000), Zhang and Karathanasis (1997), and Szymański and Skiba (2013). Table 2.4 presents the distribution of all Fe–Mn nodules (not only the coarser nodules separated by sieving) obtained from image analysis of micrographs.

Most nodules occur in illuvial horizons (Btx or Btg), coinciding with the higher content of iron (oxyhydr)oxides in lower soil horizons (Table 2.2). Iron (oxyhydr) oxides from upper soil horizons are eluviated to lower horizons, or accumulate as coarse nodules, or form organo-mineral complexes. In illuvial horizons, iron (oxyhydr)oxides form fine, soft nodules that dissolve easily, or they occur as irregular iron hypocoatings. This suggests that illuvial horizons are saturated with water for longer periods than eluvial horizons. This is related to their greater clay content (Table 2.2) and, also, drying of the upper layers by tree roots. Thus, an

Table 2.2 Physical and chemical properties of the studied soils

Horizon	Depth (cm)	Sand (%) (1–0.05 mm)	Silt (%) (0.05–0.002 mm)	Clay (%) (<0.002 mm)	Db[a] (Mg/m^3	P[b] (%)	SOC[c] (%)	pH (H_2O)	Fe_o[d] (%)	Al_o[e] (%)
Storozhynets										
O	0–5	n.a.[f]	n.a.	n.a.	n.a.	n.a.	n.a.	n.a.	n.a.	n.a.
AEg	5–22	17.0	66.0	17.0	1.29	48.4	2.4	4.5	1.49	0.04
Eg	22–33	14.0	63.0	23.0	1.39	44.2	0.6	4.9	1.59	0.04
Btx	33–60	12.0	52.0	36.0	1.53	39.5	0.6	5.3	1.96	0.07
2Btg	60–100	19.0	34.0	47.0	1.58	37.3	0.5	5.6	2.21	0.07
2BC	100–140	13.0	33.0	54.0	1.55	37.6	n.a.	n.a.	n.a.	n.a.
Ispas										
O	0–3	n.a.	n.a.	n.a.	n.a.	n.a.	n.a.	n.a.	n.a.	n.a.
AEg	3–21	26.0	54.0	20.0	1.01	58.4	2.6	4.7	1.39	0.05
Eg	21–35	21.0	56.0	23.0	1.37	44.8	0.9	4.7	1.41	0.05
Btx1	35–52	19.0	49.0	32.0	1.41	42.2	0.6	4.7	1.66	0.06
Btx2	52–110	22.0	49.0	29.0	1.46	40.3	0.6	4.7	1.88	0.07
Btg	110–140	22.0	50.0	28.0	1.46	39.9	0.5	5.1	1.92	0.06
Mysliv										
O	0–2	n.a.	n.a.	n.a.	n.a.	n.a.	n.a.	n.a.	n.a.	n.a.
A	2–14	14.0	71.0	15.0	1.30	49.4	2.5	4.8	1.22	0.04
Eg	14–30	11.0	68.0	21.0	1.30	46.1	1.3	5.4	1.49	0.05
Btx1	30–49	10.0	64.0	26.0	1.37	42.3	1.0	5.3	1.74	0.05
Btx2	49–57	11.0	64.0	25.0	1.46	40.2	0.6	5.4	1.73	0.06

(continued)

Table 2.2 (continued)

Horizon	Depth (cm)	Sand (%) (1–0.05 mm)	Silt (%) (0.05–0.002 mm)	Clay (%) (<0.002 mm)	Db[a] (Mg/m^3	P[b] (%)	SOC[c] (%)	pH (H_2O)	Fe_o[d] (%)	Al_o[e] (%)
Btg	57–120	10.0	73.0	17.0	1.67	35.6	0.4	5.0	1.98	0.06
Piilo										
O	0–6	n.a.	n.a.	n.a.	n.a.	n.a.	n.a.	n.a.	n.a.	n.a.
A	6–16	17.0	63.0	20.0	n.a.	n.a.	3.7	4.8	0.81	0.03
AE	16–31	n.a.	n.a.	n.a.	n.a.	n.a.	n.a.	n.a.	n.a.	n.a.
Eg	31–43	16.0	66.0	18.0	n.a.	n.a.	0.8	4.9	1.55	0.05
Btx1	43–72	16.0	62.0	22.0	n.a.	n.a.	0.5	5.1	1.64	0.07
Btx2	72–110	17.0	58.0	25.0	n.a.	n.a.	n.a.	5.2	1.57	0.05
BC	110–123	n.a.	n.a.	n.a.	n.a.	n.a.	n.a.	n.a.	n.a.	n.a.

[a] *Db* Bulk density
[b] *P* Total porosity
[c] *SOC* Soil organic carbon
[d] Fe_o Ammonium oxalate extractable Fe
[e] Al_o Ammonium oxalate extractable Al
[n.a.] Not analysed

Table 2.3 Distribution of hard Fe–Mn nodules

Horizon	Depth (cm)	1.0–0.5 mm	0.5–0.25 mm	0.25–0.10 mm	Sum
		$g\ kg^{-1}$			
Storozhynets					
O	0–5	n.a.[a]	n.a.	n.a.	n.a.
AEg	5–22	6.30	3.35	3.33	12.98
Eg	22–33	3.14	2.40	2.63	8.17
Btx	33–60	2.07	1.95	1.85	5.87
2Btg	60–100	7.16	3.95	5.30	16.41
2BC	100–140	1.01	0.75	1.75	3.51
Ispas					
O	0–3	n.a.	n.a.	n.a.	n.a.
AEg	3–21	24.42	8.05	5.20	37.67
Eg	21–35	9.46	6.65	4.03	20.14
Btx1	35–52	4.06	1.73	2.08	7.86
Btx2	52–110	2.18	2.40	2.17	6.75
Btg	110–140	1.66	1.43	2.65	5.73
Mysliv					
O	0–2	n.a.	n.a.	n.a.	n.a.
A	2–14	2.93	4.70	10.00	17.63
Eg	14–30	2.85	1.73	2.05	6.63
Btx1	30–49	2.74	1.98	2.58	7.29
Btx2	49–57	1.03	1.73	1.28	4.03
Btg	57–120	5.65	4.85	3.63	14.13
Piilo					
O	0–6	n.a.	n.a.	n.a.	n.a.
A	6–16	4.59	2.98	2.58	10.14
AE	16–31	n.a.	n.a.	n.a.	n.a.
Eg	31–43	11.38	3.50	2.53	17.41
Btx1	43–72	6.79	2.50	1.95	11.24
Btx2	72–110	3.55	2.40	3.18	9.12
BC	110–123	1.60	0.95	2.00	4.55

[a]*n.a.* Not analysed

analysis of the content, size, and shape of Fe–Mn nodules in soils provides information on the duration of waterlogging.

Micromorphology

Using Stoops' terminology (2003), Albeluvisols are characterized by intrusive redox pedofeatures (Fe and Fe–Mn coatings on the walls of voids and ped faces),

Table 2.4 Content of Fe–Mn pedofeatures according to percentage area in thin section

Horizon	Depth (cm)	Fe–Mn nodules and hypo- and quasicoatings (%)	Fe–Mn nodules (%)	Hypo- and quasicoatings (%)
Storozhynets				
O	0–5	n.a.[a]	n.a.	n.a.
AEg	5–22	19.27	7.87	11.40
Eg	22–33	9.50	6.50	3.01
Btx	33–60	36.19	17.35	18.84
2Btg	60–100	37.25	19.79	17.47
2BC	100–140	n.a.	n.a.	n.a.
Ispas				
O	0–3	n.a.	n.a.	n.a.
AEg	3–21	13.47	8.47	5.00
Eg	21–35	20.33	11.59	8.74
Btx1	35–52	18.83	8.10	10.73
Btx2	52–110	26.55	16.54	10.01
Btg	110–140	22.18	15.71	6.47
Mysliv				
O	0–2	n.a.	n.a.	n.a.
A	2–14	9.86	2.93	6.93
Eg	14–30	10.98	6.70	4.28
Btx1	30–49	33.91	23.15	10.75
Btx2	49–57	4.80	4.17	0.63
Btg	57–120	n.a.	n.a.	n.a.
Piilo				
O	0–6	n.a.	n.a.	n.a.
A	6–16	18.08	7.21	11.46
Eg	31–43	15.49	5.65	9.84
Btx1	43–72	18.03	13.72	4.31
Btx2	72–110	14.19	8.24	5.95
BC	110–123	12.06	6.86	5.30

[a]*n.a.* Not analysed

impregnative pedofeatures (Fe–Mn nodules and iron and iron–manganese hypo- and quasicoatings), and depletion pedofeatures (zones lighter in colour due to removal of Fe (oxyhydr)oxides and Mn oxides). Iron and iron–manganese coatings occur only in illuvial horizons (fragipan and argillic, Fig. 2.2a, b). Some may also be related to migration of iron and manganese within illuvial horizons when the soil is saturated long enough to reduce the iron (oxyhydr)oxides and manganese oxides (Vepraskas et al. 2006). The Fe–Mn coatings do not show Fe-enriched and

Mn-enriched layers which might indicate a rapid shift from reduced to oxidized conditions (Huang et al. 2008). And Fe-rich clay coatings in illuvial horizons suggest simultaneous migration of iron (oxyhydr)oxides and clay particles (Nikorych et al. 2014).

Two main types of Fe–Mn nodules are observed: (1) orthic nodules showing undifferentiated internal fabric and gradual boundaries (Fig. 2.3a, b) and (2) orthic nodules showing undifferentiated internal fabric and sharp boundaries (Fig. 2.3c–f).

The orthic nodules consist of iron (oxyhydr)oxides and manganese oxide, quartz grains, humus (especially in upper soil horizons), and clay minerals (especially in illuvial horizons). They are similar to Fe–Mn nodules found in Albeluvisols in the Carpathian foothills of southern Poland described by Szymański and Skiba (2013). Orthic, irregular nodules are the most common in all of the studied soil profiles (Fig. 2.3a, b); they are most abundant in illuvial horizons (Btx and Btg) but are also present in upper horizons (A, AEg, Eg). Undifferentiated internal fabric (the same as the matrix) and gradual boundaries indicate that the nodules formed in situ—presumably by periodic water saturation leading to the reduction of Fe and Mn compounds and subsequent desaturation, oxidation, and precipitation as Fe (oxyhydr)oxides and Mn oxides. This implies that the nodules are not relict features but are forming now (Hseu and Chen 1996; Vepraskas 2004, Szymański and Skiba 2013). Disorthic nodules also occur in the upper part of the studied soils (A, AEg,

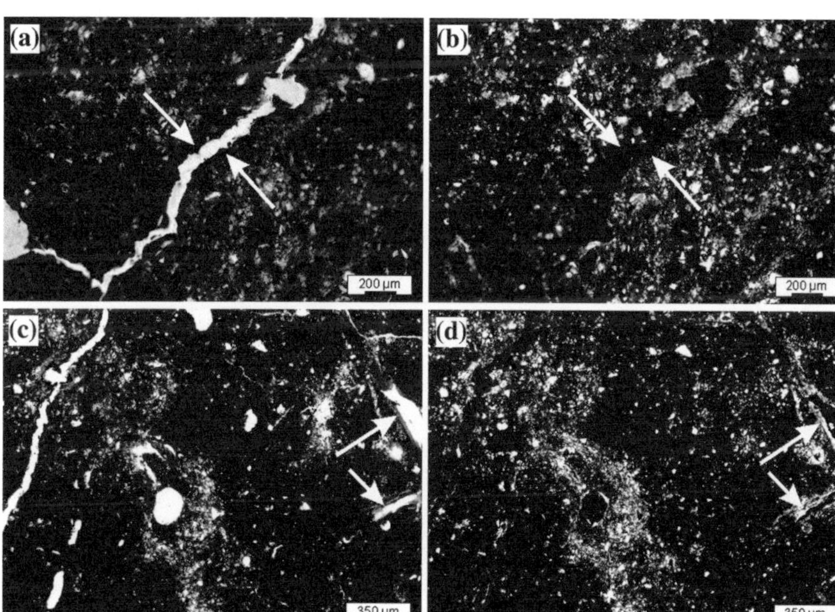

Fig. 2.2 Storozhynets profile: Fe–Mn coatings (*white arrows*) in fragipan (Btx) (**a**, **b**). Iron depletion and iron impregnative hypocoatings and clay coatings (*white arrows*) in fragipan (Btx) (**c**, **d**). Plane polarized light (**a**, **c**) and crossed polarized light (**b**, **d**)

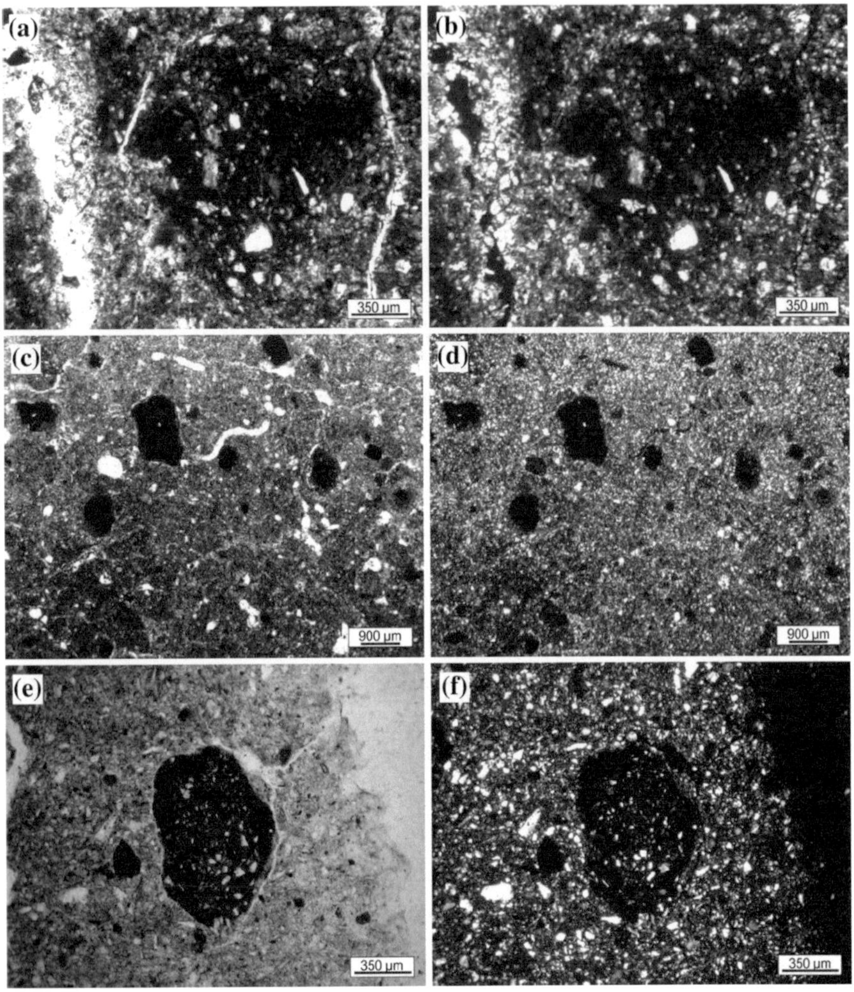

Fig. 2.3 Micromorphology of orthic Fe–Mn nodules: with gradual boundary in fragipan from the Ispas profile (**a**, **b**); with sharp boundary in eluvial horizon from the Piilo profile (**c**, **d**); and with a sharp boundary in eluvial horizon from the Storozhynets profile (**e**, **f**). Plane polarized light (**a**, **c**, **e**) and crossed polarized light (**b**, **d**, **f**)

and Eg horizons); they might form from orthic nodules that have been displaced by faunal mixing of soil material.

In addition to Fe–Mn nodules, impregnative and depletion hypocoatings occur in the illuvial horizons (Fig. 2.2c, d). These features are also related to cyclic reduction and oxidation (Vepraskas 2004; Lindbo et al. 2010; Nikorych et al. 2014). Iron and iron–manganese depletion hypocoatings occur mainly along vertical cracks and channels, indicating that they relate to flowing water containing dissolved organic matter (DOM) that is a source of energy for the microorganisms responsible for reduction. In many cases, clay depletion hypocoatings are associated with Fe and Fe–Mn depletion hypocoatings; the reduction and eluviation of iron (oxyhydr)oxides and manganese oxides facilitates dispersion and translocation of clay. In turn, eluvial horizons are characterized by bleaching that indicates depletion of iron (oxyhydr)oxides and manganese oxides. The process is linked with illuvial horizons of low hydraulic conductivity and periodic stagnation of water table above them. Reduction and depletion of iron (oxyhydr)oxides and manganese oxides is also related to the concentration of organic matter in the upper soil and the greater porosity of the upper solum, compared with the illuvial horizons, which facilitates the migration of water containing DOM (Vepraskas 2004).

Chemical Composition of Fe–Mn Nodules

The main constituents of Fe–Mn nodules are SiO_2 (55.0–62.5%), Fe_2O_3 (20.1–27.8%), Al_2O_3 (7.1–10.1%), MnO (1.1–4.0%), K_2O (1.2–1.9%), and P_2O_5 (0.6–1.9%); other constituents rarely exceed 1% (Table 2.5). The data are similar to those obtained by Zhang and Karathanasis (1997), Tan et al. (2006), and Szymański and Skiba (2013) for soils in Kentucky (USA), various parts of China, and the Carpathian foothills in Poland, respectively. The content of SiO_2 in Fe–Mn nodules shows a uniform distribution throughout the profile with only small variance (2–4%) between horizons; the content of Al_2O_3 follows a similar pattern with variance from 1 to 3%, but the Al_2O_3 content of Fe–Mn nodules is slightly higher in the fragipan, most likely linked with a higher content of clay minerals in illuvial horizons (Nikorych et al. 2014).

At first sight, the distribution of Fe_2O_3 and MnO in the nodules shows no particular pattern, suggesting that there is no relationship between the quantity of the nodules and content of Fe and Mn in the soil profile. However, on summation of the Fe and Mn content, their joint distribution becomes uniform—variances do not exceed 2%, with the exception of the Piilo profile (coefficient of variation 5%). The chemical data for the nodules show that if the nodules from the eluvial horizon contain a larger amount of Fe in comparison with nodules obtained from the illuvial horizon, then the content of Mn in nodules obtained from the eluvial horizon is lower than in nodules obtained from the illuvial horizon, and vice versa (except the Mysliv profile). However, this is only a tendency and not a regular pattern, as indicated by relatively high coefficients of variation (10–50% for Fe_2O_3 and almost 100% for MnO).

Enrichment factors (EF) vary according to the element. The content of Si and Al in the nodules is lower than in the surrounding soil material (EF < 1) likewise for Mg, K, and Na. On the other hand, the nodules are enriched in Fe (4–14 times) and especially in Mn (7–40 times) in comparison with the surrounding soil material (Table 2.6). This agrees with data presented by Tan et al. (2006), for Chinese soils, and Palumbo et al. (2001), for Sicilian soils.

Table 2.5 Chemical composition of Fe–Mn nodules obtained using SEM–EDS

Horizon	Depth (cm)	SiO_2 (%)	Al_2O_3 (%)	Fe_2O_3 (%)	MnO (%)	CaO (%)	MgO (%)	K_2O (%)	Na_2O (%)	P_2O_5 (%)	SO_3 (%)	TiO_2 (%)	Cr_2O_3 (%)	CoO (%)	NiO (%)	CuO (%)	ZnO (%)
Storozhynets																	
Eg	22–33	55.82	7.44	27.68	1.30	1.27	0.36	1.43	0.73	1.31	1.47	0.52	0.12	0.12	0.17	0.07	0.18
Btx	33–60	57.08	9.87	23.75	3.31	1.02	0.60	1.71	0.62	0.57	0.41	0.46	0.04	0.16	0.16	0.04	0.19
2Btg	60–100	58.34	7.85	27.08	1.82	0.72	0.31	1.24	0.50	0.81	0.36	0.54	0.05	0.08	0.11	0.07	0.13
Ispas																	
Eg	21–35	58.55	8.55	26.00	1.56	0.49	0.42	1.70	0.61	0.66	0.38	0.67	0.08	0.15	0.06	0.05	0.08
Btx1	35–52	56.85	10.40	23.28	4.01	0.53	0.58	1.74	0.65	0.78	0.37	0.49	0.07	0.08	0.07	0.07	0.04
Mysliv																	
Eg	14–30	57.39	8.70	26.10	1.12	1.21	0.51	1.75	0.75	1.07	0.43	0.46	0.10	0.09	0.07	0.07	0.15
Btx1	30–49	53.99	9.58	27.65	1.84	1.26	0.61	1.91	0.79	0.98	0.40	0.55	0.05	0.16	0.08	0.06	0.10
Btg	57–120	58.67	7.13	26.19	1.10	0.77	0.37	1.76	1.01	1.59	0.33	0.46	0.02	0.12	0.13	0.16	0.19
Piilo																	
Eg	31–43	62.52	7.20	20.12	3.21	0.63	0.29	1.75	0.65	1.91	0.43	0.69	0.27	0.13	0.04	0.02	0.14
Btx1	43–72	56.13	8.05	27.78	1.16	0.78	0.40	1.79	0.56	1.74	0.81	0.41	0.11	0.01	0.06	0.13	0.08

Table 2.6 EF for selected elements in Fe–Mn nodules in relation to the elements in surrounding soil material

Horizon	Depth (cm)	SiO_2	Al_2O_3	Fe_2O_3	MnO	CaO	MgO	K_2O	Na_2O
Storozhynets									
Eg	22–33	0.70	0.93	10.48	7.22	3.63	0.33	0.79	0.63
Btx	33–60	0.83	0.78	4.91	33.1	2.22	0.44	0.95	0.89
2Btg	60–100	0.92	0.54	4.26	10.11	1.38	0.25	0.59	0.45
Ispas									
Eg	21–35	0.80	0.66	14.29	9.75	0.68	0.49	0.89	0.60
Btx1	35–52	0.79	0.74	5.91	33.42	0.63	0.57	0.95	0.83
Mysliv									
Eg	14–30	0.72	1.10	10.70	22.40	5.04	0.78	0.85	0.76
Btx1	30–49	0.72	0.97	6.39	15.33	3.60	0.62	0.89	0.88
Btg	57–120	0.81	0.69	6.40	7.33	1.31	0.38	0.77	1.23
Piilo									
Eg	31–43	0.79	0.97	6.27	40.13	2.63	0.45	0.88	0.79
Btx1	43–72	0.73	0.93	7.91	14.50	1.70	0.32	0.85	0.66

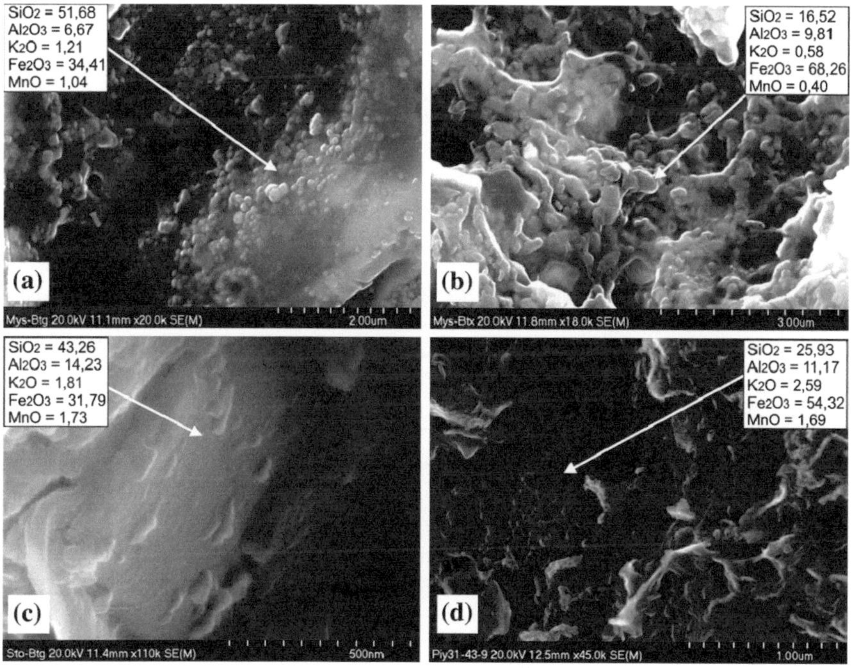

Fig. 2.4 Surface of Fe–Mn nodules from Btx1 and Btg horizons in the Mysliv profile (**a**, **b**), from Btg horizon in the Storozhynets profile (**c**), and from Eg horizon in the Piilo profile (**d**) with microbial cells and content of selected elements (SEM–EDS)

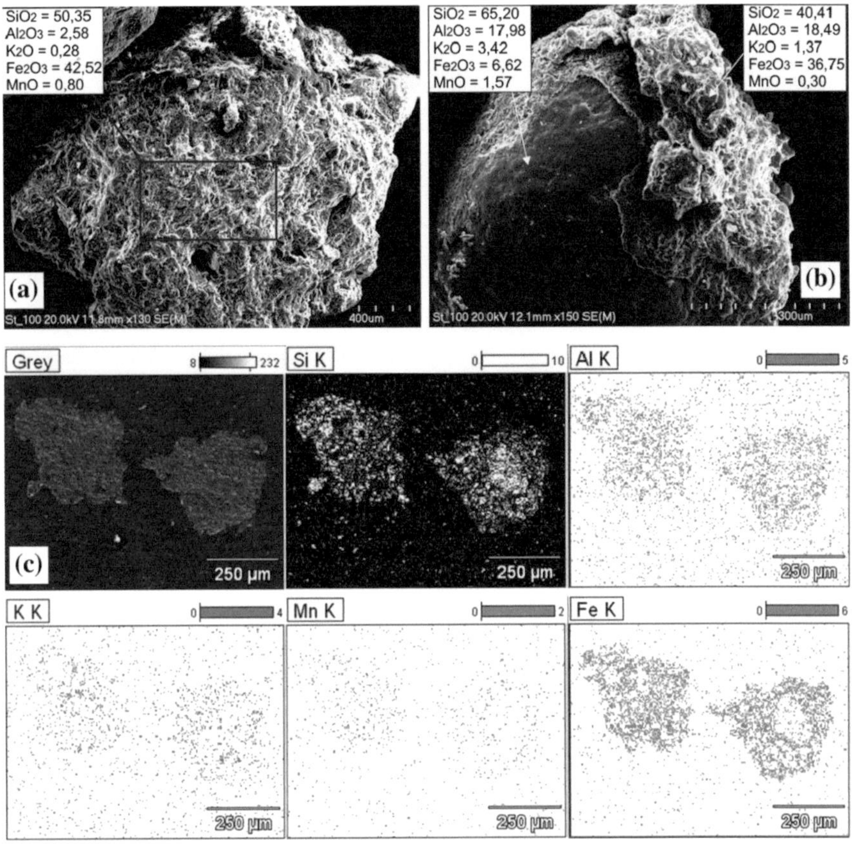

Fig. 2.5 Typic, orthic Fe–Mn nodule (**a**) and concretion showing concentric crusts (**b**), and content of selected elements in both pedofeatures from 2BC horizon of the Storozhynets profile; section of typic, orthic Fe–Mn nodule (**c** *left*) and nucleic Fe–Mn nodule (**c** *right*), and spatial distribution of selected elements in the both nodules from Eg horizon of the Piilo profile (SEM–EDS)

The mechanism of formation of Fe–Mn nodules might be biological and/or chemical (Timofeeva and Golov 2010). The weak correlation between the quantity of nodules and their content of Fe and Mn (for Fe_2O_3, r = –0.22; for MnO, r = –0.09) suggests a biological origin. Chemotrophic bacteria form local colonies or films, depending on micro-conditions, and the different quality and quantity of soil microorganism communities lead to variability in the content of Fe and Mn in nodules. The physiological manner of deposition of these elements in microbial cells—depending on specific local conditions—may also play a role (Timofeeva and Golov 2010). Specific local conditions may occur within the same soil horizon; for example, during strong anaerobiosis in any particular horizon, there are always micro-sites containing oxygen, and vice versa (Aristovskaya 1980).

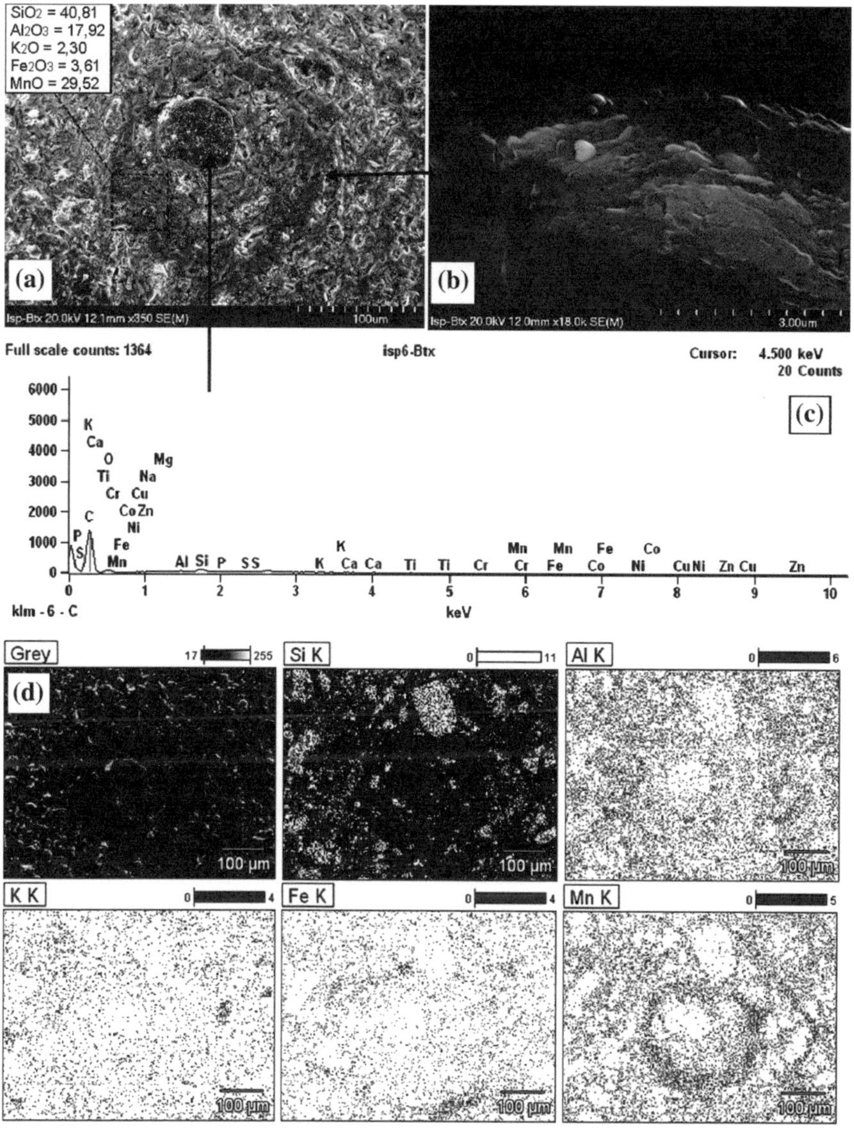

Fig. 2.6 Internal structure of Fe–Mn concretion showing concentric band enriched in Mn (**a**), presence of microorganisms in the band (**b**), chemical composition of centre (probably fungal hyphae) of the pedofeature (**c**), and spatial distribution of selected elements in the pedofeature (**d**) from Btx1 horizon of the Ispas profile (SEM–EDS)

Figure 2.4 shows the porous, pitted surface of an Fe–Mn nodule covered with microbial cells. The surface of the nodule in contact with these microorganisms has a high content of Fe and Mn; in other places, the content of Fe and Mn is lower

($\sim$50%), which suggests that micro-organisms play a crucial role in the accumulation of Fe and Mn, leading to the formation of nodules.

Two main types of Fe–Mn nodules are observed. The internal fabric of almost all of the studied nodules indicates that the nodules are micro-aggregates (Fig. 2.5a) with a high content of relatively uniformly distributed Fe and Mn (Fig. 2.5c left nodule).

However, some of the nodules (concretions) are characterized by concentric layers that have larger quantities of Fe and Mn (Fig. 2.5b, c right nodule). Such a concentric internal structure indicates cyclic formation of the nodule, which we may attribute to periodic wet and dry periods. During the wet season (and reducing conditions because of lack of oxygen), both Fe and Mn shift to a soluble, reduced state and migrate with the soil solution. When the soil begins to dry (and oxidation predominates), both Fe and Mn are oxidized and deposited onto a variety of morphological features such as the walls of pores, faces of aggregates, and mineral grains (Huang et al. 2008). Repetition of these cycles leads to the formation of a concentric internal structure of Fe–Mn nodules and their seasonal growth (Fig. 2.6) (Manceau et al. 2003).

Conclusions

The studied Albeluvisols are characterized by intrusive redox pedofeatures, impregnative redox pedofeatures, and depletion redox pedofeatures. Orthic nodules of undifferentiated internal fabric and gradual boundaries are most common; orthic nodules showing undifferentiated internal fabric and sharp boundaries also occur. Additionally, illuvial horizons exhibit Fe–Mn coatings as well as impregnative and depletion hypocoatings—the latter mainly along vertical cracks and channels.

The concentration of coarse, hard Fe–Mn nodules in the upper part of the soil profiles indicates that cyclic redox processes occur most often above illuvial horizons. This is linked with the low permeability of the fragipan, leading to periodic formation of a perched water table, as well as the greater content of organic matter in the upper layers. However, the highest content of all nodules (not only coarse nodules) occurs in illuvial horizons (Btx or Btg). This is related to a higher content of iron (oxyhydr)oxides in lower soil horizons and their longer periods of waterlogging compared with eluvial horizons.

Fe–Mn nodules are composed mainly of SiO_2, Fe_2O_3, Al_2O_3, MnO, K_2O, and P_2O_5. The content of Si, Al, Mg, K, and Na in the nodules is lower than that in the surrounding soil material. On the other hand, the nodules are enriched in Fe (4–14 times) and especially in Mn (7–40 times) compared with the surrounding soil material. The chemical composition and spatial distribution of the main components of Fe–Mn pedofeatures indicate the occurrence of two types of pedofeatures: nodules and concretions, nodules being much more common.

The crucial role of microorganisms in the accumulation of Fe and Mn in nodules is demonstrated by the distinctive chemical composition of the nodule surface in contact with microorganisms.

References

Aristovskaya TV (1980) Microbiology of soil formation processes. Nauka, Leningrad (Russian)
Brewer R (1964) Fabric and mineral analysis of soils. Wiley, New York
Cornu S, Cattle JA, Samouëlian A et al (2009) Impact of redox cycles on manganese, iron, cobalt, and lead in nodules. Soil Sci Soc Am J 73:1231–1241
Dawson BSW, Ferguson JE, Campbell AS, Cutler EJB (1985) Distribution of elements in some Fe–Mn nodules and an iron-pan in some gley soils of New Zealand. Geoderma 35:127–143
Gasparatos D, Tarenidis D, Haidouti D, Oikonomou G (2005) Microscopic structure of soil Fe–Mn nodules: environmental implications. Environ Chem Lett 2:175–178
Hickey PJ, McDaniel PA, Strawn DG (2008) Characterization of iron–manganese cemented redoximorphic aggregates on wetland soils contaminated with mine wastes. J Environ Qual 37:2375–2385
Hseu ZY, Chen ZS (1996) Saturation, reduction and redox morphology of seasonally-flooded Alfisols in Taiwan. Soil Sci Soc Am J 60:941–949
Huang L, Hong J, Tan W et al (2008) Characteristics of micromorphology and element distribution of iron-manganese cutans in typical soils of subtropical China. Geoderma 146:40–47
IUSS Working Group WRB (2006) World reference base for soil resources 2006. World Soil Resources Reports 103, FAO, Rome
IUSS Working Group WRB (2014) World reference base for soil resources 2014. World Soil Resources Reports 106, FAO, Rome
Latrille C, Elsass F, van Oort F, Denaix L (2001) Physical speciation of trace metals in Fe-Mn concretions from a rendzic lithosol developed on Sinemurian limestones (France). Geoderma 100:127–146
Lindbo DL, Rhoton FE, Hudnall WH et al (2000) Fragipan degradation and nodule formation in Glossic Fragiudalfs of the Lower Mississippi River Valley. Soil Sci Soc Am J 64:1713–1722
Lindbo DL, Stolt MH, Vepraskas MJ (2010) Redoximorphic features. In: Stoops G, Marcelin V, Mees F (eds) Interpretation of micromorphological features of soils and regoliths. Elsevier, Amsterdam, pp 129–147
Manceau A, Tamura N, Celestre RS et al (2003) Molecular-scale speciation of Zn and Ni in soil ferromanganese nodules from loess soils of the Mississippi basin. Environ Sci Technol 37:75–80
Nelson DW, Sommers LE (1996) Total carbon, organic carbon, and organic matter. In: Sparks DL et al. (eds) Methods of soil analysis. Part 3. Chemical methods. SSSA Book Series, vol 5. SSSA and ASA, Madison, pp 961–1010
Nikorych V, Szymański W, Polchyna S, Skiba M (2014) Genesis and evolution of the fragipan in Albeluvisols in the Precarpathians in Ukraine. Catena 119:154–165
Palumbo B, Bellanca A, Neri R, Roe MJ (2001) Trace metal partitioning in Fe–Mn nodules from Sicilian soils, Italy. Chem Geol 173:257–269
Quénard L, Samouëlian A, Laroche B, Cornu S (2011) Lessivage as a major process of soil formation: a revisition of existing data. Geoderma 167–168:135–147
Schwertmann U, Fanning DS (1976) Iron-manganese concretions in hydrosequences of soils in loess in Bavaria. Soil Sci Soc Am J 40:731–773
Sipos P, Németh T, May Z, Szalai Z (2011) Accumulation of trace elements in Fe-rich nodules in a neutral-slightly alkaline floodplain soil. Carpathian J Earth Environ Sci 6(1):13–22

Stoops G (2003) Guidelines for analysis and description of soil and regolith thin sections. SSSA, Madison

Suarez DL, Langmuir D (1976) Heavy metals in a Pennsylvania soil. Geochim Cosmochim Acta 40:589–598

Szendrei G, Kovács-Pálffy P, Földvari M, Gál-Sólymos K (2012) Mineralogical study of ferruginous and manganiferous nodules separated from characteristic profiles of hydromorphic soils in Hungary. Carpathian J Earth Environ Sci 7(1):59–70

Szymański W, Skiba M (2013) Distribution, morphology and chemical composition of Fe–Mn nodules in Albeluvisols of the Carpathian Foothills, Poland. Pedosphere 23(4):445–454

Tan WF, Liu F, Li YH et al (2006) Elemental composition and geochemical characteristics of iron-manganese nodules in main soils of China. Pedosphere 16:72–81

Timofeeva YO, Golov VI (2010) Accumulation of microelements in iron nodules in concretions in soils: a review. Eurasian Soil Sci 43(4):434–440

Vepraskas MJ (2001) Morphological features of seasonally reduced soils. In: Richardson JL, Vepraskas MJ (eds) Wetland soils: genesis, hydrology, landscapes and classification. Lewis Publishers, Boca Raton, pp 163–182

Vepraskas MJ (2004) Redoximorphic features for identifying aquic conditions. Technical Bulletin 301, North Carolina Agricultural Research Service, Raleigh

Vepraskas MJ, Richardson JL, Tandarich JP (2006) Dynamics of redoximorphic feature formation under controlled ponding in a created riverine wetland. Wetlands 26:486–496

Zaidelman FR, Nikiforova AS (2001) Genesis and diagnostic meaning of soil neoformations of forest and forest-steppe zones. Moscow University Press (Russian)

Zaidelman FR, Nikiforova AS (2010) Ferromanganese concretionary neoformations: a review. Eura Soil Sci 43:248–258

Zhang M, Karathanasis AD (1997) Characterization of iron-manganese concretions in Kentucky Alfisols with perched water tables. Clays Clay Miner 45:428–439

Chapter 3
Fractal Properties of Coarse/Fine-Related Distribution in Forest Soils on Colluvium

Volodymyr Yakovenko

Abstract Establishment of lithological homogeneity of the soil profile is a key to interpreting the genesis of soils, especially soils developed in colluvium. Study of fractal properties allows step-by-step establishment of lithological homogeneity, or lithological breaks, without time-consuming determination of particle size distribution and mineralogy. The genetic profile of forest soils in the gullied Dnieper Prysamarya evolved alongside transport and depositional processes in erosional elements of the landscape. Micromorphology reveals a three-tier fractal structure of microstructural elements related to the distribution of coarse and fine particles (c/f-related distribution). *Calcic Chernozem* near the edges of gullies are characterized by lithological homogeneity of the solum and underlying loess parent material; *Luvic Phaeozem* on the slopes of gullies are not lithologically homogeneous—the layers below the solum differ in morphometric parameters at the second and third levels of c/f-related distribution. However, the sola of all soils in the catena have similar fractal properties and morphometric characteristics of the c/f-related distribution because of the colluvial processes operating along the slope.

Keywords Forest soils · Micromorphology · Fractal properties · Coarse/fine-related distribution · Lithological breaks

Introduction

In the gullied landscape of the Pridneprovye Steppe, forest soils have been formed under dynamic conditions and under the influence of a complex set of factors. Transport and deposition of colluvium on erosional elements of the landscape is a major process shaping the soil profile, building up and erasing layers of soil parent material; textural differentiation by lessivage further complicates the picture. To

V. Yakovenko (✉)
Department of Geo-Botany, Soil Science and Ecology, Oles Honchar Dnipropetrovsk National University, 72 Gagarin Av, Dnipro 49010, Ukraine
e-mail: yakovenko_v@i.ua

D. Dent and Y. Dmytruk (eds.), *Soil Science Working for a Living*,
DOI 10.1007/978-3-319-45417-7_3

understand the genesis of soils in colluvium, it is important to establish the lithological homogeneity of the soil profile according to criteria of transition between lithologically different parts of the profile and the contrast of such transitions.

At the micromorphological level, measurable criteria of homogeneity are provided by characteristics of the related distribution of coarse and fine particles (c/f-related distribution): c/f limit, c/f ratio, the proportions of various particle size classes, and their spatial distribution. Further, in the forest soils, a multi-level microstructure may be observed that offers the possibility of identifying several c/f limits, each with specific characteristics. Where the overall picture of the microstructure at each level of the c/f-related distribution is similar to the picture at other levels, such features of micromorphological organization meet the definition of fractal structures (Mandelbrot 1982; Feder 1991; Morozov 2002), well-described by Bozhokin and Parshin (2001):

> Self-similarity – an hierarchic organizing principle of fractal structures will not undergo significant changes when viewing them through the microscope with different magnifications. As a result, the structure at a small scale looks, on average, the same as at a large scale. Self-similarity is the main characteristic of a fractal; it means that it is more-or-less identically constructed across a wide range of scales. So with increasing magnification, small fragments of the fractal are observed to be very similar to the large.

With this in mind, we have investigated the lithological homogeneity of the profiles of forest soils in colluvium at a micromorphological level using the fractal properties of the c/f-related distribution.

Materials and Methods

Study Area

The study area lies in Novomoskovsk District of Dnipropetrovsk region, part of the Pridneprovye Steppe of southeastern Ukraine—within the natural steppe zone (Lavrenko 1980) and the northern humid cold temperate region (Semenyuta 1948) with a mean annual temperature of 8 °C and annual precipitation more than 500 mm. Table 3.1 shows climatic data for the period from 1964 to 1993.

The soil parent materials are mainly Quaternary loess and sands but also Tertiary clays (Sobolev 1939) eroded by a network of gullies and ravines (Belova and Travleev 1999). Most of the land is intensively farmed so steppe vegetation remains only in areas unsuitable for agriculture, such as the slopes of gullies and ravines; and forest is represented by gullied and floodplain forests (Belgard 1950) with *Quercus robur* L., *Tilia cordata* Mill., *Fraxinus exelsior* L., *Acer campestre* L., *Ulmus minor* Mill., *Acer platanoides* L, and *Pinus sylvestris* L.

Table 3.1 Air temperature and precipitation 1964–1993 (after Lazarenko 1995)

Met. station	Jan	Feb	Mar	April	May	June	July	Aug	Sept	Oct	Nov	Dec	Year
Mean air temperature (°C)													
Gubiniha	−5.7	−4.8	0.6	9.3	16.0	19.1	20.5	19.9	14.1	7.8	1.5	−2.5	8.1
Pavlograd	−5.3	−3.9	1.1	8.7	15.7	19.5	21.3	20.3	14.7	8.1	2.0	−1.8	8.4
Dnipropetrovsk	−5.1	−4.2	2.1	9.4	15.9	18.9	21.5	20.5	15.2	8.5	2.0	−2.0	8.5
Mean precipitation (mm)													
Gubiniha	47	33	31	39	45	69	61	45	41	37	44	50	542
Pavlograd	42	30	27	36	49	68	52	41	41	30	40	45	501
Dnipropetrovsk	43	38	34	38	45	62	56	39	37	33	42	47	514

Table 3.2 Soil profile data

Profile	Location	Landform	Vegetation	Classification (IUSS 2014)
PP 115-B	48°47′19.1″N 035°27′19.5″E 114 m asl	Slope 3°, aspect S	*Festuca valesiaca* Schleich *Poa angustifolia* L. *Poa nemoralis* L. *Lathyrus tuberosus* L. *Achillea millefolium* L. *Salvia nemorosa* L.	Calcic Chernozem
PP 111-B	48°47′19.1″N 035°27′19.5″E 88 m asl	Slope 14°, aspect S	*Fraxinus exelsior* L. *Acer campestre* L. *Ulmus minor* Mill. *Tilia cordata* Mill, *Acer platanoides* L.	Luvic Phaeozem
PP 109-B	48°47′17.5″N 035°27′16.5″E 76 m asl	Bottomland (ravine floor)	*Quercus robur* L. *Fraxinus exelsior* L. *Acer campestre* L. *Ulmus minor* Mill. *Tilia cordata* Mill, *Acer platanoides* L.	Luvic Gleyic Phaeozem

Field Observations and Sampling

Soil pits were dug into a south-facing slope in the upper part of the Hlyboky ravine, in the upland gullied landscape (Belgard 1971), 2 km north of Andriyivka village. The catena is represented by three soil profiles: on virgin steppe between the fields and forest edge (PP 115-B), on the mid-slope (PP 111-B) and in the bottom (PP 109-B) of the ravine (Table 3.2). Full field descriptions are presented by Yakovenko (2014), soil water conditions by Travleev (1977), and soil profile structure by Rozanov (2004).

Undisturbed samples for micromorphological study were collected from each genetic soil horizon (those from subsoil layers by soil tube), dried, and impregnated with resin (Dobrovolsky 1983). Micromorphology was described following Parfyonova and Yarilova (1977) and Bullock et al. (1985). Micro-morphometric measurements of the size and content of the mineral grains of the skeletal microstructure were carried out by Gagarina (2004). Photographs were taken with an eTREK DCM320 digital camera.

Results and Discussion

Soil Macro-morphology

Water erosion is widespread across the arable around the gullies, particularly during summer rainstorms (Fig. 3.1a, b) so the gullies themselves receive and accumulate considerable amounts of colluvium.

Fig. 3.1 Slope processes in and around the Hlyboky ravine: **a** sheet erosion of adjacent arable; **b** colluvium intercepted by fallen tree trunks and branches

Fig. 3.2 Profiles of: **a** *Calcic Chernozem*, **b** *Luvic Phaeozem*, **c** *Luvic Gleyic Phaeozem*

As a result, *Calcic Chernozems* near the edges of the gullies (PP 115-B) are a soil transit zone. The soil profile is normal, freely drained, but with the signs of erosion in the surface horizon (Fig. 3.2a): O (3.5–0 cm), Ak (0–6 cm), Bk1 (6–27 cm), Bk2 (27–40 cm), Ck—calcareous loess (40–120 cm+).

Forest soils on the slopes and in the bottom of the ravine are characterized by strongly disturbed profiles in colluvium:

Luvic Phaeozems on the slopes (PP 111-B), freely drained in fine loamy colluvium, have a polycyclic genetic profile (Fig. 3.2b): O (2–0 cm), A1 (0–9 cm), A2 (9–46 cm), A3 (46–88 cm), At (88–138 cm), ABt (138–160 cm), Btk (160–187 cm), Ctk (187–230 cm), 2Ck (230–350 cm).

Luvic Gleyic Phaeozems in the bottom of the ravine (PP 109-B), developed in fine loamy, humose colluvium with impeded drainage in the subsoil, also have a polycyclic profile (Fig. 3.2c): O (3–0 cm), A1 (0–8 cm), A2 (8–34 cm), A3 (34–60 cm), At (60–118 cm), ABt (118–132 cm), Btg (132–166 cm+).

Fractal Properties of the c/f-Related Distribution

The algorithm to study fractal properties has three stages: (1) identification of the fractal structure of the c/f-related distribution, (2) determination the level of similarity, (3) determination of morphometric parameters of the related distribution of each identified level of similarity. These parameters include c/f limit, dimensions of the coarse fraction, and c/f ratio. An additional option is specificity of the spatial distribution of coarse particles at each level.

Observations were made at low (×35), middle (×90), and high (×200) magnifications. Low magnification revealed the first level of related distribution of coarse and fine particles, and morphometric measurements were made; at middle magnification, we studied material distinguished as fine particles at low magnification, determined by the c/f limit and other morphometric indices; and so on at high magnification. As a model of multi-level similarity of c/f-related distribution in different scales, we may construct a geometric fractal known as a Sierpinski carpet (Fig. 3.3).

The related distribution of the fractal structure is manifested as a visual image of micromorphological organization that combines quantitative and qualitative characteristics. The complex characteristic thus obtained enables us to diagnose the lithological homogeneity or breaks in the structure of the profile and, in the latter case, the degree of contrast (Table 3.3).

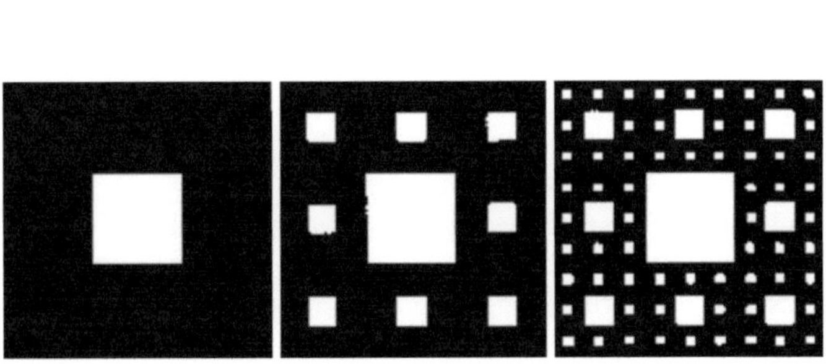

Fig. 3.3 Steps of building the Sierpinski carpet (Feder 1991) are similar to the allocation of levels of similarity in the related distribution of micromorphological study

Table 3.3 Characteristics of c/f-related distribution of forest soils

Genetic horizons	c/f limit (mm)	Coarse particles (mm)	c/f ratio (%)	Spatial distribution of coarse particles
Calcic Chernozem				
Ak (0–6 cm)	0.15	0.15–0.45	1:99	Not homogeneous
	0.02	0.02–0.1	14:86	Homogeneous
	0.005	0.005–0.01	n.a.[a]	Homogeneous
Bk1 (6–27 cm)	0.15	0.15–0.29	Coarse grains < 1	Not homogeneous
	0.02	0.02–0.1	14:86	Homogeneous
	0.005	0.005–0.01	n.a.	Homogeneous
Bk2 (27–40 cm)	0.15	0.15–0.21	Coarse grains < 1	Not homogeneous
	0.02	0.02–0.1	15:85	Homogeneous
	0.005	0.005–0.01	n.a.	Homogeneous
Ck (40–120 cm)	0.15	0.15–0.3	Coarse grains < 1	Not homogeneous
	0.02	0.02–0.09	10:90	Homogeneous
	0.005	0.005–0.01	n.a.	Homogeneous
Luvic Phaeozem				
A1 (0–9 cm	0.15	0.15–0.52	6:94	Not homogeneous
	0.02	0.02–0.09	3:97	Homogeneous
	0.005	0.005–0.01	n.a.	Homogeneous
A2 (9–46 cm)	0.15	0.15–0.59	3:97	Not homogeneous
	0.02	0.02–0.09	7:93	Homogeneous
	0.005	0.005–0.01	n.a.	Homogeneous
A3 (46–88 cm)	0.1	0.1–0.78	2:98	Not homogeneous
	0.02	0.02–0.06	8:92	Homogeneous
	0.005	0.005–0.01	n.a.	Homogeneous
At (88–138 cm)	0.1	0.1–0.53	3:97	Not homogeneous
	0.02	0.02–0.06	8:92	Homogeneous
	0.005	0.005–0.01	n.a.	Homogeneous
ABt (138–160 cm)	0.1	0.1–0.51	2:98	Not homogeneous
	0.02	0.02–0.06	8:92	Homogeneous
	0.005	0.005–0.01	n.a.	Homogeneous
Btk (160–187 cm)	0.1	0.1–0.45	2:98	Not homogeneous
	0.02	0.02–0.06	8:92	Homogeneous
	0.005	0.005–0.01	n.a.	Homogeneous
Ctk (187–230 cm)	0.1	0.1–0.4	3:97	Not homogeneous
	0.02	0.02–0.06	7:93	Homogeneous
	0.005	0.005–0.01	n.a.	Homogeneous

(continued)

Table 3.3 (continued)

Genetic horizons	c/f limit (mm)	Coarse particles (mm)	c/f ratio (%)	Spatial distribution of coarse particles
2Ck (230–250 cm)	0.1	0.1–0.45	2:98	n.a.
	0.02	0.02–0.03	8:92	n.a.
	0.003	0.003–0.005	n.a.	n.a.
2Ck (250–300 cm)	0.1	0.1–0.6	1:99	n.a.
	0.02	0.02–0.03	8:92	n.a.
	0.003	0.003–0.005	n.a.	n.a.
2Ck (300–350 cm)	0.1	0.1–0.45	1:99	n.a.
	0.02	0.02–0.03	10:90	n.a.
	0.003	0.003–0.005	n.a.	n.a.
Luvic Gleyic Phaeozem				
A1 (0–8 cm)	0.1	0.1–0.45	1:99	Not homogeneous
	0.02	0.02–0.06	10:90	Homogeneous
	0.005	0.005–0.01	n.a.	Homogeneous
A2 (8–34 cm)	0.1	0.1–0.45	5:95	Not homogeneous
	0.02	0.02–0.06	5:95	Homogeneous
	0.005	0.005–0.01	n.a.	Homogeneous
A3 (34–60 cm)	0.1	0.1–0.57	5:95	Not homogeneous
	0.02	0.02–0.06	5:95	Homogeneous
	0.005	0.005–0.01	n.a.	Homogeneous
At (60–118 cm)	0.1	0.1–0.7	7:93	Not homogeneous
	0.02	0.02–0.06	6:94	Homogeneous
	0.005	0.005–0.01	n.a.	Homogeneous
ABt (118–132 cm)	0.1	0.1–0.38	2:98	Not homogeneous
	0.02	0.02–0.06	8:92	Homogeneous
	0.005	0.005–0.01	n.a.	Homogeneous

[a]Not analyzed

Calcic Chernozem

The c/f-related distribution of the soil has a three-tier fractal structure (Fig. 3.4). The lower size limit of the first level is 0.15 mm for the whole profile. The upper limit of the dimension of the coarse fraction varies at different levels from 0.21 to 0.45 mm. The content of the coarse fraction is small, less than 1%, and its spatial distribution is not homogeneous but disordered.

The limit of the second level of c/f-related distribution is 0.02 mm for the entire profile. In the solum, the coarse fraction comprises grains of 0.02–0.1 mm; and in the loess parent material, grains of 0.02–0.09 mm. The content of coarse particles in the solum is 14–15, and 10% in the parent material. The spatial distribution of coarse particles at this level of similarity is homogeneous.

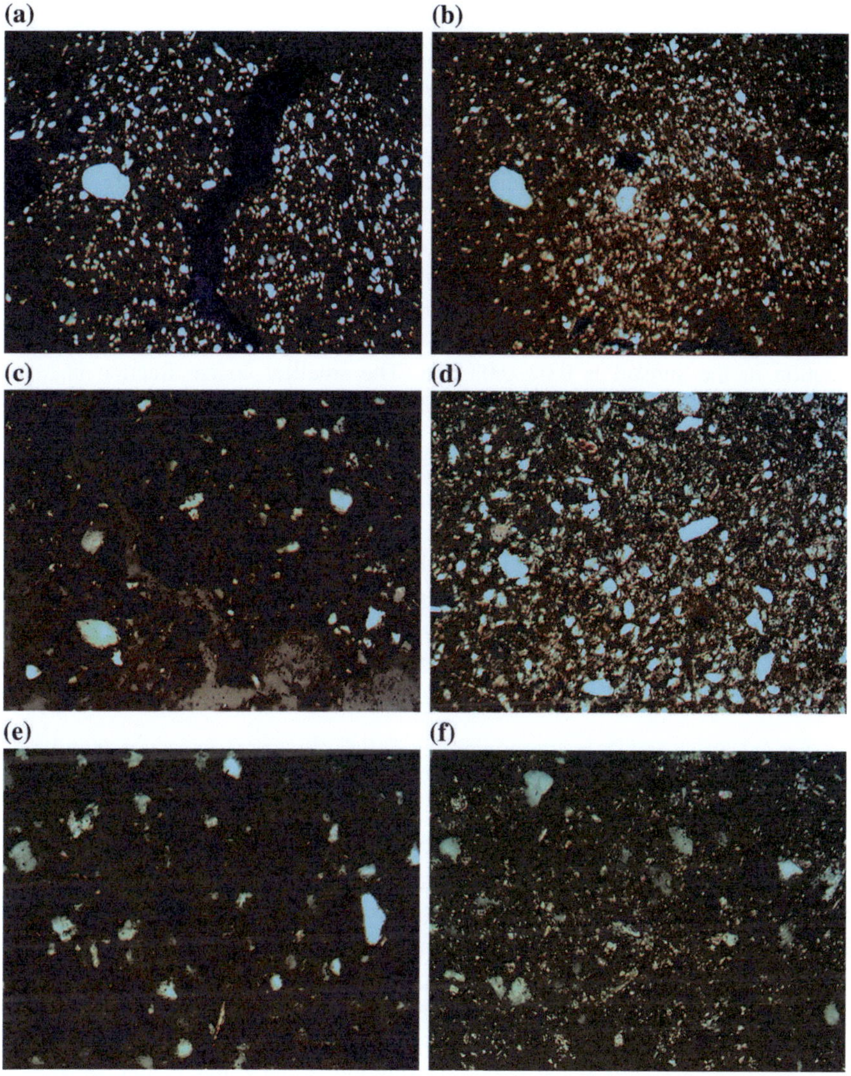

Fig. 3.4 *Calcic Chernozem* fractal structure of the c/f-related distribution of Bk1 horizons (**a**, **c**, **e**) and Ck horizons (**b**, **d**, **f**). The similarity levels have the same values of c/f limit. Microstructure of the first level is shown in **a**, **b** (XPL = with crossed Nicols, width of photograph: 2.47 mm); of the second level in **c**, **d** (XPL, width of photograph: 1 mm); and of the third in **e**, **f** (XPL, width of photograph: 0.46 mm)

The size limit of the third level of c/f-related distribution is 0.005 mm. The coarse fraction consists of particles of 0.005–0.01 mm, both in the solum and in the parent material; the spatial distribution of coarse particles is homogeneous.

These characteristics of the c/f-related distribution of the solum and parent material establish the lithological homogeneity of the whole soil profile.

Luvic Phaeozem

The soil on the slopes of the ravine is characterized by a three-level structure of the related distribution of coarse and fine particles (Fig. 3.5). The limit of the first level is 0.15 mm throughout the profile. The upper limit of the dimension of the coarse fraction varies from 0.4 to 0.78 mm; the content of coarse particles ranges from 2 to 6% in the solum and 1–2% in the subsoil (2Ck); the spatial distribution of the grains of the coarse fraction is not homogeneous but disordered.

The size limit of the second level is 0.02 mm for the entire profile. In surface horizons A1 and A2, the coarse fraction comprises grains of 0.02–0.09 mm—in the underlying horizons including the Pilk—0.02–0.06 mm. The size of the coarse fraction in the subsoil is 0.02–0.03 mm. The smallest coarse fraction is in the surface layer—3%; the biggest is in the 2Ck—10%. The spatial distribution of the grains is homogeneous.

The size limit of the third level is 0.005 mm for the solum including the Ctk, and 0.003 mm for the subsoil. The coarse fraction comprises particles of size 0.005–0.01 mm in the solum, together with the Ctk, and 0.003–0.005 mm in the subsoil 2Ck. The spatial distribution of the particles of the coarse fraction is homogeneous.

The results indicate lithological homogeneity within the solum and a change in the characteristics of the subsoil 2Ck according to indicators of the second and third levels of similarity of c/f-related distribution.

Luvic Gleyic Phaeozem

The microstructure of the *Luvic Gleyic Phaeozem* also shows three levels of similarity (Fig. 3.6). The limit of the first level of the c/f-related distribution is 0.1 mm throughout the profile. The upper limit of the coarse fraction varies from 0.38 to 0.7 mm; the content of coarse grains varies from 1 to 7%; and the spatial distribution of grains is not homogeneous but disordered.

The limit of the second level is 0.02 mm. Most of the coarse grains at this level are in the fraction of 0.02–0.06 mm; the content of coarse grains ranges from 5 to 10% in different horizons; the spatial distribution of the grains is homogeneous.

The limit of the third level is 0.005 mm. The coarse portion is composed of the particles sized 0.005–0.01 mm; the spatial distribution of the coarse material is homogeneous.

The fractal structure indicates lithological homogeneity of the soil profile in the bottom of the ravine.

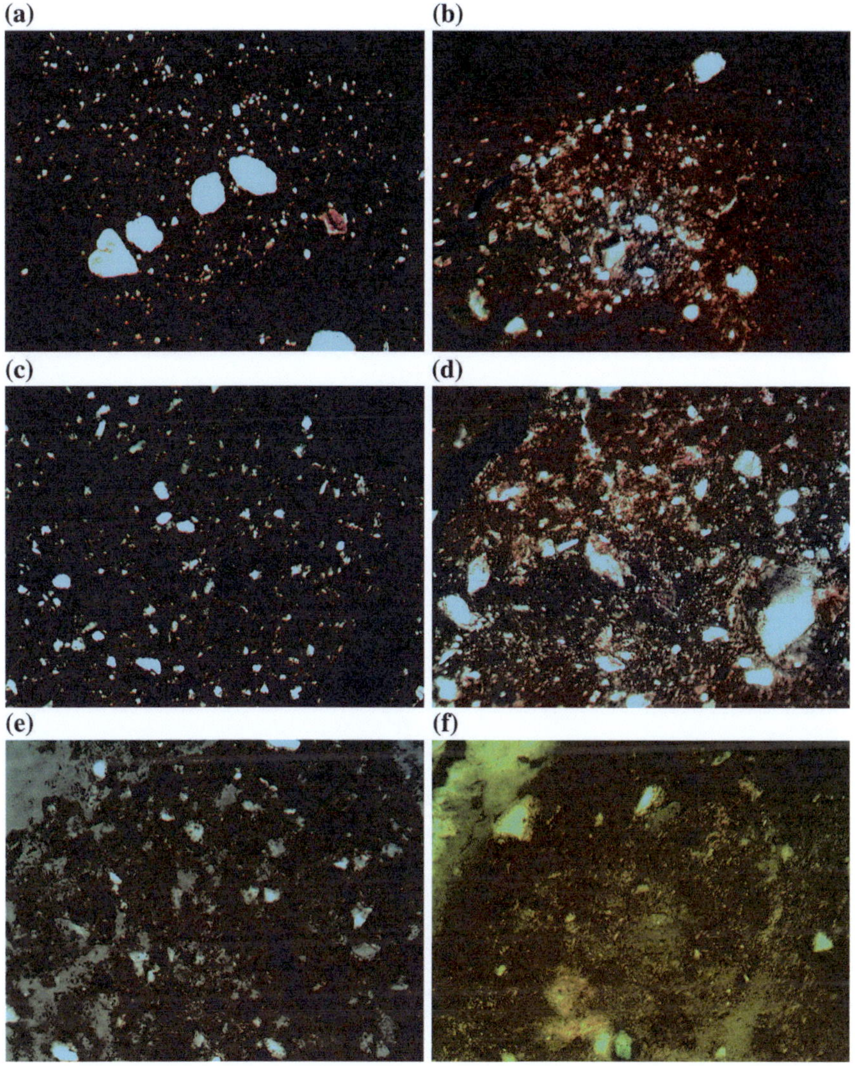

Fig. 3.5 *Luvic Phaeozem* fractal structure of the c/f-related distribution of the horizons A2 (**a**, **c**, **e**) and 2Ck (**b**, **d**, **f**). Compared with the solum, subsoil has different indicators at the second and third levels of the related distribution. The microstructure at the first level is illustrated by **a**, **b** (XPL, width of photograph: 2.47 mm); at the second—**c**, **d** (XPL, width of photograph: 1 mm); and at the third level—**e**, **f** (XPL, width of photograph: 0.46 mm)

(a) (b)

(c) (d)

(e) (f)

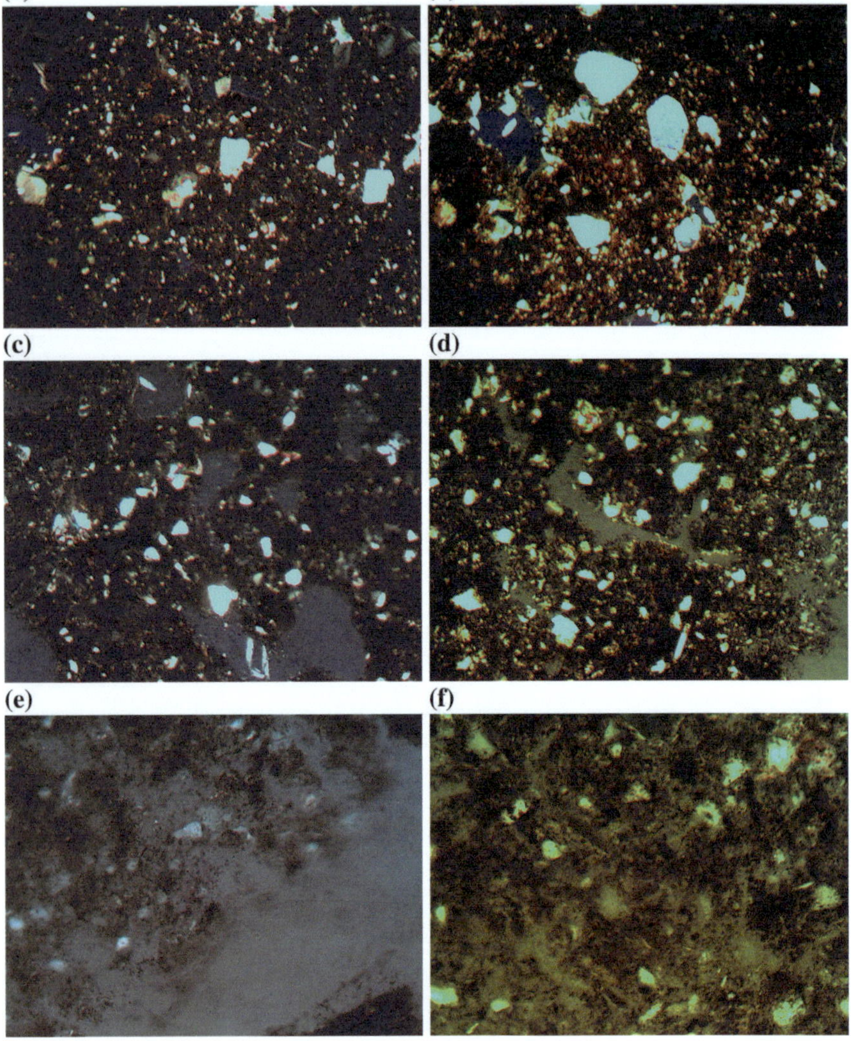

Fig. 3.6 *Luvic Gleyic Phaeozem* fractal structure of the c/f-related distribution of horizons A2 (**a**, **c**, **e**) and ABt (**b**, **d**, **f**). The parameters of selection and the microstructure of the similarity levels are the same for all the powers of the profile. First level—**a**, **b** (XPL, width of photograph: 2.47 mm); the second—**a**, **d** (XPL, width of photograph: 1 mm); and the third—**e**, **f** (XPL, width of photograph: 0.46 mm)

Conclusions

- The studied forest soils of Prysamarya exhibit a three-level fractal organization of the related distribution of microstructure elements. At each level of the relative distribution, the micromorphological picture is similar to the levels selected at a different scale.
- *Calcic Chernozems* near the edge of the gullied land are characterized by lithological homogeneity of the solum and underlying loess parent material. However, *Luvic Phaeozems* situated on the sloping sides of the Hlyboky ravine exhibit a lithological break; the subsoil beneath the solum differs in morphometric parameters at the second and third levels of the related distribution of coarse and fine particles.
- The soil genetic horizons of *Calcic Chernozems, Luvic Phaeozems,* and *Luvic Gleyic Phaeozems* have the similar fractal properties and morphometric characteristics of c/f-related distribution. Lithological similarities of micromorphological organization along the catena may be attributed to the colluvial genesis of their parent material.
- The study of fractal properties allows a step-by-step establishment of lithological homogeneity and lithological breaks within soil profiles without the need for time-consuming determination of particle size distribution and mineralogy—unless these are the aims of the research.

References

Belgard AL (1950) Forest vegetation of south-east of Ukraine. KSU, Kiev (Russian)
Belgard AL (1971) Steppe forestry. Forest Industry, Moscow (Russian)
Belova NA, Travleev AP (1999) Natural forest and steppe soils. Dnepropetrovsk State University, Dnepropetrovsk (Russian)
Bozhokin SV, Parshin DA (2001) Fractals and multifractals. Regular and chaotic dynamics, Izhevsk (Russian)
Bullock P, N Fedoroff, A Jongerius et al (1985) Handbook for soil thin section description. Waine Research Publications, Wolverhampton
Dobrovolsky GV (1983) Methodological guide to soil micromorphology. MSU, Moscow (Russian)
Feder E (1991) Fractals. Mir, Moscow (Russian)
Gagarina EI (2004) Micromorphological method for studying soil. St. Petersburg University Press, St. Petersburg (Russian)
IUSS Working Group WRB (2014) World reference base for soil resources 2014. World Soil Resources Report 106, FAO, Rome
Lavrenko EM (1980) Steppe. Vegetation of the European part of USSR. Nauka, Leningrad (Russian)
Lazarenko PI (1995) Ecological and biological basis of agricultural districts of the territories. Porogy, Kiev (Russian)
Mandelbrot BB (1982) The fractal geometry of nature. Freeman, New York

Morozov AD (2002) Introduction to the theory of fractals. Institute of Computer Science, Izhevsk (Russian)
Parfyonova EI, Yarilova EA (1977) Guide to micromorphological studies in soil science. Nauka, Moscow (Russian)
Rozanov BG (2004) Soil morphology. Academic project. Moscow University Press, Moscow (Russian)
Semenyuta AN (1948) Climate of the south-east of USSR. Sci Bull DSU Dnepropetrovsk 30:181–185 (Russian)
Sobolev SS (1939) Soils of Ukraine and steppe Crimea. Soils of USSR, vol 3. Academy of Sciences of the USSR Publishing House, Moscow-Leningrad (Russian)
Travleev LP (1977) Features of local soil moisture in wooded gullies and their geological and hydrological characteristics. Quest Forest-Steppe Sci Environ 7:31–39 (Russian)
Yakovenko VM (2014) The influence of colluvial processes on macro- and micromorphology of gullied forest soil. Gruntoznavstvo 15(3–4): 74–88 (Ukrainian)

Part II
Assessment of Resources and Risks

Chapter 4
The Last Steppes: New Perspectives on an Old Challenge

David Dent and Zhanguo Bai

Abstract Proxy global assessment of land degradation using satellite AVHRR measurements of NDVI since 1981 reveals that about 22 per cent of the land surface has been degrading over the last thirty years. The areas hardest hit are sub-Equatorial Africa with outliers in the Ethiopian highlands and the Sahel; the Gran Chaco, Pampas, and Patagonia; the steppes from Moldova eastwards through Ukraine, Russia, and Kazakhstan; the Russian far east and northeast China; and swaths of high-latitude forest. All kinds of land use are afflicted. The data reveal both long-term trends and, also, many reversals; most parts of the world have experienced both greening and browning. Land degradation across the steppes corresponds with the best arable soil in the world—*Chernozem*, and its dryland cousin—*Kastanozem*. Both AVHRR and the more detailed MODIS data show that something dramatic happened in the mid-1990s which is not explained by rainfall variability but is probably related to the dismantling of the industrial-scale farming system of the former Soviet Union.

Keywords Land degradation · Global and regional assessment · NDVI time series · Steppe · Chernozem

A Perfect Storm

The land provides 95 per cent of our food and clothing as well as timber, bio-fuel, fresh water, and environmental services that underpin our economy and society. And it's taken for granted. Between 1965 and 1980, the *green revolution* increased crop yields threefold; for a generation, global food production outpaced population growth, world food prices fell in real terms (Fig. 4.1), and political attention turned

D. Dent (✉)
Chestnut Tree Farm, Forncett End, Norfolk NR16 1HT Norwich, UK
e-mail: dentsinengland@hotmail.com

Z. Bai
ISRIC-World Soil Information, 6700 AJ Wageningen, The Netherlands

D. Dent and Y. Dmytruk (eds.), *Soil Science Working for a Living*,
DOI 10.1007/978-3-319-45417-7_4

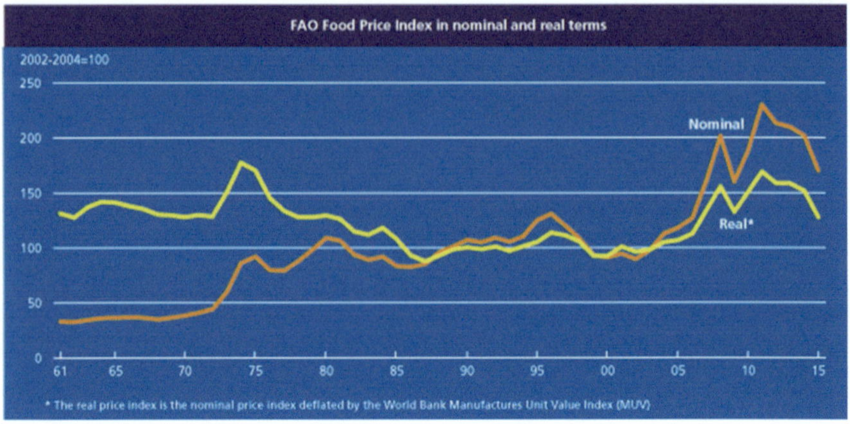

Fig. 4.1 World food prices 1961–2015, after FAO 2016

elsewhere. It is too early to say whether the recent volatility of food prices marks the end of an era but it has concentrated minds, once again, on food and water security.

Today's policy makers face all the old challenges writ large:

- *Burgeoning demand.* By 2050, 75% more food will be needed than now, more than double in developing countries.
- *We are not making any more land.* The area under cereals peaked in 1981, total arable in 1991; the last thirty years have witnessed degradation of one-quarter of the land surface; and tracts of the best land are being lost to urban development and infrastructure.
- *The food system is unsustainable.* The green revolution depended on cheap fuel, fertilizer and irrigation applied to new, responsive crop varieties. Fuel and fertilizer are no longer cheap—notwithstanding the recent drop in oil prices; water resources are overcommitted; yields have levelled off—in some places, they are declining and there is compelling evidence that present management is driving land degradation; climate change is bringing more extreme weather and a rising sea level that threatens great cities and prime farmland.

We are sailing into a perfect storm of ever-growing demand for food, energy and fresh water; the stresses of land degradation and climate change; destabilization of governments that cannot provide for their people's basic needs; and mass migration from poor countries to the rich. And there are no charts—over the last 30 years, knowledge of the land and capacity to face up to land resources issues have atrophied (Dent and Dalal-Clayton 2014).

Land degradation is just one facet of the challenge but a contentious one. Its extent, severity, costs, and impact on food security and the environment are uncertain so it has rarely been at the top of anyone's policy agenda—the only exception was America's response to the Dust Bowl of the 1920s and 1930s. So the investments needed to arrest land degradation are not being made—they are not even known. Uncertainty stems from the lack of a well-accepted definition of land degradation and the kinds of degradation accounted for, uncertainties in the data, and questionable science. We each have our own definition: a loss of capacity to deliver the goods and services that we most value—but these are many and various so it is hard to reach agreement; and whereas landslides, gullies cutting through farmland, dust storms, silted rivers, and salinity are obvious signs of degradation, others like depletion of humus and biodiversity are invisible. Questions that need to be answered in a scientifically justifiable way include the following:

- Is land degradation a global issue or just a collection of local problems?
- Which regions are hardest hit?
- Is it mainly a problem of drylands?
- Is it associated mainly with farming?
- Is it driven by population pressure—or poverty?

New Perspectives

A Totally Unexpected Success

Natural resource surveys have always made good use of new technology that was originally developed for other purposes. The title of a paper in *Advances in Space Science* says it all: '*The exciting and totally unexpected success of AVHRR in applications for which it was never intended*' (Cracknell 2001). AVHRR is the Advanced Very High Resolution Radiometer carried by NOAA weather satellites since 1981—actually very low resolution (4 km), even compared with the Landsat satellite data already being collected at the time, but its large field of vision and daily global coverage make it ideal for global monitoring. And, by chance, the ratio of red to near-infrared radiation measured by the radiometer, the Normalized Difference Vegetation Index (NDVI = (NIR − RED)/(NIR + RED)), is a good measure of photosynthetic capacity and an indicator of vegetation dynamics.

All sorts of applications are possible thanks to consistent, time-series data at spatial resolutions from 20 m to 8 km—the people who maintain these data sets are unsung heroes. The GIMMS data set (Tucker et al. 2004) now comprises 35 years' data corrected for instrument calibration, variations in solar and view zenith angle, and volcanic eruptions. Cloud and haze effects are filtered by taking the highest fortnightly value within composite 8 km^2 blocks of pixels. The derived map of global photosynthetic capacity (Fig. 4.2) shows where things grow and where they do not.

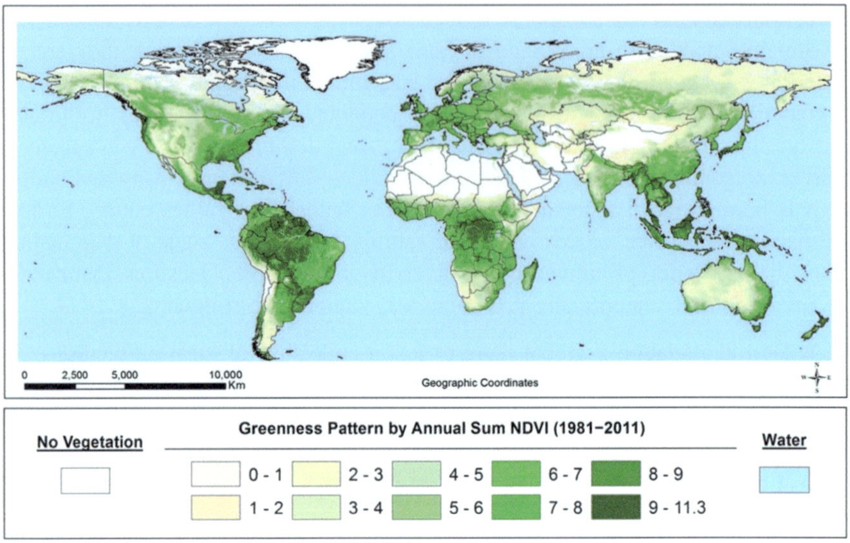

Fig. 4.2 Global photosynthetic capacity by annual sum NDVI 1981–2011

Global Assessment of Land Degradation

If we define land degradation as a long-term loss of ecosystem function and productivity from which the system cannot recover unaided (UNEP 2008), then it may be assessed by proxy using NDVI. But a decreasing trend does not necessarily mean land degradation, or an increasing trend necessarily improvement—vegetation depends on climate, land use and management, large-scale ecosystem disturbances such as fires, and atmospheric fertilization. To interpret NDVI in terms of land degradation, we have to eliminate false alarms.

We can do this for climatic variability because we have a century's compatible data. As a rule of thumb: in drylands where vegetation dynamics are driven by rainfall, declining rain-use efficiency (RUE, calculated as NDVI divided by rainfall) is correlated with land degradation; in humid areas, where vegetation is not as strongly driven by rainfall, NDVI itself is strongly correlated with vegetation dynamics. However, there are no global time series for land use, which has to be examined case by case. And there is an issue of credibility; leaps of deduction (some might say imagination) are required to translate measurements of reflected solar radiation into the information that policy makers want—such as loss of production and environmental services, action needed to arrest these losses, and the economic and political payback for this action.

The first global assessment of land degradation using actual measurements was made in 2007 by analysing the greening and browning revealed by climate-adjusted NDVI (Bai et al. 2008). The aim was to identify black spots where significant biological change is happening—so these could be investigated and attended to on the ground. To provide a more tangible measure of land degradation, NDVI was translated into net primary productivity (NPP) by correlation with NPP measurements made by the satellite-borne moderate-resolution imaging spectroradiometer (MODIS) launched in 2000 (Zhao et al. 2005).

What we saw was quite different from received wisdom that reckoned degradation was worst in the Amazon rainforest, in drylands generally, and the Sahel in particular. In fact, nearly all the usual suspects showed improvement! This intelligence was met with the usual public reaction to a new truth: '*It's not true*'. Then, grudgingly: '*It's against scripture*'. We are still waiting for the final accolade: '*We knew it all along*'.

Figure 4.3 is an updated analysis of land degradation at the 90 per cent significance level (Bai and Dent 2015) using GIMMS NDVI data for 1981–2011 (Pinzon and Tucker 2014). The picture differs from our earlier analysis—not just because of the longer run of data but because of changes in GIMMS processing to better correct for changeovers between successive AVHRR sensors. Importantly, the new calibration does not assume stationarity (i.e. that there is no overall trend in NDVI) but reveals the underlying trends. Procedural changes like this do nothing for our credibility, but we should be right from now on.

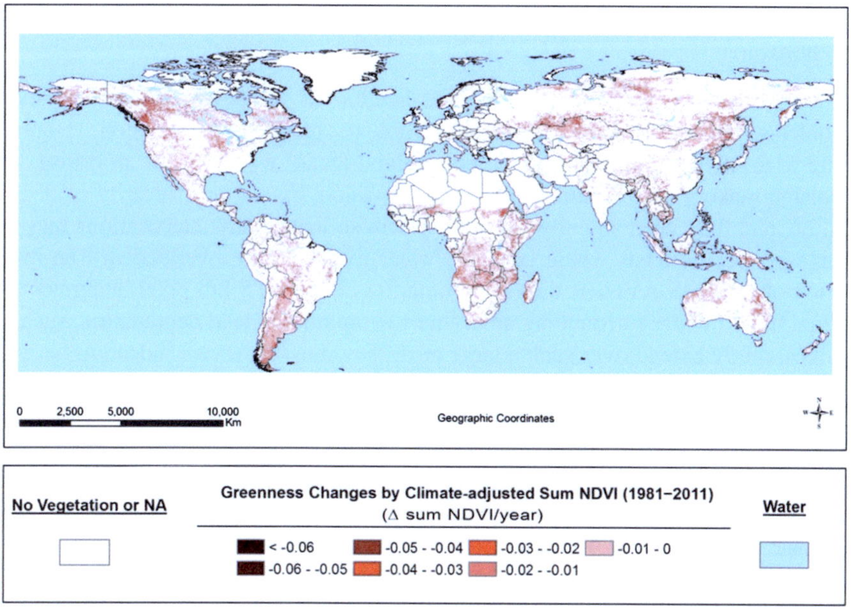

Fig. 4.3 Global climate-adjusted annual sum NDVI 1981–2011

We can now give some answers to our original questions:

- *Land degradation is a global issue.* Our calculations indicate that about 22% of the land has been degrading over the last thirty years, corresponding to a loss of carbon sequestration of about 150 million tonnes but an order of magnitude greater loss of soil organic carbon.
- *The areas hardest hit* are sub-Equatorial Africa with outliers in the Ethiopian highlands and the Sahel; the Gran Chaco, Pampas, and Patagonia; the steppes from Moldova eastwards through Ukraine and Russia to Kazakhstan; the Russian far east and northeast China; and swaths of high-latitude forest.
- *All kinds of land use are afflicted.* Cropland comprises 13% of the global land area but 15% of degrading land; rangeland occupies 29% of the land area but 42% of degrading land; forest occupies 23% of land but 37% of degrading land.
- *Fourteen per cent of land shows an increasing NDVI trend, which may indicate improving conditions.*
- *Comparison of rural population density with land degradation shows no simple pattern. Taking infant mortality and the percentage of young children who are underweight as proxies for poverty, there is some correlation but we need a more rigorous analysis.*
- *Most parts of the world have experienced both greening and browning* (de Jong et al. 2012). With more than 30 years of consistent NDVI data we see both the long-term trends and, also, many reversals. The data reveal both natural, cyclical changes and, also, changes of trend beyond individual drought cycles, and these deviations from the local or regional trend may be a warning of problems on the horizon.

So we have a global issue *and* a collection of pressing regional and local problems. Arresting land degradation, not to mention remediation, requires long-term investment. Budgetary constraints mean we have to prioritise; so decision-makers need to know the precise location, extent, and severity of degradation, and they need an early warning system so that timely interventions may be made in specific areas where land and water resources are overloaded. The resolution of GIMMS AVHRR data is a limitation in that its 8-km pixel integrates the signal from a wider surrounding area. Many symptoms of land degradation, such as gullies, rarely extend over such a large area; they must be severe indeed to be seen against the signal of surrounding unaffected areas. Annually aggregated data also lose much of the detail but use of the Harmonic Analysis of NDVI Time Series algorithm to eliminate the effects of haze, snow, and cloud cover (de Jong et al. 2011) instead of taking the highest fortnightly value of the aggregate 8-km pixel, and integration with climate, soils and land use information, opens the door to more forensic analysis of the wealth of information within the data set. And, for the last fifteen years, newer sensors offer sub-km resolution.

The State of the Steppes

Nowhere highlights misguided policies more dramatically than the grain belt that stretches across the steppes from Moldova eastwards, through the Ukraine and Russia, to Kazakhstan. Figure 4.4 shows the negative changes in climate-adjusted NDVI over 31 years at 90% probability. Between 1981 and 2011, greenness increased by 4.7% across Russia and Kazakhstan and by 4.5% across Moldova and Ukraine—but not across the grain belt.

Comparing the NDVI decline with regional soil patterns, it is clear that the swath of degradation across the steppe corresponds with *Chernozem* soils and their dry-land cousins *Kastanozem*. It is also clear that *Chernozem* elsewhere are faring better than those across the Eurasian steppe (Fig. 4.5). This (Fig. 4.6) is the best arable soil in the world. Surely, extinction in its home range is more than a parochial matter.

In the case of Moldova, both AVHRR and the more detailed MODIS data show that something dramatic happened in the mid-1990s (Fig. 4.7).

This change in vegetation dynamics is not explained by rainfall variability. Taking winter wheat as a representative crop and comparing the national average yield with wheat grown in rotation after an early-harvested predecessor on *Typical chernozem* at the Selectia Experimental Station in the north of the country, we see no downward trend in experimental yields but a dramatic fall in national averages.

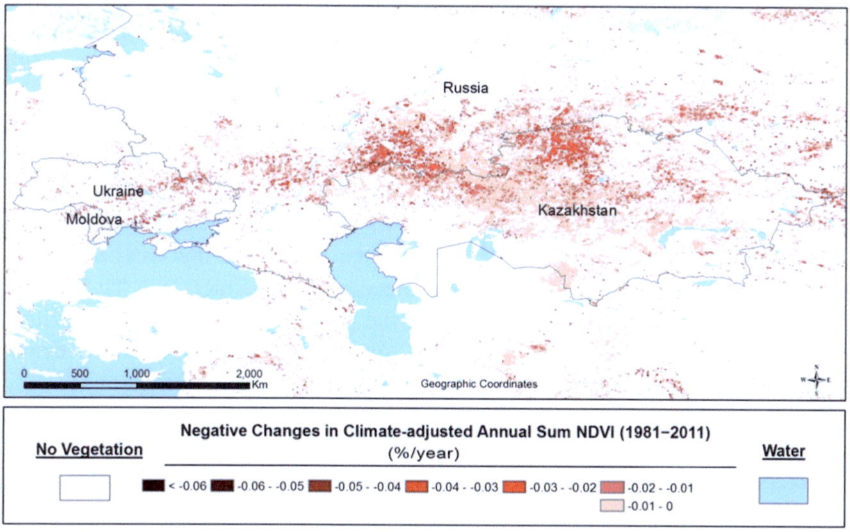

Fig. 4.4 Negative changes in climate-adjusted annual sum NDVI across the steppes, 1981–2011

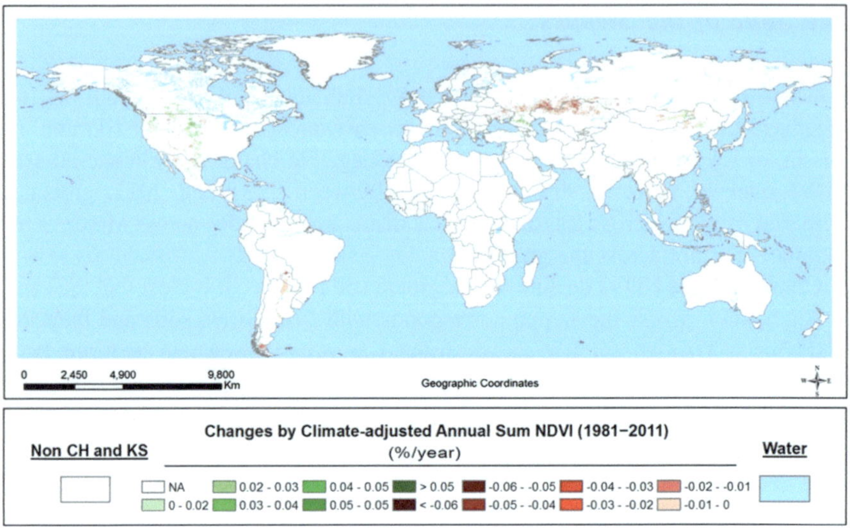

Fig. 4.5 Changes in climate-adjusted annual sum NDVI in the areas of *Chernozem* and *Kastanozem* worldwide, 1981–2011

The national yield was 78 per cent of the experimental station yield over the period 1981–1992: excluding the severe drought years 1996 and 2003, national yields were only 51% of the experimental station yield from 1994 to 2012. The downturn shown in Fig. 4.8 may be related to the dismantling of the industrial-scale farming system of the former Soviet Union and, beginning in 1992, redistribution of the land in small plots to the entire rural population without providing the capital and expertise to farm the land sustainably. Land policies in Ukraine, Russia, and Kazakhstan also appear to be driving the best soil in the world to extinction.

By the end of the twentieth century, under increasingly industrial farming, *Chernozem* had lost 30–40% of the humus which binds the granular structure that renders the soil drought-proof and resilient against erosion by wind and water, which fuels the extraordinary biodiversity that creates soil structure in the first place and accomplishes nutrient cycling and internal pest control, and which constituted in itself a huge reserve of plant nutrients. According to Kovda's (1983) figures, losses of soil organic carbon amounted to c40 t/ha over a century of farming; losses under bare fallow amounted to 50 t/ha in 50 years (Krupenikov et al. 2011). That was three decades ago and things have not improved. The NDVI data show an inexorable decline in productivity over the last twenty years or so. As others have indicated in his symposium, many a field has seen no fertilizer in many a year, let alone farmyard manure. The best soil in the world continues to yield by running down its humus reserves, but a tipping point is approaching when its resilience will be lost. Then, the black earth will wash, or blow, away—there are historical precedents in the dust bowl of the American mid-west and across the steppes themselves.

Fig. 4.6 2 m profile of *Typical chernozem* with its topsoil of unrivalled thickness and humus content, photographed by the senior author in 1974. *Chernozem* is now an endangered species

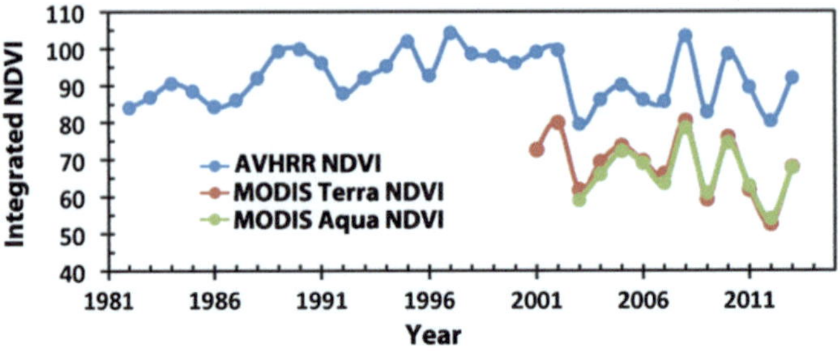

Fig. 4.7 Integrated NDVI over the growing season for part of Moldova

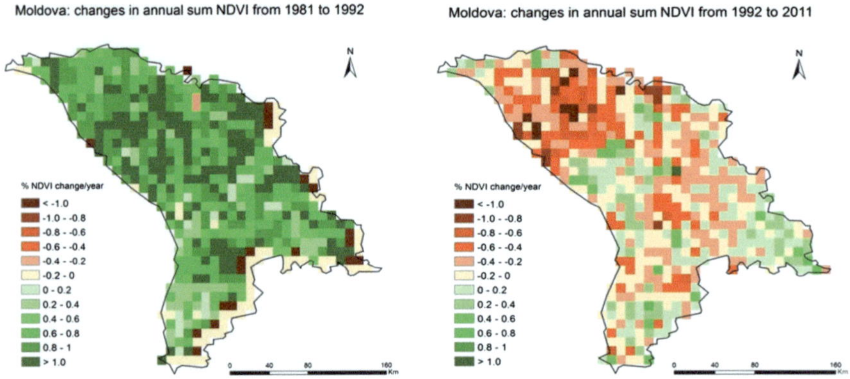

Fig. 4.8 Greening and browning in Moldova pre- and post-1992

Acknowledgments We are indebted to Prof. Boris Boincean for the yield data for Moldova.

References

Bai Z, Dent DL, Olsson L, Schaepman ME (2008) Proxy global assessment of land degradation. Soil Use Manag 24:223–234

Bai Z, Dent DL (2015) Mapping soil degradation by NDVI. In: Lal R (ed) Encyclopedia of soil science, 2nd edn. Taylor and Francis, New York

Cracknell A (2001) The exciting and totally unexpected success of AVHRR in applications for which it was never intended. Adv Space Res 28:23–24

Dent DL, Dalal-Clayton DB (2014) Securing land resources: information needs today and tomorrow. Environmental Governance Series 8, IIED, London

de Jong, R, de Bruin S, Wit A, Schaepman MA, Dent D (2011) Analysis of monotonic greening and browning trends from global NDVI time-series. Remote Sens Environ 115:692–702

de Jong R, Verbesselt J, Schaepman ME, de Bruin S (2012) Trend changes in global greening and browning: contribution of short-term trends to longer term change. Global Change Biol 18:642–655

Kovda VA (ed) (1983) Russian chernozem 100 years after Dokuchaev. Nauka, Moscow (Russian)

Krupenikov IA, Boincean BP, Dent DL (2011) The black earth. Ecological principles for sustainable agriculture on chernozem soils. Springer, Dordrecht

Pinzon J, Tucker CJ (2014) A non-stationary 1981-2012 AVHRR NDVI 3 g time series. Remote Sensing 6:6929–6960. doi:10.3390/rs6086929

Tucker CJ, Pinzon J, Brown M (2004) Global Inventory Modeling and Mapping Studies (GIMMS) satellite drift corrected and NOAA-16 incorporated normalized difference vegetation index (NDVI), monthly 1981–2002. University of Maryland Global Land Cover Facility Data Distribution, Baltimore

UNEP (2008) The land. Ch 3. In: Global Environment Outlook GEO4. UNEP, Nairobi, pp 81–114

Zhao M, Heinsch F, Nemani R, Running S (2005) Improvements of the MODIS terrestrial gross and net primary production global data set. Remote Sens Environ 95:164–176

Chapter 5
Using Multispectral Satellite Imagery for Parameterisation of Eroded Chernozem

Tatiana Byndych

Abstract Multispectral data from the *Sich-2* satellite are used for mapping and parametric description of *Chernozem* soils in the North-Steppe province of Ukraine in order to: (1) delineate soil mapping units and evaluate their spatial structure; (2) quantitatively assess specific soil properties using geo-statistics and multidimensional data processing. Field description of soil profiles from the mapped soil classes demonstrated the efficacy of *K*-means cluster analysis for determining different elements of the soil cover. Maps of humus content, created by statistical and mathematical modelling, enabled the parameterisation of eroded *Chernozem* and further detailed the analysis of the structure of the humus scalar field.

Keywords Soil cover · Chernozem · Geo-information technologies · Remote sensing · Multispectral scanning · Cluster analysis

Introduction

Soil degradation is driven by a combination of natural and socio-economic factors. Its adverse effects on agricultural production, water supply and the environment pose severe problems for sustainable development worldwide. Ukraine is no exception; according to the National report on the state of soil fertility, 38 per cent of farmland (some 15.9 million ha) suffers from erosion, of which 4.5 million ha is moderately or severely eroded (Dept Agricultural Policy and Food 2010). By some estimates, the topsoil has been completely lost from 68 thousand ha (Bulygin and Nearing 1999). Since the main soils concerned are *Chernozem*, an operational system for diagnosis and monitoring of eroding *Chernozem* is an imperative.

Every facet of the landscape can be accurately characterised by conventional soil survey but this is time-consuming, some areas are hard to reach, and it is not

T. Byndych (✉)
National Scientific Centre, Sokolovskyi Institute for Soil Science and Agrochemistry Research, 4 Tchaikovsky St, Kharkiv 61024, Ukraine
e-mail: tanyabyndych@mail.ru

D. Dent and Y. Dmytruk (eds.), *Soil Science Working for a Living*,
DOI 10.1007/978-3-319-45417-7_5

practicable to analyse long-term soil change using field data alone. We can meet the challenge by making use of remote sensing (RS) and geographic information systems (GIS), thanks to continual instrumental improvement and computerised image interpretation in GIS, elaboration of theoretical foundations for assessing the structure of soil cover, and application of geo-statistics to provide testable estimates of reliability. In particular, high-resolution multispectral satellite imagery offers objective data of the detail and geometric and geographic accuracy that we need to diagnose and map soil erosion.

In soil erosion studies, RS data have been used mainly to describe the cover-management factor by means of land-cover classification (Millward and Mersey 1999; Ma et al. 2003), while GIS has been used to derive the topographic factor from DEMs, data interpolation, and calculation of soil erosion loss (Cerri et al. 2001; Bartsch et al. 2002; Wang et al. 2003). Recently, Goldshleger et al. (2010) combined hyperspectral (narrow-band) imagery with ground-penetrating radar and frequency-domain electromagnetic induction to produce detailed, three-dimensional maps of soil degradation. The Remote Sensing of Soil Cover Laboratory of the Sokolovskyi Institute has been applying modern approaches to the delineation and description of degraded soils since 2001, but we have been held back without access to much of the latest RS data—so we are keen to investigate new data from the Ukrainian space industry.

Materials and Methods

We have tested modern technologies for mapping eroded *Chernozem* using Ukrainian *Sich-2* satellite data that provide digital images of the Earth's surface in the panchromatic and multispectral bands with a resolution of 8.2 m, and in the mid-infrared range with a resolution of 41.4 m (Table 5.1). The imagery may be used to monitor land use and crop rotation compliance, calculate the total area of crops, identify stages of crop development, detect soil erosion and salinity, study natural conditions that affect agricultural activities (waterlogging, sharp changes in relief) and identify agricultural land that has suffered losses due to adverse natural conditions.

The study area is the Rozovka polygon which occupies 300 ha in Yasinovatskiy District of Donetsk Region. It lies within the Donetsk physiographic region of the left-bank Dnieper Northern Steppe province of the Ukraine (Popov et al. 1968)—an upland comprising Carboniferous, Triassic, Permian, Jurassic, Cretaceous, Paleogene and Quaternary sediments. The Quaternary cover includes loess, gravelly loam and clay colluvium, and sand and clay alluvium.

Imagery was acquired on 19 July 2012 from the bare, dry soil surface, after harvest and a month after the last rainfall. Research included statistical analysis of the image, creation of a provisional soil map and system of soil sampling, field investigation of the soil pattern and laboratory analysis of soil samples, expert assessment of image complexity and analytical results as the basis for image

Table 5.1 Critical parameters of the Sich-2 satellite (http://www.nkau.gov.ua)

Parameters	Units
Mass	176 kg
Orbit: height, inclination	700 km, 98.24°
Stabilisation	Triaxial active
Maximum deviation angle from nadir	30°
Multispectral scanner—spectral ranges	
Panchromatic	0.51–0.90 μm
Multispectral	0.51–0.59; 0.61–0.68; 0.80–0.89 μm
Pixel step projection in nadir	~8.2 m
Swath width in nadir	~48.8 km
Mid-infrared scanner	
Width of spectral ranges	1.51–1.70 μm
Pixel step projection in nadir	41.4 m
Swath width in nadir	58 km

classification and soil cover models, derivation of mathematical models describing the relationship between optical characteristics of soil and other soil attributes, parameterisation and geo-statistical analysis of the spatial variation of soil indicators, and extrapolation procedures based on interpretation of spectral signatures. Earlier research had revealed regional changes in the optical characteristics of *Chernozem* related to physical and chemical attributes that allowed us to delineate areas suffering from soil erosion or salinisation. We take it that the contrast within the imagery is linked with the heterogeneity of soil parent material and the intensity of degradation processes such as soil erosion—for instance by visual observation of high-contrast areas, elongated downslope.

The imagery afforded contrast enough to delineate soil mapping units and, further, to attempt a parametric description of selected properties of degraded soils within these delineations. Preliminary data analysis included collecting thematic information about the study area and schematic linkage of thematic data with major interpretive features of the satellite imagery: brightness and structure of the image (the size and shape of features of the land surface and the nature of brightness distribution within them).

With the aid of a GPS, 75 soil samples were collected from the 0–10 cm layer, and 7 soil pits were dug to characterise the soils in the field. Samples were collected according to Soil Survey Standards of Ukraine (ISO 10694-1995, DSTU 4728-2007, DSTU 4730-2007). Analytical data were compiled in a regional database. Statistical and data processing methods used GIS TNT-lite for geo-referencing of space images, NDVI calculation, primary image processing, transformation, general statistical analysis and image classification; and STATISTICA 10 for variance, correlation and regression analysis.

Results and Discussion

The first stage in decoding the imagery was definition of common statistical indicators and analysis of distribution curves of its optical brightness, which showed that the variation has several modal values and, thus, the image of the bare soil surface encompasses diverse objects. The long dry summer and early harvest of the crops provide a large window of opportunity for use of the optical brightness of the soil surface as an indicator of soil attributes, unaffected by vegetation cover and variation in soil moisture content. Analysis of plots of the optical brightness of the image suggested that the soil cover of the Rozovska polygon could best be depicted by four distinct mapping units.

The provisional soil map (Fig. 5.1) was created in the final stage of image processing by the *K*-means method of cluster analysis. This map was the basis for soil sampling and positioning of soil pits for field observations. Further, we observed correspondence between soil delineations and individual elements of the micro-relief (micro-watershed, breaks in slope, bottom of a drainage hollow, etc.) so the map reflects a certain orderliness of the soil cover; its boundaries separate soil bodies with specific internal structure and variability characteristics.

Following the field examination, analytical determinations of more than 50 indicators of soil properties were conducted. The results, compiled in the regional database, revealed correlation between the brightness of multispectral image in different bands and various soil attributes, including humus content and soil texture (for the infrared range, correlation coefficients (r^2) between the brightness of the soil surface and total humus content and clay content were −0.72 and −0.68, respectively).

Approximating functions were used during regression analysis to express the relationship between the optical brightness of bare soil and its humus content. Point representation of the data showed a nonlinear relationship and an exponential model proved to be the most successful (Fig. 5.2).

This model was characterised by the lowest value of the regression error as calculated by Eq. 5.1 (Kutsenko 2008):

$$\delta = \sqrt{\frac{\sum_{i=1}^{n} (y_i - y_{ai})^2}{n}} \tag{5.1}$$

δ Error of regression
y_i Experimental data on total soil humus content
y_{ai} Estimated parameter using exponential model
n Number of samples

Verification of the derived model, using data that were not used for its calculation, showed good convergence in the main range of values and some understating of calculated data in the range of the minimum and maximum values

Fig. 5.1 Soil map derived from classification of Sich-2 satellite data

(Fig. 5.3); deviations from the calculated values are within the experimental error of the analytical method for determining humus content.

Field description of soil profiles confirmed the effectiveness of the multispectral imagery for the identification of *Chernozem* formed on different parent materials (loess, colluvium derived from shale) and for distinguishing degrees of erosion (weakly, moderately and strongly eroded soils, which are characterised by progressively lower humus contents). The taxonomic units represented by the mapping units in Fig. 5.1 are listed below, according to the Ukrainian soil nomenclature and the World Reference Base (WRB), using the scheme for harmonisation drawn up by Medvedev et al. (2003):

1. *Ordinary chernozem*, medium humus on loess and loess-like parent material, weakly eroded (*Chernic chernozem* in WRB);
2. *Meadow-chernozemic* soil, medium humus depth on loess and loess-like parent material (*Haplic phaeozem*);
3. *Chernozem*, fine clay, moderately and strongly eroded (*Luvic chernozem*);

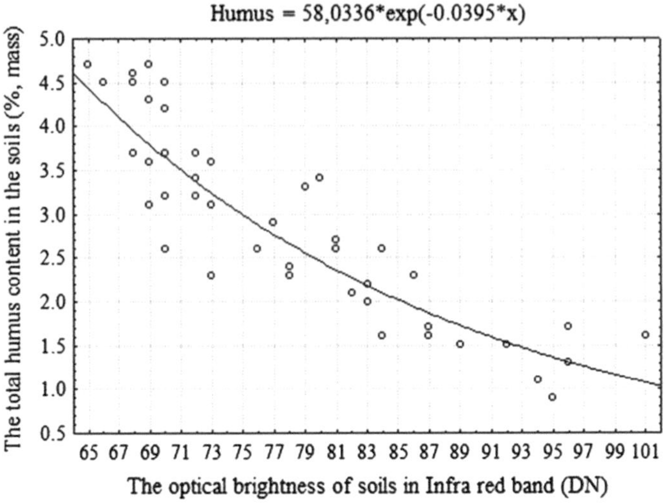

Fig. 5.2 Exponential model of dependence of soil humus content on its optical brightness in the infrared band

4. Rubbley *Chernozem* on non-calcareous rock (colluvium from sandstone or slate) and strongly eroded *(Chernic chernozem)*.

The digital map (Fig. 5.4) was then used to analyse the scalar field structure of this soil indicator: visual field analysis and testing stationarity of the expected value; smoothing functions, determination of the nature of a trend and random function approximation; finding an analytical expression of regionally correlated component and removing the trend; correlation analysis and identification of significant periodic components; analysis of the spectral density of the dispersion; and verification of the allocation and estimates of the indicator distribution parameters.

General conclusions about differences between the selected classes may be drawn from graphs comparing the approximate empirical distribution of total humus content performed for different classes of soils. Data for these graphs were obtained by extraction of the indicator values from the digital map. Figure 5.5 compares the estimated distribution of humus vales with comparable normal distributions. The difference between average values for these classes was about 1.0 per cent of humus. However, the graphs also show a fairly wide parametric range in all of these classes (from 1.8 to 3.5%); distribution of humus content in the soils of the third class departs most from a normal distribution but is characterised by a smaller range of variation.

This information was used to produce robust estimates of the variation in humus content across the study area. We may conclude that classification of the image into four classes is not optimal for depicting the pattern of soil cover in this area. However, this cartographic model may be taken as a starting point.

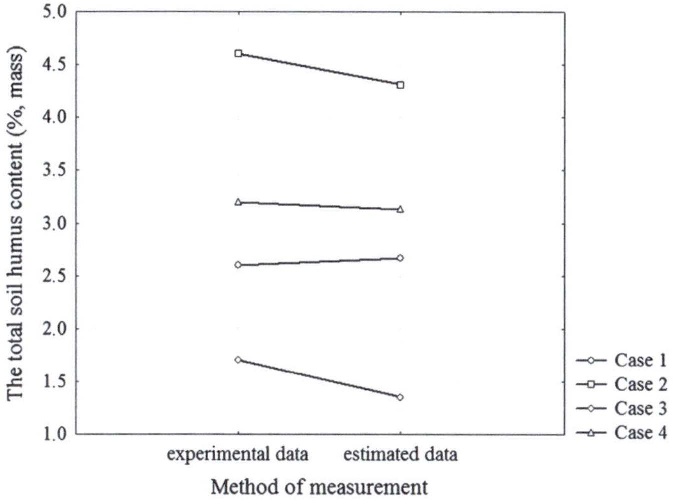

Fig. 5.3 Graphical representation of verification results

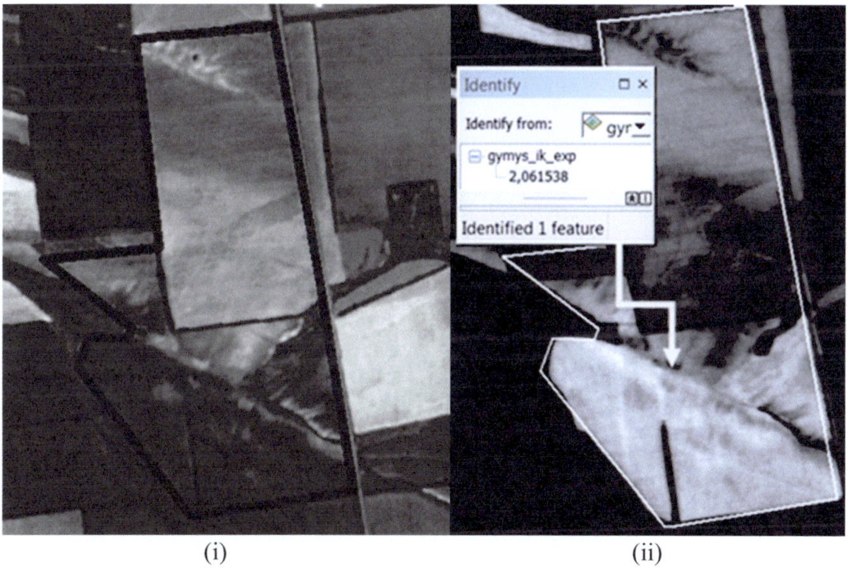

Fig. 5.4 Images of the Rozovka polygon **i** in infrared band and **ii** on the derived digital map of total soil humus content

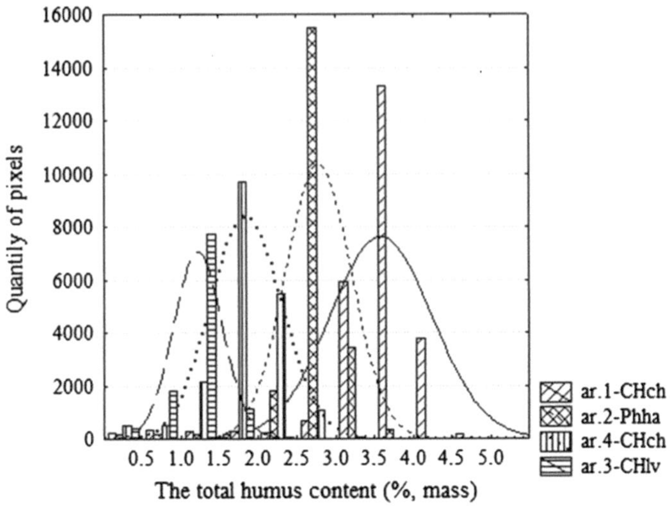

Fig. 5.5 Distribution of total humus content in the mapped soils

Conclusions

- Multispectral data from the Sich-2 satellite may be used with advantage for parametric description of soil attributes in the northern steppe province of Ukraine.
- *K*-means cluster analysis is effective for determination of soil types and degree of soil erosion.
- The strong correlation between the optical brightness of soil in the infrared range and the total content of soil humus was used in simulating the soil cover heterogeneity. Cartograms of total soil humus content provided a basis for quantitative analysis of the structure of the humus scalar field.
- Directions for further research include the following:
  - Systematisation of equations which explain the relationship of optical properties of the soil surface with fundamental physical and chemical properties, so as to make real and useful differentiation of soil quality from satellite imagery;
  - Methods, techniques and rules for the use of remote sensing data for diagnosis and quantitative assessment and mapping of the soil cover in our region.

Acknowledgments I gratefully acknowledge the specialists of Dniprocosmos for providing the satellite imagery free of charge and all the staff of the Remote Sensing of Soil Cover Laboratory of NSC ISSAR for their assistance during the field and analytical researches. I also thank Academician VV Medvedev for his invaluable comments on the manuscript.

References

Bartsch KP, van Miegroet H, Boettinger J, Dobrwolski JP (2002) Using empirical erosion models and GIS to determine erosion risk at Camp Williams. J Soil Water Conserv 57:29–37

Bulygin SY, Nearing M (1999) Formation of ecologically balanced landscapes: the problem of soil erosion. University Publishing House, Kharkov (Russian, English abstract)

Cerri CEP, Dematte JAM, Ballester MVR et al (2001) GIS erosion risk assessment of the Piracicaba River Basin, southeastern Brazil. Mapp Sci Remote Sens 38(3):157–171

Dept Agricultural Policy and Food (2010) National report on the state of soil fertility in Ukraine. Sokolovskyi Institute for Soil Science and Agrochemistry Research, Kharkiv (Ukrainian)

Goldshleger N, Ben-Dor E, Lugassi R, Eshel G (2010) Soil degradation monitoring by remote sensing: examples with three degradation processes. Soil Sci Soc Am J 74(5):1433–1445

Kutsenko MV (2008) Introduction to geographic information systems and environmental modelling. University Publishing House, Kharkiv (Ukrainian)

Ma JW, Xue Y, Ma CF, Wang ZG (2003) A data fusion approach for soil erosion monitoring in the Upper Yangtze River Basin of China based on the USLE model. Int J Remote Sens 24:4777–4789

Medvedev VV, Laktionova TM, Kanash OP (2003) Soils of Ukraine (genesis and agronomic characteristics). University Publishing House, Kharkiv

Millward AA, Mersey JE (1999) Adapting the RUSLE to model soil erosion potential in a mountainous tropical watershed. Catena 38:109–129

Popov VP, Marynich AM, Lan'ko AI (1968) Physical geographic regions of the Ukrainian SSR. Kiev University Publishing House, Kiev (Russian)

Wang G, Gertner G, Fang S, Anderson AB (2003) Mapping multiple variables for predicting soil loss by geostatistical methods with TM images and a slope map. Photogramm Eng Remote Sens 69:889–898

Chapter 6
Pedo-geochemical Assessment of a Holsteinian Occupation Site

Yuriy Dmytruk and Vadim Stepanchuk

Abstract A preliminary geochemical examination of materials from the archaeological sequence at Medzhibozh A reveals evidence of the use of fire. Irregular, dark ochre, ashy circles that look like the remains of hearths were unearthed in 2012 and 2013 in the uppermost artifact-bearing layer associated with episode zv-1 of the Zavadovka stratigraphic layer (Holsteinian MIS 11). We present the methodology, results, and interpretation of a geochemical study of these remains to determine whether they are natural or anthropogenic. The main conclusion is that we are dealing with hearths which, at the moment, are the oldest probable remains of controlled fireplaces in Ukraine.

Keywords Ukraine · Lower Paleolithic · Holsteinian · Early hearth · Geochemical analysis · Soil · Microelements

Introduction

Identification of places of early human activity presents important and unresolved problems, especially the geochemical investigation of the environmental context and its influence on settlement and human activities. This is certainly true of Ukraine where, at present, there are few data on geochemical aspects of Paleolithic sites. One promising direction of geochemical studies is the investigation of early hearths. Until recently, the most ancient reliable evidence of controlled use of fire in Ukraine was from the Middle Paleolithic, mostly since the last interglacial, and this fits with the established picture from other areas of Europe and the Near East (Goldberg et al. 2012). Earlier evidence of the use of fire is scarce; the earliest outside Africa, at Gesher Benot Ya'aqov in Israel, is dated *circa* 800 Kya

Y. Dmytruk (✉)
Yuriy Fedkovych Chernivtsi National University, Chernivtsi, Ukraine
e-mail: yuri.dmy@gmail.com

V. Stepanchuk
Institute of Archaeology, National Academy of Sciences of Ukraine, Kyiv, Ukraine

D. Dent and Y. Dmytruk (eds.), *Soil Science Working for a Living*,
DOI 10.1007/978-3-319-45417-7_6

(Alperson-Afil 2008), but many believe the onset of habitual use should be set between 300 and 400 Kya (Shimelmitz et al. 2014). Recently, evidence of early use of fire in Oldowan sites not far from Ukraine has been unearthed in Dagestan and, likely, in Moldova. However, the use of fire in the Ukrainian Lower Paleolithic remained contentious until recent finds in Medzhibozh where remains of several fires have been recovered in different lithological and cultural horizons, sometimes accompanied by burnt bones and flints.

A systematic approach to establishing the intentional character of a fire includes tests of sediments and biological materials such as ash, charcoal, fragments of bones, charred plants, phytoliths, hair, and skin or, even, unusual materials such as ochre, flint, and rubble (Barbetti 1986). Microelement analysis has been used in archaeology for a long time, especially in the analysis of soil in settlement areas (Wilson et al. 2009; Davidson et al. 2007; Entwistle et al. 2007). Several authors give weight to the increase of K, Zn, Cu, Mn, Pb and Zn in old fire sites (Wilson et al. 2009; Dirix et al. 2013). Cu, Zn, Pb and P are generally indicative of hearths (Wilson et al. 2008; Davidson et al. 2007). Characteristically, burnt wood and bones increase the content of these elements, while a maximum of zinc is often found in charcoal. Peat, turf and dung, used as fuel, significantly affect the content of microelements. At the same time, elements such as Ti, Ca, Na, Sr, Ni, Fe, Al, V and Zr are often geogenic and may not be indicative of human activity (Wilson et al. 2007, 2008).

A priori, compilation of data on the chemical composition of materials from hearths from different places is problematic for several reasons (Barbetti 1986): first, because different fuels were used (trees, bushes, grass, turf, anything else that could burn); secondly, various other materials enter the fire (bones of mammals, birds and fish, as well as shellfish); and thirdly, different periods of time have elapsed since the fire (from tens to hundreds of thousands of years) and so changes in fired material might occur in different ways and with varying intensity, depending on natural conditions of the specific area. The conditions of diagenesis are various so the processes of leaching and general removal of material dominated in some places, while elsewhere either accumulation or alternating processes of accumulation and migration were dominant; weathering and diagenesis of fire-affected materials continue even after the remains of the fire are covered by sediments; the underlying surfaces are also diverse in structure, grain size, and other physical properties and, therefore, react differently to high temperatures; besides, these surfaces were different in chemical composition in the first place.

Charcoal and carbonized particles of organic material are quite stable. Rubification of haematite at high concentrations also helps identify hearths, along with estimation of contents of clay, iron and calcite in soils and sediments (Butler and Dawson 2013). Microelement analysis of materials from ancient settlements should also consider the background amounts of these same chemical elements, their variability, capacity for migration or accumulation, and suitability as indicators. Definition of the local geochemical background is critical; otherwise, it is impossible to distinguish the anthropogenic component of changes in chemical composition of sediments and soils because of interaction between natural factors and anthropogenic impacts (Matschullat et al. 2000; Wells et al. 2000). However, some researchers eschew calculations of geochemical background, focusing on the spatial structure and not on absolute values (Entwistle and Abrahams 1997; Linderholm and Lundberg 1994).

The Lower Paleolithic sites near Medzhibozh are located on the left bank of Southern Bug River, next to Medzhibozh city in the Letichiv District of Khmelnitsky Region (Fig. 6.1). Systematic archaeological surveys have been conducted since 2008 by the Medzhibozh Paleolithic (since 2013—Lower Paleolithic) Expedition of the Institute of Archaeology of the National Academy of Sciences of Ukraine, directed by VN Stepanchuk, in collaboration with SN Ryzhov and OG Pogorilets of the Kyiv National University of Taras Shevchenko and the Mezhibizh State Historical and Cultural Reserve. At present, two localities are known: Medzhibozh 1 and Medzhibozh A. The sequences include Middle and Lower Pleistocene sediments containing faunal remains, stone artifacts, and likely traces of hearths. Our research was conducted at Medzhibozh A (49°25′48″N, 27°23′08″E) where artifacts and other important archaeological remains were found in sediments and soils of Zavadovka (MIS-11), Lubny (MIS-13-15), Martonosha (MIS-73-19), and Shyrokino (MIS-21-35) stratigraphic horizons encompassing sod-podzolic, meadow and marsh soils, and lake-alluvial floodplain sediments (Fig. 6.2). For the first time in Ukraine, inhabited surfaces and evidence of early use of fire dated to c. 400,000 years were recovered. Samples for geochemical analysis were selected from the main horizons as well as archaeological materials (Fig. 6.3; Table 6.1).

Fig. 6.1 Location of Medzhibozh A, Ukraine

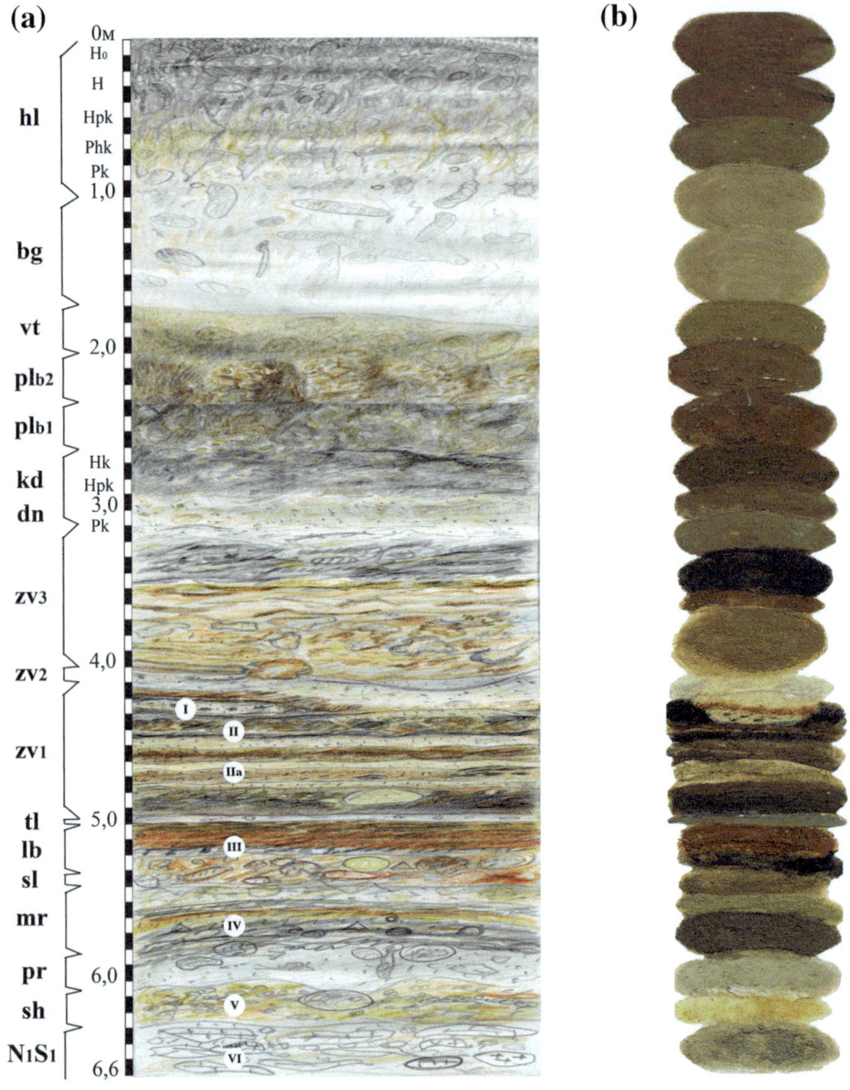

Fig. 6.2 Medzhibozh A. Sketch (**a**) and smears (**b**) of the natural material of Pleistocene deposits (after Matviishina and Karmazynenko 2014). I–VI—position of cultural horizons

Archaeological Context

Medzhibozh A is one of only a few stratified Lower Paleolithic sites discovered in Ukraine. The well-known Korolevo and Maly Rakovets sites are located west of the Carpathians, while Medzhibozh 1 and Medzhibozh A are the most informative stratified Lower Paleolithic localities discovered east of the Carpathian arc

Fig. 6.3 Medzhibozh A, 2012: cultural horizons I (**a**) and II (**b**) of Holsteinian age. *Top left* is the hearth site from the uppermost horizon

(Stepanchuk et al. 2010). Both sites are multilayered and their uppermost cultural horizons are dated to the early phases of the Mindel-Riss interglacial, ca. 380–420 Kya. Paleo-pedology and geomorphology suggest both sites represent remains of repeated hominid occupation of periodically flooded, riparian sites. Medzhibozh 1 has a longer history of investigation and the age of culture-bearing sediments is supported by a wide spectrum of biostratigraphy and geostratigraphy indications (Rekovets et al. 2007). Recent studies provide additional arguments for Mindel-Rissian (MIS 9-11) age of the uppermost horizon and, also, reveal the presence of an earlier assemblage attributed to the Lubny stage (MIS 13-15).

Medzhibozh A was discovered only in 2011. During 2012–2013, the lower part of the sediment sequence was investigated on the steep left bank of the Southern Bug, 30 m above the flood plain. In a 6m-thick lower portion of the sequence, seven distinct horizons of artifact occurrences were identified, separated by sterile layers of gravel and clay. In sediments of Zavadoka (MIS 11), Lubny (MIS 13),

Table 6.1 Position of samples in Medzhibozh A sequence and correlation with archaeological and paleo-pedological data

No.	Sample number (level)	Archaeological finds	Depth from surface (cm)	Basic sediment	Ecological features	Stratigraphic horizon	MIS
Samples of 2012 (*Stepanchuk* 2014)							
1	1	–	20–30	Silty light loamy soil	Modern soil	A	MIS 1
2	2	–	140–160	Sand and silty light loam	Periglacial loess	C	MIS 3
3	3	–	210–220	Sand and silty loam	Buried brown steppe soil	vt	
4	5	–	380	Laminated loamy material	Floodplain deposits	zv-3	MIS 9
5	8	–	405	Light-coloured sand	Alluvium of cold phase	zv-2	MIS 10
6	7	+ (layer 1)	410–415	Dark-coloured sandy clay loam	Warm-phase floodplain deposits interbedded with sands	zv-1	MIS 11
7	6	− (beneath layer 1)	420	Loam			
8	9	− (beneath layer 1)	415	Loam			
9	4	+ (layer 2)	490	Red-brown loam			
10	10		490				
11	11	–	510–515	Meadow lime	Hydromorphic soil	tl?	MIS 12
12	12	+	565	Dark-coloured sandy material	Floodplain soil deposits of warm phase	mr?	MIS 17-19

(continued)

Table 6.1 (continued)

No.	Sample number (level)	Archaeological finds	Depth from surface (cm)	Basic sediment	Ecological features	Stratigraphic horizon	MIS
Samples of 2013							
13	10A	+ (the top of layer 1)	410	Dark-coloured loam	Floodplain deposits of warm phase	zv-1	MIS 11
14	11A	+ (the top of layer 1)	410	Sandy material with charcoal (?)			
15	11B	+ (layer 1)	415	Dark-coloured clay loam			
16	11C	+ (layer 1)	415	Burnt (?) bone			
17	4A	+ (layer 2)	490	Dark-coloured loam mixed with ash (?)	Floodplain deposits of warm phase		
18	12A	–	560	Ferruginous sandy material		mr?	MIS 17-19
19	13A	+ (~ layer 4)	570	Brownish black loam			
20	13B	–	575	Meadow lime			

Martonosha (MIS 17-19), and Shirokino (MIS 21-35) stratigraphic horizons (represented by sod-podzolic, meadow and marshy soils, and by lacustrine-fluvial floodplain materials), stone artifacts have been found, sometimes accompanied by remains of Bovidae, horses, deer, and wild boar.

Evidence of fire at Medzhibozh A is of real interest—an irregular, dark ochre-coloured circle representing remains of a likely hearth was unearthed in 2012 in the uppermost artifact-bearing layer associated with episode zv-1 of Zavadovka stratigraphic horizon, analogous to Holstein (MIS 11) (Fig. 6.3). A similar spot that looks like a hearth with ashy material was discovered at the same level in 2013. It is worth noting that some flint artifacts recovered from this horizon show clear marks of fire. XRF analysis of Medzhibozh artifacts and some burnt archaeological flints from the later Paleolithic context reveals similarity of their chemical composition. Preliminary data on the chemical composition of the soil (abundance of Ca and low content of Mn and Fe) support a firey interpretation of the hearth spots (SP Karmazinenko personal communication 2014), and we should mention the discovery of a similar likely hearth in the Holstein culture-bearing sediments at Medzhibozh 1 (Stepanchuk 2014) where there have also been finds of burnt bones (AM Moigne personal communication 2012). Evidence of fire, particularly controlled use of fire, is unique in the context of the Ukrainian Lower Paleolithic and deserves further substantial argumentation.

Most of the stone artifacts of Medzhibozh A are knapped flint pebbles (Fig. 6.4): microchopping tools and microchoppers with isolated removals, as well as small flakes with intentional retouch or retouch of use; assemblages include no cores, regular products of their knapping, or corresponding tools. There also are products made of quartz, quartzite, limonite, fragments of granite, and other coarse-grained rocks. Some artifact-bearing horizons include many sharp-edged, multifaceted fragments and pieces of vein quartz, representing likely products of intentional crushing and fragmentation. In a typological sense, Medzhibozh A industry is poor; it includes almost no retouched forms and, although there are some more or less stable types of flake tools such as points, billhooks, raclettes, and end-scraper-like forms, these are isolated occurrences. The main purpose of processing was to obtain a sharp-angled edge but not a flake. This was achieved by simple fragmentation, dissection, or crushing on an anvil. It cannot be excluded that such archaism was determined (or significantly intensified) by the available raw materials.

All assemblages of Medzhibozh A, even the most recent with a relatively late age of about 400 Kya, belong to technological Mode 1. Medzhibozh A's Oldowan industry provides a perspective on the cultural and behaviour of the earliest pebble-tool traditions in Europe. And we now know that fire was employed! In this context, the pilot data from the geochemical study of Medzhibozh A is particularly valuable. Sampling was conducted in 2012–2013 with isolated samples from different levels of profile and little series were sampled either directly from the Lower Paleolithic culture-bearing sediments or from underlying or overlying deposits (Table 6.1).

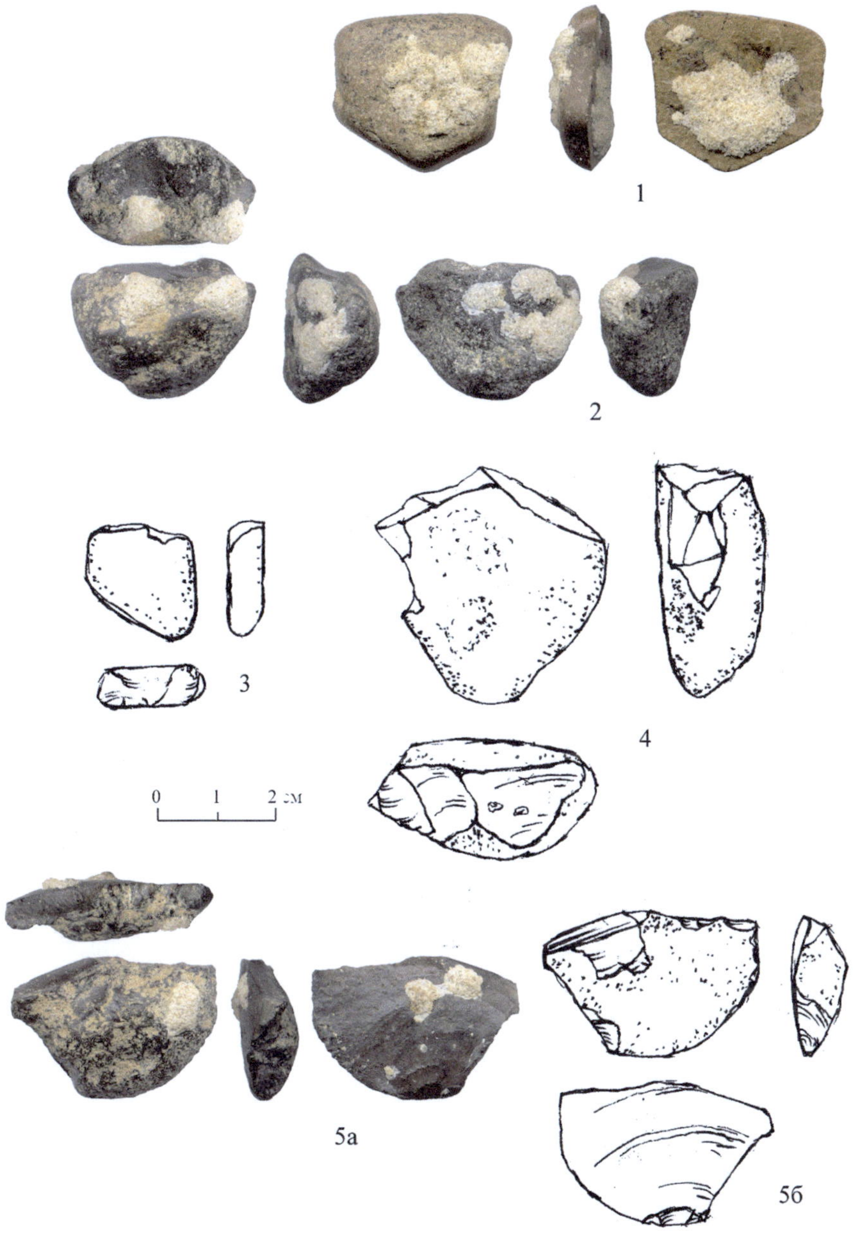

Fig. 6.4 Medzhibozh A, 2012: cultural horizon II. Artifacts: 1—quartzite, 2–5—flint

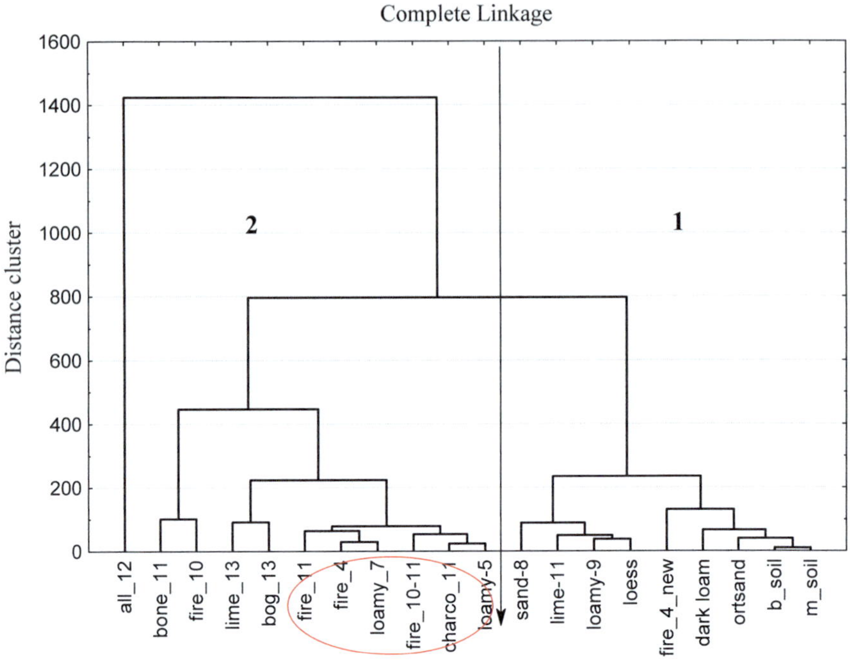

Fig. 6.5 Tree of relationships between the horizons of Medzhibozh A sequence based on total content of microelements

Materials and Methods

The geochemical status of various materials, including soils, has been used to distinguish between possible artifacts and other signs and areas of human activity and natural objects and processes. Accordingly, we assessed the geochemical status of natural objects: soil and its parent material, marl, turf, and ortsand, comparing them with the geochemical status of materials of problematic genesis that might relate to human activity—in this case the material of hearths. The geochemical status of hearth material is compared with the microelement content of buried soils and ash-bearing sediments of mainly sandy nature. Characteristically, the sandy material appeared to be poor in most microelements compared to other sediments. However, their high content of dark-coloured material that visually corresponds to hearths may indicate peculiarities of geochemical status caused by fire. Of course, diagenetic changes that may have occurred over tens and hundreds of thousands of years should be considered in this analysis. The remains of hearths at Medzhibozh are associated with alluvial sediments. It is likely that the hearths were located on a floodplain (they are overlain by alluvium). Obviously, the floods must have been gentle so that some of the hearth could remain in situ, and although large quantities of ash were not found, hearth areas survived and demonstrate surfaces altered by heat.

The sequence of operations was as follows: (1) archaeological analysis and site description, (2) sampling-defined horizons, (3) sample preparation and laboratory tests, (4) mathematical and statistical characterization of results, and (5) generalizations and conclusions. In the last stage, we analyzed contamination of the site by modern, in particular anthropogenic, processes; background geochemical status, and the geochemical status of both natural and likely anthropogenic objects.

The first stage of research must be an assessment of contamination of the site by human activities. In the absence of such impacts, the analysis of chemical elements can be carried out in the same way as for the background territory. Assessment of pollution is performed by calculation of distinct coefficients—a comparative analysis of microelements in horizons of the studied profile with respect to certain background indicators. For this purpose, we used the geoaccumulation index of Rodriguez-Barroso et al. (2009) that gives seven grades of intensity from unpolluted to the very heavily polluted. The index of saturation of soil with chemical elements was determined as described in Dmytruk (2006, 2013) compared with a predefined background microelement content; coefficients of accumulation-dissipation were calculated and, by calculating their geometric mean, indices of accumulation were defined. Theoretically, the background horizon might be:

a. The horizon of alluvium at the base of the profile. However, this material is allochthonous, and therefore, it would not be correct to compare it with geochemical status of subaerial formations.
b. The parent material of modern soil—Bug loess. Comparison with hearth material would not be correct as this analysis will reflect differences between the two different materials. By analogy, the comparative analysis of hearth material with the soil of Vytachiv stage (MIS 3) is also inappropriate.
c. Geochemical status of modern background soil of the profile beyond the site (M1 see Table 6.2) —Again, comparison would be made between different substances formed under different ecological and landscape conditions.
d. Comparative analysis of geochemical status of various substances of undoubtedly natural character (ortsand, marl) and substances of controversial aspect (bones, turf) with each other and with the hearth material.
e. *Our proposed approach*, after a preliminary cluster analysis of microelements of all horizons and materials (Fig. 6.5): comparison of the geochemical status of hearths with those materials that are most distinctive and, also, with the most similar and, also, determination of background values of chemical elements based on individual averages. It is expedient to conduct comparative analysis according to both the absolute content of microelements and by relative indicators (coefficients of saturation of environment by chemical elements), indices (ratio of chemical elements) and conservative chemical elements more stable in environment than other microelements (in our case, only manganese is appropriate).

Results and Discussion

Table 6.2 presents geoaccumulation indices that indicate the background total content of chemical elements, demonstrating that the site is not contaminated with heavy metals. The elevated levels of certain chemical elements in some places are within the range of natural processes of accumulation. Even the surface layers of the modern soil in agricultural landscapes demonstrate metal contents within the range of background values. Because no accumulation of any microelement was identified in profile of the background soil, we may with confidence calculate accumulation factors in relation to any of genetic horizons of background soil, including the upper layer containing most organic matter and which often exhibits an increased concentration of microelements (Table 6.3).

Table 6.2 Total content of microelements (mg/kg) and the corresponding geoaccumulation indices of different horizons of Medzhibozh A sequence

Horizon	Pb	Cu	Ni	Cr	Zn	Mn
Medzhibozh A sequence						
1*. Modern soil	11.7/0.07**	15.1/−0.84	14.4/−0.82	12.0/−0.24	27.4/−0.21	361/−0.21
2. Loess	12.3/0.14	6.25/−1.12	13.6/−0.91	8.40/−0.76	23.1/−0.46	198/−1.07
3. Buried soil	10.6/−0.07	7.25/−0.90	15.2/−0.75	13.4/−0.08	25.9/−0.30	363/−0.20
4. Loam	7.10/−0.65	5.20/−1.38	9.20/−1.47	8.25/−0.78	19.1/−0.74	510/0.29
5. Sand	0.60/−4.22	1.30/−3.38	2.45/−3.38	2.25/−2.66	4.5/−2.82	150/−1.48
6. Dark loam	4.70/−1.25	15.6/0.20	49.2/0.95	8.25/−0.78	15.7/−1.02	379/−0.14
7. Sand under dark	5.30/−1.07	7.35/−0.88	18.8/−0.44	5.30/−1.42	13.7/−1.21	551/0.40
8. Sand	1.50/−2.90	1.75/−2.95	6.95/−1.88	3.25/−2.13	5.30/−2.58	227/−0.88
9. Fire-4 (2012)	12.7/0.19	9.90/−0.45	20.4/−0.32	13.1/−0.12	31.2/−0.03	571/0.45
10. Fire-10 (2012)	8.0/−0.48	27.1/1.00	**76.4/1.58**	10.4/−0.45	27.5/−0.21	**942/1.18**
11. Marl	11.2/0.01	24.2/0.84	18.7/−0.45	15.9/0.16	43.6/0.46	222/−0.91
12. Basal alluvium	9.30/−0.26	9.75/−0.48	33.5/0.39	5.40/−1.39	23.5/−0.44	**1573/1.92**
13. Fire-10–11(2013)	**54.1/2.28****	7.0/−0.95	9.20/−1.47	16.0/0.17	39.4/0.31	521/0.32
14. Charcoal 2013	6.30/−0.82	4.90/−1.47	18.6/−0.46	19.6/0.47	34.5/0.12	500/0.26
15. Fire-11	5.0/−1.16	**64.0/2.24**	10.0/−1.35	14.4/0.02	32.1/0.01	540/0.37
16. Bone-11	6.10/−0.87	10.6/−0.36	12.8/−0.99	16.8/0.24	41.8/0.39	**867/1.06**
17. Fire-4 (2013)	12.5/0.16	**89.0/2.71**	10.2/−1.32	17.3/0.29	60.9/0.94	289/−0.53
18. Ortsand	6.90/−0.69	1.85/−2.87	15.2/−0.75	12.2/−0.22	24.2−0.39	325/−0.36
19. Bog	13.9/0.32	0.55/−4.62	17.7/−0.53	15.3/0.11	24.3/−0.39	723/0.79
20. Marl	8.40/−0.41	12.2/−0.15	19.4/−0.39	25.3/0.83	34.2/0.10	634/0.60
Background soil (profile M1): Podzolized chernozem medium loam on loess						
He	14.6/0.39	7.70/−0.82	12.0/−1.09	10.4/−0.45	28.0/−0.18	284/−0.55
Hi	13.6/0.29	9.0/−0.59	13.0/−0.97	11.5/−0.30	32.6/0.04	281/−0.57
Hp	17.2/0.62	7.80/−0.80	13.4/−0.93	10.4/−0.45	33.0/0.05	276/−0.60
Ph	16.2/0.54	6.65/−1.03	12./−1.03	9.31/−0.61	27.0/−0.24	183/−1.19
Pk	17.2/0.62	8.20/−0.73	16.0/−0.67	11.5/−0.30	34.7/0.13	211/−0.98

*12 upper horizons—samples selected in 2012, 13 and below—in 2013; numbers correspond the order of position of horizons in Table 6.1

**1–2 moderately polluted; 2–3 moderate–moderately polluted (according to Rodriges-Barroso 2009)

Table 6.3 Coefficients of accumulation of microelements in the Medzhibozh A sequence in relation to the upper humus horizon of profile M1

Horizons	Pb	Cu	Ni	Cr	Zn	Mn
Medzhibozh A sequence						
1. Modern soil	1.70	0.85	0.95	1.29	1.44	1.13
2. Loess 140**	1.79	0.71	0.89	0.91	1.21	0.62
3. Buried soil 210	1.54	0.82	1.00	1.44	1.36	1.14
4. Loam 380	1.03	0.59	0.60	0.89	1.00	1.60
5. Sand 405	0.09	0.15	0.16	0.24	0.24	0.47
6. Dark loam 410	0.68	1.76	**3.23**	0.89	0.82	1.19
7. Sand under dark 420	0.77	0.83	1.23	0.57	0.72	1.72
8. Sand 415	0.22	0.20	0.46	0.35	0.28	0.71
9. Fire-4 490	1.85	1.12	1.34	1.41	1.63	1.79
10. Fire-10 490	1.16	**3.06**	**5.02**	1.12	1.44	**2.95**
11. Marl 510	1.63	**2.74**	1.23	1.71	**2.28**	0.69
12. Basal alluvium 565	1.35	1.10	**2.20**	0.58	1.23	**4.92**
13. Fire-10–11 410	**7.87***	0.79	0.60	1.72	**2.06**	1.63
14. Charcoal 410	0.92	0.55	1.22	**2.11**	1.81	1.56
15. Fire-11 415	0.73	**7.24**	0.66	1.55	1.68	1.69
16. Bone 11 415	0.89	1.20	0.84	1.81	**2.19**	**2.71**
17. Fire-4 490	1.82	**10.1**	0.67	1.86	**3.19**	0.90
18. Ortsand 560	1.00	0.21	1.00	1.31	1.27	1.02
19. Bog 570	**2.02**	0.06	1.16	1.65	1.27	**2.26**
20. Marl 575	1.22	1.38	1.27	**2.73**	1.79	1.98

*Horizons of element accumulation in bold
**Depth from surface, cm

The accumulation of some elements is not associated with modern anthropogenic processes because the horizons that showed accumulation occur at depths greater than 4 m from surface. The increased quantities of individual microelements must be the result of either natural processes or, possibly, much earlier human activity.

The homogeneity of archaeological sequences is disturbed more often than natural areas—although natural sites also exhibit variability of content of microelements because of leaching, transport, weathering, and other possible processes. The heterogeneity of natural landscapes may be estimated so that the anthropogenic contribution to the content of microelements is determined accurately. To do this, it is advisable to analyze properties such as soil organic matter, acidity, and texture. One of the statistical indices that helps distinguish between natural and anthropogenic content of microelements is the variability (V, %) of their content (Table 6.4). Analyzing the variability of microelement content of specific materials (background soil, sandy material of horizons of Medzhibozh A sequence, hearth material), this difference is especially significant for lead and copper, whose variability exceeds 100%; zinc also appears to be characteristic of hearths. Thus, the largest variability of microelements is typical for samples identified as a material of hearths (Tables 6.4 and 6.5). There are several reasons for this: (1) varying intensity of diagenetic changes; (2) hearth size and duration of their operation could be different; (3) in some cases, only ashes remained, in others only burnt material; (4) diverse composition of combustible material used for the fire. The maximum content of most microelements (except lead) is identified in the hearths.

In general, the variability of the total content of microelements in natural materials is less than in the anthropogenically altered; the Medzhibozh A sequence is characterized by increased variability of microelement content because of the inclusion of various horizons and materials, including those that we recognize as hearths (Table 6.5).

Table 6.4 Statistical indicators of the total content of microelements (mg/kg dry mass)

ME	Mean	Geometric mean	Median	Min	Max	Range	SD	Skewness*	Kurtosis	V%
All horizons, Medzhibozh A sequence										
Pb	10.4	**7.44**	8.20	0.60	54.1	53.5	10.9	3.63	15.0	105
Cu	16.0	**8.11**	8.55	0.55	89.0	88.4	22.2	2.57	6.53	138
Ni	19.6	**15.3**	15.2	2.45	76.4	74.0	16.7	2.51	7.02	85.2
Cr	12.1	10.4	12.6	2.25	25.3	23.0	5.80	0.18	0.00	47.8
Zn	27.6	23.7	26.6	4.50	60.9	56.4	13.2	0.45	1.06	48.0
Mn	522	**445**	505	150	1573	1423	328	1.85	4.70	62.8
Background soil										
Pb	15.8	15.7	16.2	13.6	17.2	3.60	1.61	−0.56	−1.99	10.21
Cu	7.87	7.83	7.80	6.65	9.00	2.35	0.85	−0.23	1.13	10.83
Ni	13.4	13.3	13.0	12.0	16.0	4.00	1.56	1.62	2.95	11.63
Cr	10.6	10.6	10.4	9.31	11.5	2.19	0.92	−0.50	−0.64	8.63
Zn	31.1	30.9	32.6	27.0	34.7	7.70	3.36	−0.40	−2.58	10.83
Mn	247	243	276	183	284	101	46.8	−0.79	−2.17	18.94
Sandy material (horizons 5, 6, 8, 9)										
Pb	3.63	2.41	3.40	0.60	7.10	6.50	3.08	0.21	−4.01	85.10
Cu	3.90	3.05	3.48	1.30	7.35	6.05	2.89	0.44	−3.32	73.98
Ni	9.35	7.37	8.08	2.45	18.8	16.4	6.90	1.02	1.64	73.76
Cr	4.76	4.23	4.28	2.25	8.25	6.00	2.65	0.83	−0.43	55.62
Zn	10.6	8.89	9.50	4.50	19.1	14.6	7.00	0.48	−3.20	65.76
Mn	360	313	368	150	551	401	201	−0.09	−5.08	55.81
Background content by the cluster of natural materials										
Pb	7.44	5.15	8.75	0.60	12.30	11.70	4.72	−0.49	−1.66	63.4
Cu	9.04	5.58	6.25	1.30	24.20	22.90	8.37	0.89	−0.33	92.6
Ni	16.96	12.77	14.80	2.45	49.20	46.75	14.03	2.03	5.15	82.7
Cr	9.46	7.94	10.20	2.25	15.90	13.65	4.84	−0.42	−1.00	51.2
Zn	21.21	16.84	23.6	4.50	43.6	39.1	12.7	0.25	0.25	60.1
Mn	278	265	276	150	379	229	88.7	−0.17	−1.93	31.9

*Skewness error—0.51; kurtosis error—0.95

Table 6.5 Statistical indicators of the total content of microelements, mg/kg dry mass

Microelements	Entire sequence of Medzhibozh A	Background soil	Sandy loams	Hearths
Pb	7.92 ± 4.14/12.1*	15.8 ± 1.61/3.60	3.62 ± 3.08/6.50	14.8 ± 17.7/**49.4**
Cu	10.3 ± 8.12/25.8	7.87 ± 0.85/2.35	3.90 ± 2.88/6.05	31.1 ± 32.7/**84.1**
Ni	23.2 ± 20.8/**74.0**	13.4 ± 1.56/4.0	9.35 ± 6.90/16.4	27.7 ± 25.6/67.2
Cr	8.82 ± 4.28/**13.6**	10.6 ± 0.92/2.19	4.76 ± 2.65/6.0	14.2 ± 3.94/11.4
Zn	21.7 ± 11.0/39.1	31.1 ± 3.36/7.70	10.6 ± 7.0/14.6	34.5 ± 13.8/**45.2**
Mn	504 ± 401/**1423**	247 ± 46.8/101	360 ± 201/401	535 ± 205/653

*Arithmetic mean ± standard deviation, after forward stroke–value range

Figure 6.5 presents a cluster analysis of analytical data on the content of microelements. The result shows two almost equal clusters (11 horizons on the left and 9 on the right) and, separated from them, the horizon of the base of the sequence (alluvial cobbles, gravel and sand). The two clusters represent two characteristic groups of materials—in the right branch mostly natural; in the left branch brought about by complex processes including, probably, human activity.

The right branch cleaves into two further clusters. The lesser contains loamy soil from horizons 8 and 9, Bug loess, and marl from the base of the sequence (12th horizon)—all substances of natural genesis. The larger cluster includes dark loamy soil of the horizon 6, ortsand at a transition between horizons 12 and 13, and both modern and buried Vytachiv soil. Somewhat resembling an ancient hearth, the dark-coloured material of horizon 10 (sampled in 2013) with inclusions of material visually similar to ashes, is linked to the above-mentioned natural materials. Probably, living organisms played a critical role in their formation and, by analogy with soil material, they obviously contain organic compounds. These compounds are complexing agents, and therefore, an increased microelement status can be caused by absorption and maintenance of organic matter.

The left branch is more complicated. It includes the most remote cluster of material that resembles burnt bone and substance visually similar to hearth material (sample of horizon 10, collected in 2012); it is likely that a special part of this hearth is linked with a concentration of remains of bones that were converted in the fire. Placed in the middle is a cluster of natural materials (marl and turf, both samples from the horizon 13). The largest cluster consists of two branches, both of which include substances of both anthropogenic (in our opinion) and also natural genesis. The latter correspond with areas spatially associated with hearths—in particular, the left branch includes material that resembles a hearth (collected in 2012 from the horizon 10, second cultural layer), and a sample of a hearth from the first cultural layer (collected in 2013), alongside sandy material of horizon 7. The left cluster contains material of a sample resembling a hearth (new, discovered in 2013 on the boundary between horizons 10 and 11); and charcoal in sandy material (sample collected in horizon 11 in 2013) which is associated with sandy material of horizon 5.

Setting aside the cluster of marl–turf, where concentration of certain microelements is explained by natural processes of accumulation, most of the materials (bounded by the ellipse) appeared to be anthropogenic. Thus, the geochemical status of this cluster can be considered as a fact of certain past processes, accompanied by the accumulation of chemical elements. There is no obvious reason to regard these processes as pedogenic. The main thing is that these materials, morphologically, lie beyond the range of properties inherent to Holocene-Pleistocene soil formations; and geochemical characteristics confirm their exclusiveness.

It would be quite wrong to take for comparison, as a background, the quantities of microelements found in horizon 12 that lies at the base of the sequence. This is a layer of alluvium influenced by extraneous factors. Application of coefficients of radial differentiation is also impossible, because their values are quite local and have meaning only for a specific sequence. Therefore, for the background content of elements, we took their quantities in materials of horizons represented in the right branch of cluster (average of all 8 horizons, save for material that resembles a hearth).

We tested several ways of defining horizons of scattering–accumulation based upon different approaches to determination of background microelements: (A) their arithmetic mean (or geometric in a case of deviation from the normal distribution); (B) median value; (C) with respect to the conservative element; and (D) with respect to the background content, calculated on average for eight horizons forming the cluster of natural substances (see Fig. 6.5). The indices of saturation (IB) of materials by microelements (Table 6.6) are calculated by determining the geometric mean after measuring coefficients of concentration–dispersion (ratio of content of microelement in given horizon to its background content).

The results obtained with respect to the supposedly conservative element manganese (column C) most often drop out of the general regularity. There are caveats with the use of standard elements as conservative indicators of natural processes: first, the reference element is found in certain proportions with other elements; secondly, the reference element should not respond to inputs from anthropogenic sources; and thirdly, the reference element should be stable—not susceptible to redox, adsorption-desorption, and other diagenetic processes that may affect the composition of sediments. Reference elements should predictably and stably enter with flows from the earth's crust; if there are anthropogenic sources, they can be identified by changes in ratio microelement/reference element. Obviously, for complicated geochemical environments, it is advisable to select as reference element another conservative element (e.g. aluminum, titanium, or zirconium). Manganese is mobile under the reducing conditions that we might expect close to the water table, and sensitive to the action of fire (high content of calcium under reduced iron and manganese is reported for actual hearths).

The greatest similarity appears in the data from calculation of IB in relation to the mean values (and only slight deviation for the median). According to the background content (column D), the results are generally similar to the averages, but the number of deviations increased to four horizons. If we consider the results when all methods for calculating indices provide very similar saturation values,

Table 6.6 Indices of saturation of horizons of Medzhibozh A sequence by microelements

Horizons	A. Mean content*	B. Median	C. With respect to conservative Mn	D. Background for eight horizons**
Modern soil	**1.17**	1.09	**1.66**	**1.15**
Loess	0.83	0.78	**2.28**	0.93
Buried soil	1.04	0.97	**1.43**	**1.15**
Loam	*0.78*	*0.73*	*0.68*	*0.87*
Sand	*0.17*	*0.16*	*0.47*	*0.19*
Dark loam	1.07	1.00	**1.40**	**1.19**
Sand under dark	*0.79*	*0.74*	*0.63*	*0.88*
Sand	*0.29*	*0.27*	*0.54*	*0.32*
Fire-4 (2012)	**1.31**	**1.22**	**1.10**	**1.45**
Fire-10 (2012)	**1.83**	**1.71**	0.90	**2.03**
Marl	**1.37**	**1.28**	**3.59**	**1.52**
Basal alluvium	**1.31**	**1.22**	0.33	**1.46**
Fire 10–11 (2013)	**1.46**	**1.36**	**1.39**	**1.62**
Charcoal 2013	1.08	1.01	1.03	**1.20**
Fire-11	**1.37**	**1.28**	**1.24**	**1.52**
Bone-11	**1.27**	**1.19**	0.64	**1.41**
Fire-4 (2013)	**1.75**	**1.63**	**3.52**	**1.95**
Ortsand	*0.73*	*0.68*	*1.08*	*0.81*
Bog	*0.82*	*0.76*	*0.47*	*0.91*
Marl	**1.45**	**1.35**	1.09	**1.61**

*Mean content and median of all data of Medzhibozh A sequence
**According to the results of cluster analysis (Fig. 6.5)

then accumulation is observed for five horizons, four of them hypothesized as hearth materials: two discovered in 2012 in horizon 10 (at a depth of 490 cm, the second cultural layer); the two discovered in 2013 (at a depth of 410 cm, the first cultural layer); as well as a new case at a depth of 490 cm (second cultural layer).

For the small series and signs that are not normally distributed, nonparametric indicators such as the Spearman's rank correlation coefficient may be employed. Calculated ratios (Table 6.7) show that the most prominent by the total content of microelements are: a hearth at the boundary between horizons 10 and 11, a hearth in horizon 11 and, also, a hearth in horizon 4 (all samples are taken in 2013). Another stable horizon of accumulation is represented by horizon 11, the marl sample collected in 2012; also the marl horizon sampled in 2013 (save for the calculations on option C, assessed as a background content, but at the upper limit and close to the accumulation, IB = 1.09). The formation of marl occurs via the groundwater by precipitation and accompanied by accumulation of microelements. Similarly, the uppermost horizon of the sequence (humus-enriched soil at the modern ground surface) is dominated by accumulative processes, though of a different kind.

Dissipation is inherent in coarse textured, alluvium (horizons 4, 5, 7, and 8). Low content of microelements is also characteristic for ortsand under redox conditions at the boundary between horizons 12 and 13, and material recovered from

Table 6.7 Spearman's rank correlation coefficients for horizons of Medzhibozh A sequence ($p < 0.05$)

	1	2	3	4	5	6	7	8	9	10	11	12	13	14	15	16	17	18	19	20
1		0.6	0.6	0.6	**0.83**	**0.83**	**0.81**	0.7	0.6	**0.83**	**1.0**	0.7	0.2	0.6	**0.89**	0.7	0.7	0.6	0.6	0.7
2	0.6		**0.94**	**0.94**	**0.83**	0.7	0.7	0.7	**0.94**	0.7	0.6	0.7	0.7	**0.83**	0.3	0.7	0.2	**0.94**	**0.94**	0.7
3	0.6	0.9		**1.0**	**0.94**	0.7	0.7	**0.89**	**1.0**	0.7	0.6	0.7	0.6	**0.94**	0.3	**0.89**	0.2	**1.0**	**1.0**	**0.89**
4	0.6	0.9	1.0		**0.94**	0.7	0.7	**0.89**	**1.0**	0.7	0.6	0.7	0.6	**0.94**	0.3	**0.89**	0.2	**1.0**	**1.0**	**0.89**
5	0.8	0.8	0.9	0.9		**0.89**	**0.84**	**0.94**	**0.94**	**0.89**	**0.83**	0.7	0.3	**0.89**	0.6	**0.94**	0.4	**0.94**	**0.94**	**0.94**
6	0.8	0.7	0.7	0.7	0.8		**0.99**	**0.94**	0.7	**1.0**	**0.83**	**0.94**	0.1	0.6	0.6	0.7	0.3	0.7	0.7	0.7
7	0.8	0.7	0.7	0.7	0.8	0.9		**0.90**	0.7	**0.99**	0.8	**0.99**	0.2	0.5	0.5	0.6	0.3	0.7	0.7	0.6
8	0.7	0.7	0.8	0.8	0.9	0.9	0.9		**0.89**	**0.94**	0.7	**0.83**	0.2	0.7	0.5	**0.83**	0.2	**0.89**	**0.89**	**0.83**
9	0.6	0.9	1.0	1.0	0.9	0.7	0.7	0.8		0.7	0.6	0.7	0.6	**0.94**	0.3	**0.89**	0.2	**1.0**	**1.0**	**0.89**
10	0.8	0.7	0.7	0.7	0.8	1.0	0.9	0.9	0.7		**0.83**	**0.94**	0.1	0.6	0.6	0.7	0.3	0.7	0.7	0.7
11	1.0	0.6	0.6	0.6	0.8	0.8	0.8	0.7	0.6	0.8		0.7	0.2	0.6	**0.89**	0.7	0.7	0.6	0.6	0.7
12	0.7	0.7	0.7	0.7	0.7	0.9	0.9	0.8	0.7	0.9	0.7		0.2	0.5	0.4	0.5	0.3	0.7	0.7	0.5
13	0.2	0.7	0.6	0.6	0.3	0.1	0.2	0.2	0.6	0.1	0.2	0.2		0.6	0.1	0.4	0.2	0.6	0.6	0.4
14	0.6	0.8	0.9	0.9	0.8	0.6	0.5	0.7	0.9	0.6	0.6	0.4	0.6		0.4	**0.94**	0.3	**0.94**	**0.94**	**0.94**
15	0.8	0.2	0.3	0.3	0.6	0.6	0.5	0.4	0.3	0.6	0.8	0.4	0.1	0.4		0.6	**0.94**	0.3	0.3	0.6
16	0.7	0.7	0.8	0.8	0.9	0.7	0.6	0.8	0.8	0.7	0.7	0.5	0.4	0.9	0.6		0.5	**0.89**	**0.89**	**1.0**
17	0.7	0.2	0.2	0.2	0.4	0.3	0.3	0.2	0.2	0.3	0.7	0.3	0.2	0.3	0.9	0.5		0.2	0.2	0.5
18	0.6	0.9	1.0	1.0	0.9	0.7	0.7	0.8	1.0	0.7	0.6	0.7	0.6	0.9	0.3	0.8	0.2		**1.0**	**0.89**
19	0.6	0.9	1.0	1.0	0.9	0.7	0.7	0.8	1.0	0.7	0.6	0.7	0.6	0.9	0.3	0.8	0.2	1.0		**0.89**
20	0.7	0.7	0.8	0.8	0.9	0.7	0.6	0.8	0.8	0.7	0.7	0.5	0.4	0.9	0.6	1.0	0.5	0.8	0.8	

*Significantly important coefficients are in bold; numbering of horizons corresponds to the order of their position, according to Table 6.1

horizon 13—apparently buried peat. Our results confirm the patterns of depletion of microelements in peaty and sandy soils. It is noteworthy that amount of microelements in loess, the parent material of the modern soil, is significantly less than in the soil itself. Although there is good evidence that the parent rock serves as a source for chemical elements in soil, soil processes change these features, so it is advisable to assess the actual processes and possible relationships between soil horizons and parent materials for each ecotype.

The indices of saturation of materials by microelements indicate that substrates that we are calling hearths are characterized by a high content of studied elements, among which zinc and copper (biogenic elements whose content is often increased in most types of vegetation) are the most concentrated, while a high concentration of nickel and manganese is infrequent. Consequently, zinc and copper might be primarily recognized as proper indicator microelements of hearth materials.

Conclusions

There are certain problems in diagnosis or confirmation of material of fireplaces, or hearths. Under the term *hearth,* we mean materials that we distinguished by visual correspondence with real hearths. They are detectable, first, by colour and, possibly, by touch or some other sense, and attributed by us to residues of the once-worked hearths. The diagnostic problems arise because of the antiquity of the discovered horizons and probable peculiarities of human activity some 400,000 years ago and, also, due to diagenetic changes even before the hearths were buried by sediments.

Even if enough original post-fire material survived for identification, at the present time there are no definitive criteria for referring such artifacts to a number of genuine hearths. The materials burnt in the hearths are also unknown—whether the fuel was trees and bushes or grass and turf, or whether other substances (including food remains) went into the fire. All these factors can affect the chemical composition of post-fire material and, therefore, make impossible elaboration of unambiguous criteria for its diagnosis in periods that predate the Holocene.

Therefore, it is important in this context to investigate as many artifacts of the same nature and age as possible. Generalization of the database over time, accompanied by better research tools, will allow a more specific approach to the development of criteria for evaluating and diagnosing early hearths. For the present, on the basis of actual data of geochemical (multielement) analysis, we conclude that the materials identified in the sequence of Medzhibozh A most likely represent the remains of ancient hearths and evidence of the systematic, controlled use of fire during this period.

Acknowledgments Medzhibozh A archaeological projects were funded as part of state research program 0109U008921 (2010–2014) of the Stone Age Dept of IA NASU, by the State Fund for Fundamental Research, Ukraine 0113U004362 (F53.5/005-2013).

References

Alperson-Afil N (2008) Continual fire-making by hominins at Gesher Benot Ya'aqov, Israel. Quat Sci Rev 27:1733–1739

Barbetti M (1986) Traces of fire in the archaeological record before one million years ago? J Hum Evol 15:771–781

Butler DH, Dawson PC (2013) Accessing hunter-gatherer site structures using Fourier transform infra-red spectroscopy: applications at a Taltheilei settlement in the Canadian Sub-Arctic. J Archaeol Sci 40:1731–1742

Davidson DA, Wilson CA, Meharg AA et al (2007) The legacy of past manuring practices on soil contamination in remote rural areas. Environ Int 33:78–83

Dirix K, Muchez P, Degryse P et al (2013) Multi-element soil prospection aiding geophysical and archaeological survey on an archaeological site in suburban Sagalassos (SWTurkey). J Archaeol Sci 40:2961–2970

Dmytruk YM (2006) Ecological and geochemical analysis of soil cover in agroecosystems. Ruta, Chernivtsi (Ukrainian)

Dmytruk YM (2013) Evaluation of the geochemical status of soil in agroecosystems, 75–80. In: Transactions of the international scientific conference celebrating 10 years of the Faculty of Natural Sciences and Agroecology, Alecu Russo Balti State University, Republic of Moldova, Oct 10–11 2013. Balti

Entwistle JA, Abrahams PW (1997) Multi-element analysis of soils and sediments from Scottish historical sites. The potential of inductively coupled plasma-mass spectrometry for rapid site investigation. J Archaeol Sci 24:407–416

Entwistle JA, McCaffrey KJW, Dodgson RA (2007) Geostatistical and multi-elemental analysis of soils to interpret land-use history in the Hebrides, Scotland. Geoarchaeology 22(4):391–415

Goldberg P, Dibble H, Berna F et al (2012) New evidence on Neanderthal use of fire: examples from Roc de Marsal and Pech de l'Azé IV. Quat Int 247:325–340

Linderholm J, Lundberg E (1994) Chemical characterization of various archaeological soil samples using main and trace elements determined by inductively coupled plasma atomic emission spectrometry. J Archaeol Sci 21:303–314

Matschullat J, Ottenstein R, Reimann C (2000) Geochemical background: can we calculate it? Environ Geol 39:990–1000

Matviishyna ZM, Karmazynenko SP (2014) Results of paleopedological study of Quaternary deposits of Medzhibozh Palaeolithic, 49–69. In: Mestonahozhdenie "Medzhibozh" i problemy izucheniya nizhnego paleolita Vostochnoevropejskoj ravniny. Terno-graf, Ternopil (Ukrainian)

Rekovets L, Chepalyga A, Povodyrenko V (2007) Geology and mammalian fauna of the Middle Pleistocene site Medzhibozh, Ukraine. Quatern Int 160:70–80

Rodriguez-Barroso MR et al (2009) Evaluation of metal contamination in sediments from north of Morocco: geochemical and statistical approaches. Environ Monit Assess 159:169–181

Shimelmitz R, Kuhn SL, Jelinek AJ et al (2014) 'Fire at will': the emergence of habitual fire use 350,000 years ago. J Hum Evol 77:196–203

Stepanchuk VN (ed) (2014) Medzhybizh locality and problems of Lower Paleolithic studies on the East European plain. Terno-graf, Ternopil (Russian and Ukrainian)

Stepanchuk V, Ryzhov S, Rekovet I, Matviishina Z (2010) The Lower Palaeolithic of Ukraine: current evidence. Quatern Int 223–224:131–142

Wells EC, Terry RE, Parnell JJ et al (2000) Chemical analyses of ancient anthrosols in residential areas at Piedras Negras, Guatemala. J Archaeol Sci 27:449–462

Wilson CA, Davidson DA, Cresser MS (2007) Evaluating the use of multi-element soil analysis in archaeology: a study of a post-medieval croft (olligarth) in Shetland. Atti Soc Tosc Sci Nat Mem A 112:69–77

Wilson CA, Davidson DA, Cresser MS (2008) Multi-element soil analysis: an assessment of its potential as an aid to archaeological interpretation. J Archaeol Sci 35:412–424

Wilson CA, Davidson DA, Cresser MS (2009) An evaluation of the site specificity of soil elemental signatures for identifying and interpreting former functional areas. J Archaeol Sci 36:2327–2334

Chapter 7
Using Geostatistics, DEM and Remote Sensing to Clarify Soil Cover Maps of Ukraine

Vasyl Cherlinka

Abstract The basic soil maps of Ukraine were created during broadscale surveys between 1956 and 1961. There has been further examination of several areas but, even now, there are places without soil information. Comparison of the soil maps with remotely sensed data and digital elevation models shows some significant differences, especially in the definition of contours, and the extent of eroded soils is much increased—so there is a need for new research on spatial and chronological soil changes. For this task, we constructed a 1m-resolution digital elevation model for accurate delineation of all landform elements. Using geostatistics, we can associate soil mapping units with their landform elements and create contoured soil maps for areas where these are lacking. In the same way, it is possible to refine soil mapping units for the rest of the country.

Keywords DEM · Geostatistics · Multinomial logistic regression · Predictive soil mapping

Introduction

The basic soil maps of Ukraine were created by systematic, broadscale surveys over the period 1956–1961. Subsequently, further examination of the soil cover has been undertaken in several areas but, even today, there are places without any soil information. Under the present difficult political and economic situation in Ukraine, we cannot expect to be able to undertake new, large-scale soil surveys.

Comparison of existing soil maps with current data from remote sensing and digital elevation models (DEMs) shows some significant differences, especially in the delineation of contours. Moreover, there have been changes in the soil cover

V. Cherlinka (✉)
Department of Soil Science, Institute of Biology Chemistry and Bio-Resources,
Yuriy Fedkovych Chernivtsi National University, 2 Kozyubyns'ka St,
Chernivtsy 58012, Ukraine
e-mail: v.cherlinka@chnu.edu.ua

D. Dent and Y. Dmytruk (eds.), *Soil Science Working for a Living*,
DOI 10.1007/978-3-319-45417-7_7

since the original surveys—for instance, there has been a significant increase in the extent of soil erosion. Therefore, new research is needed to ascertain spatial and chronological changes in soil cover. One way to achieve this goal is spatial classification. A soil map is actually a set of well-defined classes (Leenhardt et al. 1994; Webster and Oliver 2007), and the initial conditions for the occurrence of a certain class (type of soil) can be analysed and used to predict the location of this class. The same procedure allows us to test existing maps—which often contain errors. A DEM is the foundation for predictive soil mapping, but not all existing DEMs are suitable; only a matrix or raster format can easily be employed, and the quality and type of DEM directly affects the accuracy of basic morphometric variables such as slope, aspect, profile, and plan curvature (Wood 1996; Shary et al. 2002; Florinsky 2012).

Of course, creation of predictive soil maps is no guarantee of accuracy—we also need further fieldwork.

Materials and Methods

The study area, Ridkivtsi, occupies 8.03 × 4.06 km in Northern Bukovina, in western Ukraine (Fig. 7.1a). The area is partly wooded, part agricultural but, also, includes settlement. Elevation ranges from 184 to 354 m; landforms include hill summits and shoulders, eroded slopes of small valleys, river terraces, and bottomland including an abandoned river channel. There is no conventional soil map for this area, but an analogous map of agro-industrial types of soil (hereafter

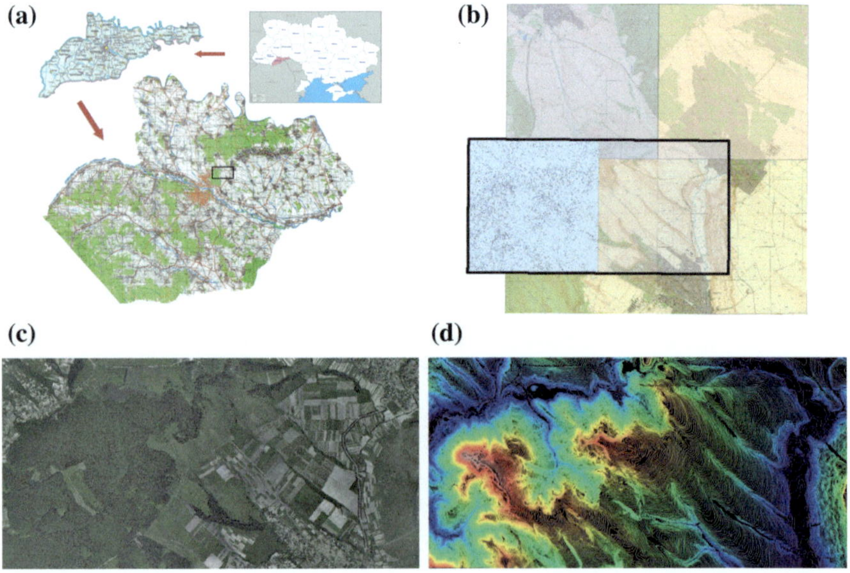

Fig. 7.1 Ridkivtsi data set: location (**a**), topographic maps (**b**), aerial photograph (**c**) and digitized contours (**d**)

referred to *the soil map*) combines some soil types of similar behaviour in *agro-industrial groups*, 18 in all. For the most part, there are no data on the soils so we explore the possibility of creating a forecast or predictive map.

Despite the increasing role of remote sensing and photogrammetry in creating DEMs, large-scale topographic maps are still the main source for terrain modelling in Ukraine—and we have to contend with a lack of modern maps (most date from the 1970/1980s), a variety of coordinate systems, and undisclosed transfer parameters between the coordinate systems (a hangover from the Soviet Union).

As a basis for maps of the study area, we used the Pulkovo 1942/CS63 Zone X2 coordinate system (code EPSG: 7826 (2016)) within the Gauss-Krueger-based coordinate system with a seven-parameter datum definition (Evenden 1990): +proj = tmerc +lat_0 = 0 +lon_0 = 26.5 +k = 1 +x_0 = 2,300,000 +y_0 = −9214.687903777511 +ellps = krass +towgs84 = 23.92, −141.27, −80.9, −0,0.35,0.82, −0.12 +units = m +no_defs. Also, one map fragment was transferred from the local Cartesian coordinate system. Maps scale 1:10,000 and 1:2000 were geo-rectified to the maximum possible number of points, removing most of the

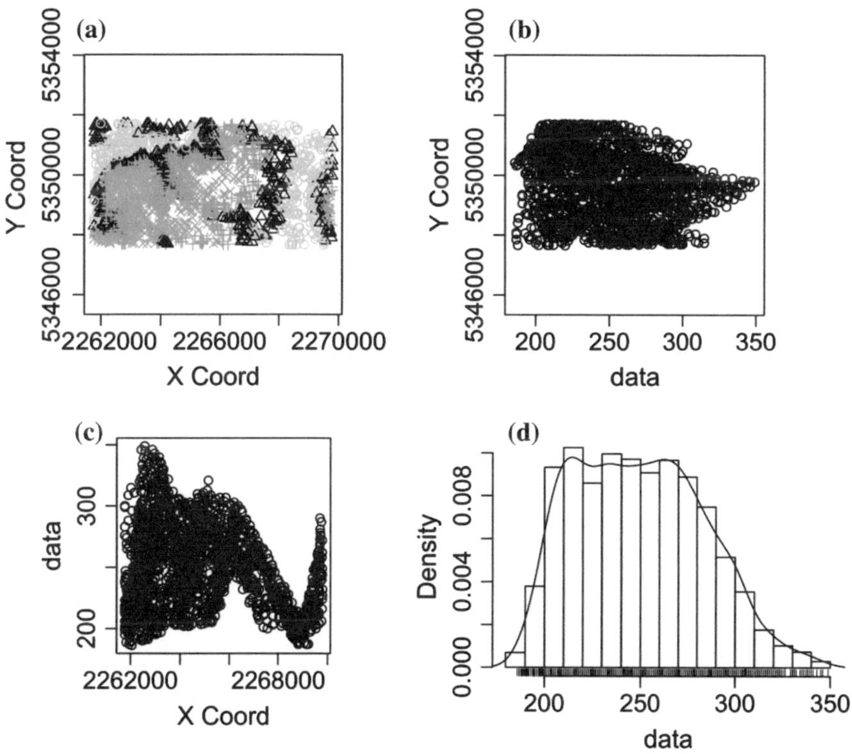

Fig. 7.2 Distribution of the target variable in geographical and feature space: **a** point only in geographical space; **b** distribution of heights of pixels in coordinates; **c** height distribution points on the x-coordinate; **d** approximately normally distributed points

geometric distortion from storage and scanning. Using this set of points and progressive transformation algorithms allowed the production of maps with qualitative geometric parameters (QGIS 2015).

For digitizing 1:10,000 scale topographic maps, we used scanned material with a resolution of 600 dpi 24 bit, and for map scale 1:2000 with 8 bit per channel. Easy Trace 7.99 software (2015) yielded contours with high precision (2910 lines, total length 3686 km). The soil map was scanned and digitized in the same way. Based on the contour map, we generated a point map (Fig. 7.2a) as an ESRI Shape file and, for comparison, we also used SRTM v4 data (2015)—so there were three sources of elevation data. The point map was used to generate multiple versions of the DEM and, then, extract drainage networks for objective selection of the grid-cell size, as described by Brenning (2008) and Smith et al. (2011). The selected DEM was then used to derive various parameters and extract landform classes using a fuzzy *k*-means algorithm and, ultimately, improve the accuracy of soil mapping units (Hengl 2009). The procedure followed methods described by Seijmonsbergen et al. (2011): for generating the DEM, we used GRASS GIS (Neteler and Mitasova 2008; GRASS Development Team 2015) and R (R Core Team 2006); for geostatistics, we used (SAGA GIS 2015) and R; and for visualization, we used QGIS, GRASS GIS, SAGA GIS and R under Debian GNU Linux 7 (Debian GNU 2015).

Results and Discussion

Variogram Modelling and Geostatistical Simulations

Before creating a smooth DEM by spline interpolation, we used more objective variogram modelling with geoR (Ribeiro and Diggle 2001) to assess anisotropy and determine the smoothness of the surface. Average height values of (*Z*) points (using a 10% sub-sample of the original data set) in coordinates showed three peaks of clustering heights (Fig. 7.2b). Comparing these data with Fig. 7.1d shows that the meridional direction of the central part of the map is the most hilly.

Two standard variograms were constructed, showing the target variable (*Z*) gradually changing in all directions (Fig. 7.3a) which means that it can be modelled using an isotropic model (Fig. 7.3b). It is relatively smooth—there is no nugget variation, and spatial autocorrelation operates over a distance of 1 km. The resulting model variogram was used to create multiple implementations of the target variable. In this way, 4 DEMs were constructed (Fig. 7.4) using the Stochastic Conditional Gaussian Simulations algorithm as implemented in the gstat package (Pebesma 2004). It chose the high kappa parameter—1 and 2 (it was originally 0.5). These DEMs were then used for comparison, modelling stream networks, and choice of optimal grid-cell size.

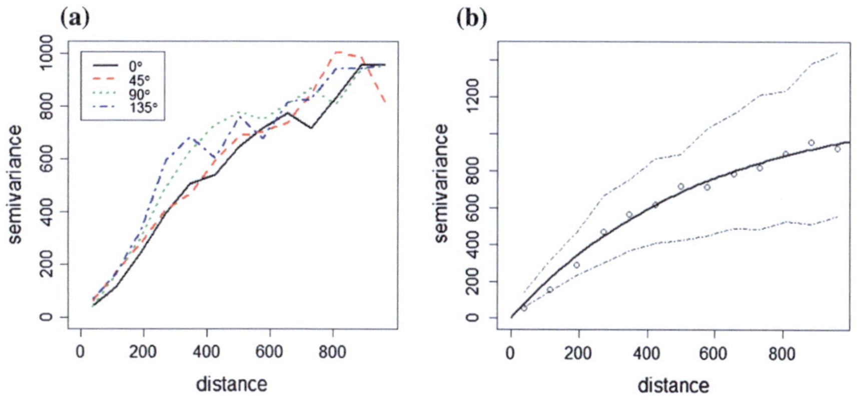

Fig. 7.3 Standard variograms fitted for elevations: **a** anisotropy in four directions; **b** isotropic variogram model fitted using the weighted least squares (WLS) and its confidence bands

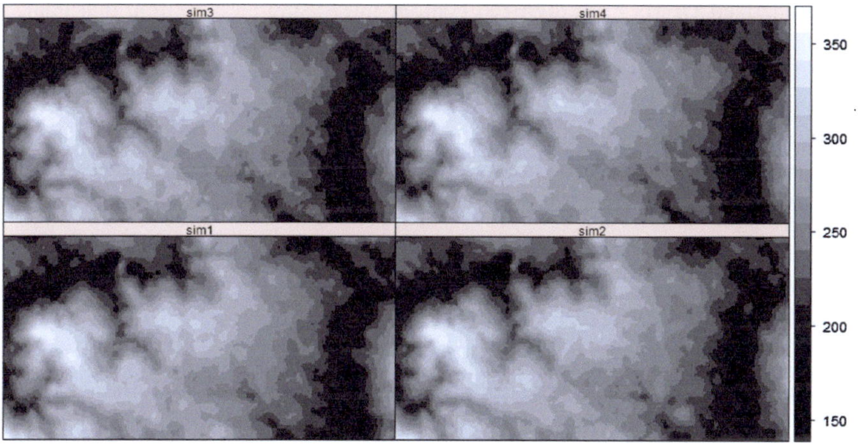

Fig. 7.4 Four realizations of the DEM generated by geostatistical simulation

Select a Grid-Cell Size

For good results, it is necessary to select an appropriate grid-cell size. Following Hutchinson (1996), the grid-cell size was selected according to the accuracy of the derived stream network. By plotting the error of mapping streams against the grid spacing, the grid-cell size that yields the maximum information in the final map may be selected; the optimal grid-cell size is the one where further refinement does not change the accuracy of derived streams.

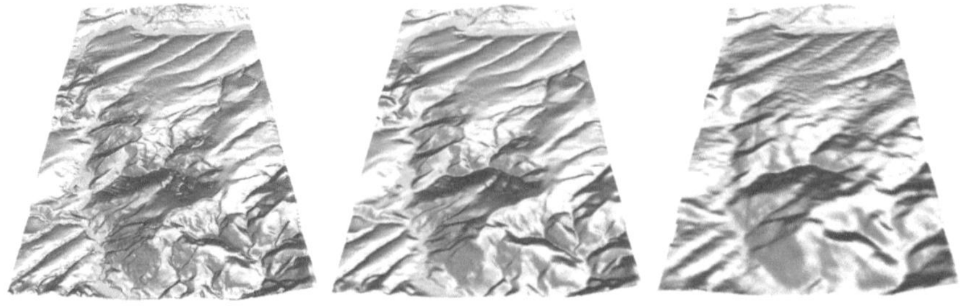

(a) DEM 1m (b) DEM 20m (c) DEM 30m (SRTM)

Fig. 7.5 DEMs of different resolutions generated directly from contours: 1 m (**a**), 20 m (**b**) and the 30 m-resolution DEM derived from SRTM data (**c**)

DEM Generation from Contour Data

We generated DEMs from digitized contour lines using an algorithm that realizes regularized splines with tension (Mitášová and Mitáš 1993; Neteler and Mitasova 2008), generating models with different spatial resolution: 1, 5, 10, 20, and 30 m. Figure 7.5 depicts DEM grid-cell size of 1 and 20 m, and for comparison, SRTM resolution of 30 m (SRTM 2015) and depicts the detail and accuracy of a 1-m resolution model for mapping geomorphic elements. This accuracy allows the use of this particular DEM as a basis for comparisons in future.

DEM Generation from Contours

Based on the DEMs, we derived drainage networks using several grid-cell sizes. We checked whether the spatial location of streams differed significantly from the one derived using the finest resolution by estimating the buffer distance, which shows distance from the channel network derived using the finest resolution and increasingly coarser resolutions (Fig. 7.6). In general, the 5 m-resolution DEM of the stream network is neat and, with ascending of resolutions (from 5 to 30 m), the spatial accuracy of mapping streams is progressively reduced; interestingly, the resolution of 30 m SRTM data is better than the 30m-resolution DEM generated from contours. We selected a 5-m grid-cell size for further study.

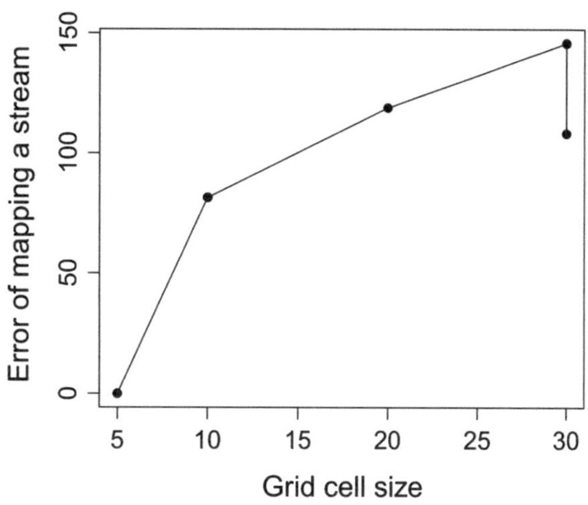

Fig. 7.6 Mean location error of stream network for varying grid-cell size (values of both coordinates in metres)

Extraction of Land Surface Parameters (LSPs) and Unsupervised Extraction of Landforms

The next step was to extract LSPs that will used to explain the distribution of soil mapping units: SAGA topographic wetness index (TWI), valley depth (VDEPTH), insolation (INSOLAT), convergence index (CONVI), SLOPE, ASPECT and plan curvature (PLANC). Then, we extracted land surface objects using the fuzzy *k*-means clustering approach implemented in a statistics package (Venables and Ripley 2002). This will optimally assign each individual pixel to an abstract class; the class centres will be selected in such a way that the within-group sum of squares is minimized. In statistical terms, this is the cluster analysis approach to extraction of features. We converted the LSPs to independent components by using principal component analysis (Fig. 7.7). From the figure, it may be concluded that the LSPs are relatively independent.

To be statistically correct, we should proceed with clustering the principal components instead of using the original predictors. We attempted, unsuccessfully, to determine the optimal number of classes for fuzzy *k*-means clustering using Venables and Ripley (2002); so, pragmatically, tried 12, 15 and 18 classes. The derived maps of predicted classes demonstrate that the size of polygons is well distributed and the polygons are spatially continuous. The question remains: What do these classes mean? The selected classes are poorly correlated with terrain and soils contours, and we will explore this further.

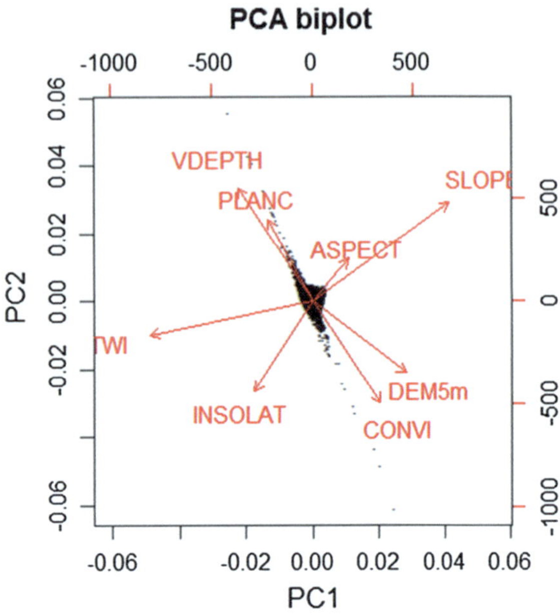

Fig. 7.7 PCA biplot showing first two components derived using eight LSPs

Spatial Prediction of Soil Mapping Units

Multinomial Logistic Regression

The LSPs were used to improve the spatial detail of the existing soil map using multinomial logistic regression which generalizes logistic regression to problems with more than two possible discrete outcomes. That is, a model is used to predict the probabilities of the different possible outcomes of a categorically distributed dependent variable, given a set of independent variables (in our case category-valued). It is implemented in the multinom method of the nnet package (Venables and Ripley 2002) which, given a set of training pixels, iteratively fits logistic models for a number of classes. The output predictions can then be evaluated against the complete geomorphological map to see how well the two maps match and where the most problematic areas are. As noted earlier, seven LSPs were chosen.

Selection of Training Pixels

To refine the existing soil map, we used training pixels from the map to fit the model. Following Hengl (2009), we place the training pixels along the medial axes for polygons of interest. The advantage of using the medial axes is that relatively small polygons will be represented in the training pixels; it is important that the algorithm will minimize the selection of transitional pixels that might well be in either of the two neighbouring soil classes. Stripped of the training pixels, the existing soil map was used as a basis for multinomial logistic regression.

Once we have allocated the training pixels, a logistic regression model was fitted using the nnet package. The next step was to predict the mapping units for Ridkivtsi. Finally, we compared the existing map (Fig. 7.8a) against the map generated using multinomial logistic regression (Fig. 7.8b) and then created a new soil map by the combination of existing and predictive maps (Fig. 7.8c).

To compare the overall fit between the two maps, we used the vcd package (Meyer et al. 2015), especially the kappa index (Li and Zhang 2007; Grinand et al. 2008; Hengl 2009). The match between the two maps is approximately 40 per cent. Given what we know about the quality of soil mapping at this location, we count this a good result; matching of soil maps typically realizes rather low correspondence (Kempen et al. 2009). Table 7.1 summarizes the relationship between actual and predicted mapping units of agro-industrial groups. Groups that occupy a large proportion of the area are generally close in actual and forecast maps but not those

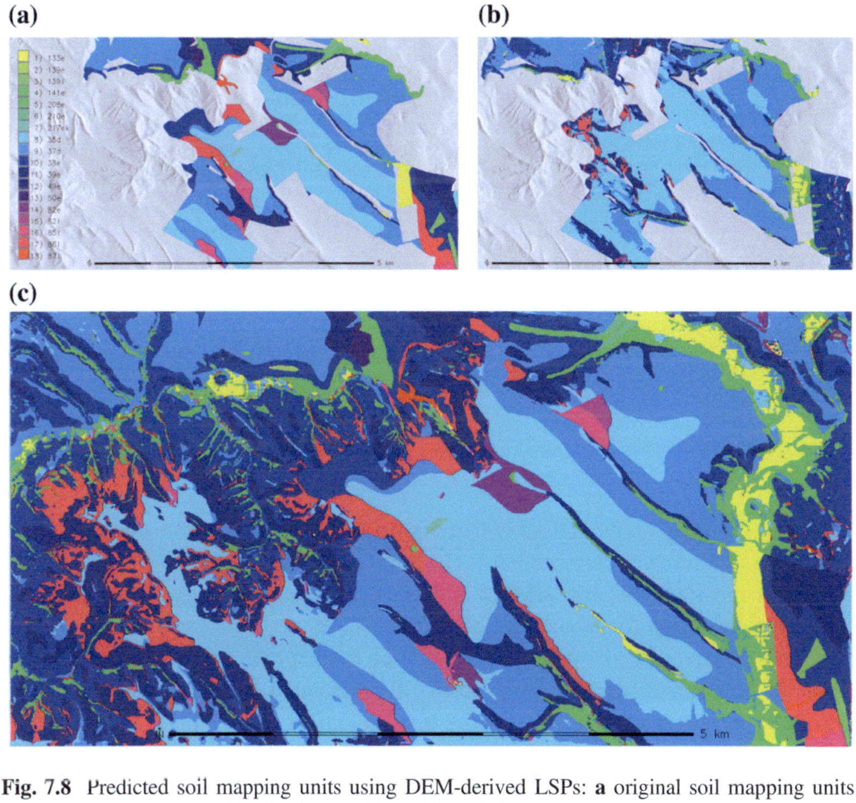

Fig. 7.8 Predicted soil mapping units using DEM-derived LSPs: **a** original soil mapping units; **b** soil mapping units predicted using multinomial logistic regression (mlr); **c** soil mapping units predicted using mlr for whole study area

Table 7.1 Original and predicted mapping units

#	Classes of agro-industrial types of soils	Original map	Predicted map	Combined whole map
		Km^2		
1	133e	0.2048	0.2351	0.9541
2	139e	0.0105	–	0.0105
3	139l	0.0880	0.0064	0.1145
4	141e	0.4657	0.6024	2.2400
5	208e	0.0099	0.0041	0.0166
6	210e	0.0378	–	0.0378
7	217ek	0.0140	0.0001	0.0164
8	35d	3.9359	4.2094	7.0124
9	37d	4.0131	4.0108	6.9790
10	38e	1.2688	1.7417	6.4720
11	39e	1.0453	1.3642	5.0267
12	49e	0.0802	0.0000	0.0809
13	50e	0.0641	0.0523	0.0640
14	82e	0.1961	0.0003	0.3261
15	82l	0.0493	0.0609	0.0510
16	85l	0.3846	0.2510	0.4899
17	86l	0.6369	0.0184	2.0747
18	87l	0.0490	–	0.2060
	No data	20.0986	20.0954	0.4800
	Total	32.65267		

occupying small areas. This might be explained by the fact that we have not included a geological/parent material factor which has a significant impact on the genesis of soils.

Conclusions

The hydrologically correct DEM with a 5m grid-cell size is accurate enough for the extraction of land surface parameters (LSPs). Geostatistics enables the association of certain areas of soil and several LSPs. This, in turn, enabled the creation of predictive soil maps for areas not yet examined.

This is of scientific and practical importance in the present difficult conditions in Ukraine. To expand research in this area, it is planned to extend the set of predictors, especially using both landform and geological parameters.

References

Brenning A (2008) Statistical geocomputing combining R and SAGA: the example of landslide susceptibility analysis with generalized additive models. SAGA Seconds Out Hamburger Beiträge zur Physischen Geographie und Landschaftsökologie 19:23–32

Debian GNU (2015) Linux, the universal operating system. http://www.debian.org

Easy Trace Group (2015) Easy Trace Pro 7.99 software package. http://www.easytrace.com

Evenden G (1990) Proj. 4—Cartographic Projections Library. Source code and documentation available from https://trac.osgeo.org/proj

Florinsky IV (2012) Digital terrain analysis in soil science and geology. Academic Press, Cambridge

GRASS Development Team (2015) Geographic Resources Analysis Support System (GRASS) software. Open Source Geospatial Foundation Project. http://grass.osgeo.org

Grinand C, Arrouays D, Laroche B, Martin MP (2008) Extrapolating regional soil landscapes from an existing soil map: sampling intensity, validation procedures, and integration of spatial context. Geoderma 143(1):180–190

Hengl T (2009) A practical guide to geostatistical mapping. University of Amsterdam

Hutchinson MF (1996) A locally adaptive approach to the interpolation of digital elevation models. In: Third international conference/workshop on integrating GIS and environmental modeling, Santa Fe, NM, 21–26 Jan 1996. National Center for Geographic Information and Analysis, University of California, Santa Barbara http://www.ncgia.ucsb.edu/conf/SANTA_FE_CD-ROM/sf_papers/hutchinson_michael_dem/local.html

Kempen B, Brus DJ, Heuvelink GB, Stoorvogel JJ (2009) Updating the 1:50,000 Dutch soil map using legacy soil data: a multinomial logistic regression approach. Geoderma 151(3):311–326

Leenhardt D, Voltz M, Bornand M, Webster R (1994) Evaluating soil maps for prediction of soil water properties. Eur J Soil Sci 45(3):293–301

Li W, Zhang C (2007) A random-path Markov chain algorithm for simulating categorical soil variables from random point samples. Soil Sci Soc Am J 71(3):656–668

Meyer D, Zeileis A, Hornik K (2015) vcd: Visualizing Categorical Data. R package version 1.4–1. https://cran.r-project.org/web/packages/vcd/

Mitášová H, Mitáš I (1993) Interpolation by regularized spline with tension: I. Theory and implementation. Math Geol 25(6):641–655

Neteler M, Mitasova H (2008) Open source GIS: a grass GIS approach-third edition. Springer, New York

Pebesma EJ (2004) Multivariable geostatistics in S: the gstat package. Comput Geosci 30(7):683–691

QGIS Development Team (2015) QGIS geographic information system. Open Source Geospatial Foundation Project. http://qgis.osgeo.org

R Development Core Team (2006) R: a language and environment for statistical computing. R Foundation for Statistical Computing, Vienna http://www.R-project.org

Ribeiro PJ, Diggle PJ (2001) geoR: a package for geostatistical analysis. R News 1(2):14–18

SAGA Development Team (2015) System for automated geoscientific analyses (SAGA GIS). http://www.saga-gis.org

Seijmonsbergen AC, Hengl T, Anders NS (2011) Semi-automated identification and extraction of geomorphological features using digital elevation data. Geomorphol Mapp: Methods Appl 15:297–335

Shary PA, Sharaya LS, Mitusov AV (2002) Fundamental quantitative methods of land surface analysis. Geoderma 107(1):1–32

Shuttle Radar Topography Mission (SRTM) (2015) http://www2.jpl.nasa.gov/srtm/

Smith MJ, Paron P, Griffiths JS (2011) Geomorphological mapping: methods and applications, vol 15. Elsevier, Amsterdam

The EPSG Geodetic Parameter Dataset (2016). http://www.epsg-registry.org/Venables WN & BD Ripley 2002 Modern applied statistics with S. Springer, New York
Webster R, Oliver MA (2007) Geostatistics for environmental scientists. Wiley, Chichester
Wood J (1996) The geomorphological characterisation of digital elevation models. Doctoral dissertation, University of Leicester

Chapter 8
Making Better Soil Maps Using Models of Tangential Curvature

Yuriy Dmytruk and Olga Stuzhuk

Abstract Modelling enables various sources of information to be integrated in a GIS. But actuality requires a methodology that can compare existing soils and ground cover information with natural descriptions of the land surface, identify errors, specify the limits of scale and definition, determine the effects of erosive processes and, finally, assemble a credible soil map. An algorithm is employed to describe the relationships between soils and landforms using models of tangential curvature that divide the land surface into two planes (bulges and depressions) which determine the distribution and concentration of materials, water and structural elements of the soil. The method places a certain soil type in a positive or negative relief element: where this correlation is broken, clarification is required.

Keywords Soil mapping · Agro-industrial types of soils · Geomorphology · DEM

Introduction

A fully functioning land registry needs the support of relevant soil maps. In the case of agricultural land, the reliability of this information has a direct impact on farm management and productivity. For urban areas, dependence is not so straightforward but soil attributes still determine the value of land, its fitness for purpose and the organization and operation of urban activities—and so have a direct impact on economic efficiency. To improve the relevance and accuracy of the cadastre, there is need for clarification of the existing procedure and correct depiction of the soil pattern in urban areas.

Y. Dmytruk · O. Stuzhuk (✉)
Department of Soil Science, Institute of Biology, Chemistry and Bio-Resources, Yuriy Fedkovych Chernivtsi National University, 2 Kotsyubynskogo St, Chernivtsy 58012, Ukraine
e-mail: stouzhouk.olga@gmail.com

D. Dent and Y. Dmytruk (eds.), *Soil Science Working for a Living*,
DOI 10.1007/978-3-319-45417-7_8

Soil mapping is of practical and scientific interest in many fields. Its methodology has been analyzed by Bulygin et al. (2003) and Krasyehy (2011). A geo-information approach to mapping has been explored by Klebanovych (2011), Stepanov et al. (2005) investigated the application of streaming models (which is the same thing as tangential curvature) to create or adjust soil maps, and Dmytruk and Cherlinka (2012, 2013) researched tangential curvature models of the land surface for soil and erosion mapping.

Ukraine has made use of agro-industrial soil groups, analogous to agro-ecological zones (Fischer et al. 2000), to provide a standard framework for the characterization of climate, soil and terrain conditions relevant to agricultural production. Existing agro-industrial soil maps of part of Chernivtsi city were used to analyze the correspondence of soil mapping units and soil boundaries with landforms and changes of slope. Soil mapping units were checked against landform units identified by analysis of a DEM, errors of location identified and a procedure elaborated to adjust their extent and boundaries by definition of tangential surface curvature. Tangential curvature is parallel to the slope and indicates the direction of maximum slope. A negative value indicates that the surface at that cell is upwardly convex; a positive profile indicates that the surface is concave; a value of zero indicates that the surface is linear. Tangential curvature affects the acceleration and deceleration of flow across the surface so we can create and edit soil maps based on the modelled surface.

Materials and Methods

We analyzed maps of agro-industrial soil groups for a part of Chernivtsi City characterized by complex landforms (Fig. 8.1) and a height difference of 170–315 m. The soil cover is represented by four mapping units (Table 8.1, in which the Ukrainian soil groups have been classified according to FAO 1988).

GrassGIS was used to derive two cartograms from the DEM:

- Slope curvature, performed by the following values of the module: v.surf.rst:—dmin = 10; tension = 40; smooth = 0.1; npmin = 300
- Tangential curvature, performed by the following values of module: v.surf.rst:—tension = 40; smooth = 0.1; npmin = 300; dmin = 10.

Soil boundaries from the extant agro-industrial soil mapping were superimposed on each terrain model.

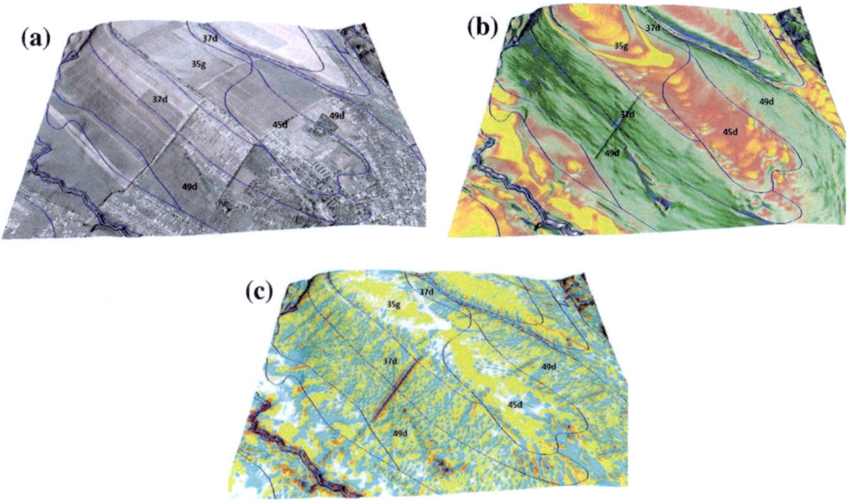

Fig. 8.1 Soil mapping units superimposed on: **a** elements of relief; **b** modelled landforms; **c** slope curvature

Table 8.1 Agro-industrial groups of soils on the research area

Code	Type of soil
35 g	*Haplic Greyzems* surface-gleyed light loam
37 d	*Haplic Greyzems* surface-gleyed medium loam
45 d	*Haplic Greyzems* and *Luvic Phaeozems* medium clay loam
49 d	*Haplic Greyzems* and *Luvic Phaeozems* medium loam

Results and Discussion

Superimposition of soil boundaries from existing soil maps allows us to visualize the relationship between soil mapping units and specific facets of the landscape; in particular the match, or mismatch, between landforms and the placement of soil mapping units (Fig. 8.1). *Haplic Greyzems* surface-gleyed light loam (mapping unit 35 g) is located where we should expect it to be but the boundary with the neighbouring *Haplic Greyzems and Luvic Phaeozems* medium clay loam (45 d) matches neither the landform boundaries nor what we would expect from our understanding of soil genesis. The mismatch between these soil delineations and the landform facets is easily seen on the model (Fig. 8.1b), and we presume that this boundary should be shifted. We should expect the *Haplic Greyzems* surface-gleyed light loam (35 g) to have a smaller footprint while increasing the range of *Haplic Greyzems and Luvic Phaeozems* medium clay loam (45 d).

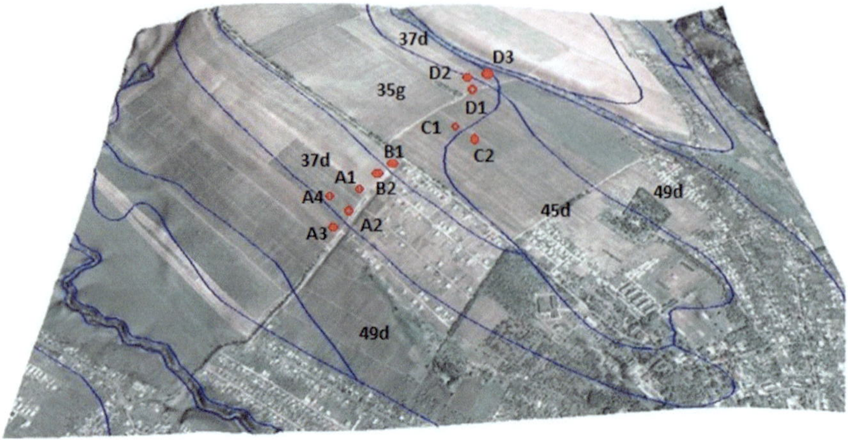

Fig. 8.2 Placement of soil inspection pits

In contrast, the area shown as *Haplic Greyzems* and *Luvic Phaeozems* medium loam (49 d) corresponds to a clearly defined slope facet (Fig. 8.1c) which suggests that its delineation is correct.

Nine soil profiles and two half-profiles were examined in the field to ascertain the actual extent of the different soils and to establish the feasibility of correcting the extant maps based on GIS modelling (Fig. 8.2; Table 8.2).

Profiles A1, A3, A4 and the A2 half-profile corresponded to mapping units 37 and 49 d. The current map (Fig. 8.1b) shows the boundary between them running across the hill where there is no break of slope—which is unlikely. Our field data (Table 8.2) also suggest a discrepancy. Its correction requires additional research but, relying on the modelled slope curvature and slope facets, we may assume that the boundary between the soils should be shifted downslope, significantly reducing the area mapped as 37 d.

Field checks (profiles B1 and B2) confirm the placement of units 35 g and 37 d on the map *(Haplic greyzems*, surface-gleyed light loams and medium loams). However, the boundaries between these soils and the *Haplic greyzems and Luvic Phaeozems* (45d) are questionable. Profiles C1 and C2 found no significant differences in the properties of the soils, in line with the cartogram of slope curvature (Fig. 8.1).

Likewise, profiles D1, D2 and D3 show the fallibility of boundaries between units 35 g and 37 d.

Table 8.2 Soil profile descriptions

B1	10–30 cm—Humus-eluvial. Grey to dark grey, powdered with silica, loam, fine subangular blocky, compact, many roots, gradual transition
	30–72 cm—Humus-eluvial. Brown loam to clay loam, finely porous, abundant worm casts and burrows, some darkening at the base associated with artificial profile inversion, clear boundary
	73–82 cm—Humus-illuvial, transition to parent material. Greyish-brownish-yellow, powdered with silica, finely porous, MnO_2 and iron (hydr)oxide mottles
B2	0–15 cm—Grey, powdered with silica, finely structured, compact, abundant worm burrows and casts, many roots, Mn and Fe mottles, brick fragments, gradual boundary
	15–43 cm—Humus-eluvial; dark, artificially inverted layer
	43–64 cm—Dark grey, powdered with silica, medium loam, fine granular, common roots, worm casts and burrows
	64–107 cm—Grey to dark grey (wet), worm casts and burrows, clear boundary
	107–117 cm—Brown clay loam, coarse subangular blocky, fragments of brick and coal
C1	0–24 cm—Grey to dark grey (wet) loam, granular, worm casts and burrows, occasional clean quartz grains, clear boundary
	24–65 cm—Humus-illuvial. Dark grey loam, medium crumb and subangular blocky, roots, clean quartz grains, clear boundary
	65–100 cm—Illuvial. Mottled greyish-yellow and brown (wet), powdered with silica, loam, medium subangular blocky to prismatic, roots, a few worm burrows and casts
C2	0–24 cm—Plough layer. Grey, powdered with silica, loam to clay loam, fine granular, worm burrows and casts, clear boundary
	24–64 cm—Humus-illuvial. Grey (wet), powdered with silica, loam, crumb and subangular blocky, compact, roots, worm burrows and prominent casts, clear boundary
	64–93 cm—Illuvial. Yellowish-brown, powdered with silica, loam, finely porous, roots, worm burrows and casts, some MnO_2 mottles
D1	0–27 cm—Plough layer. Grey, powdered with silica, loam, granular, slightly stony, roots, worm burrows and casts, abrupt boundary
	27–59 cm—Grey mottled brown and strong brown clay loam, compact, few roots, occasional worm burrows and casts, manganese nodules, vague boundary
	59–106 cm Yellowish-brown coarsely mottled dark bluish grey (wet), massive, MnO_2 nodules
D2	0–27 cm—Grey loam, fine granular structure, loose, slightly stony, roots, worm burrows and casts, clean quartz grains gradual boundary
	27–42 cm—Humus-eluvial. Grey loam, medium crumb and subangular blocky, compact, slightly stony, worm burrows and casts, clean quartz grains, abrupt boundary
	42–86 cm—Illuvial-humus. Mottled yellowish-brown clay loam, medium crumb and coarse subangular blocky, worm burrows and casts, clean quartz grains, gradual boundary
	86–96 cm—Illuvial-gley. Mottled yellowish-brown and grey clay loam, MnO_2 nodules, roots

(continued)

Table 8.2 (continued)

D3	0–27 cm—Humus-eluvial plough layer. Greyish-brown, medium subangular blocky and fine granular, slightly stony, fine roots, clear boundary
	27–48 cm—Humus-eluvial. Greyish-brown, subangular blocky, compact, fine roots, worm burrows and casts, clean quartz grains, clear boundary
	48–80 cm—Humus-eluvial. Yellowish-brown clay loam; weakly structured, patchy black MnO_2 deposition
A3	0–34 cm—Humus. Dark grey (dry) to black loam
	34–60 cm—Humus. Dark grey loam, crumb and subangular blocky, roots, worm burrows and casts, fine clean quartz grains, gradual wavy boundary
	60–107 cm—Transitional. Dark grey loam, medium crumb and subangular blocky, roots, worm burrows and casts, fine clean quartz grains
	107–122 cm—Transitional. Yellowish-brown clay loam, medium crumb and coarse subangular blocky, common clean fine quartz grains
A4	0–34 cm—Turf. Very dark greyish-brown loam, fine crumb structure, roots, worm burrows and casts, clean fine quartz grains, gradual boundary
	34–57 cm—Humus. Dark greyish-brown clay loam, crumb structure, roots, crotovinas, worm burrows and casts, clean quartz grains, clear boundary
	57–89 cm—Transitional. Mixed dark grey and yellowish-brown clay loam, crumb and coarse subangular blocky, roots, crotovinas, clear boundary
	89 cm + Brownish-yellow (wet) clay loam, coarse subangular blocky
A1	0–30 cm—Grey (dry) to dark grey (moist) loam to clay loam, coarse crumb structure, slightly stony, abundant roots, occasional clean quartz grains, uneven boundary
	30–64 cm—Brownish-yellow clay loam, coarse crumb structure, worm burrows and casts, clean fine quartz grains, irregular boundary
	64–90 cm—Transitional. Brownish-yellow clay loam, coarse subangular blocky, roots, gradual boundary
	90–120 cm—Brownish-yellow clay loam
A2	Similar to A1 but higher humus content

Conclusions

Detailed analysis of the digital elevation model, cartograms of tangential and simple curves and field observations all indicate systematic errors in the placement of soil boundaries in the extant agro-industrial soil maps. Most likely, they result from drawing boundaries by hand-and-eye, as opposed to rule-based modelling using GIS.

The outcomes affect decisions about the organization of the territory and the requirements of land legislation. The necessary adjustments can be made with the aid of modelling backed up by field checks.

Acknowledgments Publications are based on the research provided by the grant support of the State Fund for Fundamental Research (Project F 64/24 – 2015 c).

References

Bulygin SY, Shatohin A, Achasov AB, Truskavets SR (2003) The need for a new methodology for soil mapping. Pedology 4(1–2):5–10 (Russian)

Dmitruk YM, Cherlinka VR (2013) Applications of hydrologically correct and environmentally—appropriate digital terrain models. Environ Secur Sustain Resour Use 1(7):126–131

Dmitruk YM, Cherlinka VR (2012) Effect of DEM resolution on the accuracy of input data for modelling erosion processes. Phys Geogr Geomorphol Interdep Sci J 2(6):95–102

FAO (1988) Revised legend, soil map of the world. World Soil Resources Report 60, FAO-Unesco-ISRIC, Rome

Fischer G, van Velthuizen HT, Nachtergaele FO (2000) Global agro-ecological zones assessment: methodology and results. IIASA Interim Report: http://pure.iiasa.ac.at/6182/

Klebanovich NV, Bakun VV (2011) Research on possibilities of application of GIS technology to build models of relief for urabanized areas (for example, Minsk city). In: International congress on informatics: information systems and technologies: materials of the international scientific congress 30 Oct–3 Nov, 1, 2, pp 76–81 (Russian)

Krasyehy YN. Features of modern methodology of soil mapping. http://archive.nbuv.gov.ua/portal/chem_biol/Aig/2011_76/Kraseha.pdf. List of valuable soils: http://zakon2.rada.gov.ua/laws/show/z0979-03/page

Stepanov IN (2005) Theory of plasticity of relief and the new thematic map. Russia Sci Publ, Moscow (Russian)

Chapter 9
Determination of Soil-loss Tolerance for Chernozem of Right-Bank Ukraine

Sergiy Chornyy and Nataliya Poliashenko

Abstract Soil loss tolerance (*T*-value) is the maximum rate of soil erosion that may be tolerated and still allow high, sustainable crop yields. For modelling soil loss tolerance for *Chernozem* of the Right-Bank Steppe of Ukraine, a modified productivity index (MPI) is developed by summing productivity values of each 10 cm layer over the upper metre thickness of the soil. Productivity values depend on the content of humus, available phosphorus and potassium, bulk density, and pH. The equation is defined by the size of change in MPI over a given time and, also, the rate of decline of MPI in the topsoil caused by soil erosion. Calculations for eroded and not-eroded *Ordinary chernozem* and *Southern chernozem* under the condition of losing not more than 5% of soil fertility over 100 years indicate a soil loss tolerance of 5–7 t/ha/year.

Keywords Soil loss tolerance · Soil erosion · Productivity index · Chernozem

Introduction

For long-term management of soil resources and the design of erosion control measures, it is helpful to compare actual rates of soil erosion with a defined, tolerable value. The concept of soil loss tolerance (*T*-value, SLT) refers to the maximum level of soil erosion that still allows the land to sustain an economic level of crop productivity. SLT has been variously defined in the literature—for instance, the State Standards of Ukraine devoted to scientific soil erosion terminology (State Standard of Ukraine 2009) define it as follows: *the maximum soil loss from erosion which doesn't lead to degradation of the soil cover, and is established taking into account the existing or prospective soil control resources and/or rate of topsoil formation*. Definitions take one of the three approaches: first, conservation of

S. Chornyy (✉) · N. Poliashenko
Department of Soil Science and Agrochemistry, Mykolayiv National Agrarian University, 9 Paryzka Komuna, Mykolayiv 54020, Ukraine
e-mail: s.g.chornyy@gmail.com

D. Dent and Y. Dmytruk (eds.), *Soil Science Working for a Living*,
DOI 10.1007/978-3-319-45417-7_9

important soil attributes such as the status and humus content of the topsoil or rooting layer (Wischmeier and Smith 1978; USDA 2014), secondly, rates of soil loss compared with the rate of soil formation (Alexander 1988; Chornyy 1999; Svetlichny et al. 2004), and thirdly, change of productivity caused by soil erosion (Pierce et al. 1984; Duan et al. 2012) where SLT (t/ha/year) is calculated as a function of a permissible value of reduced soil productivity taking account of initial soil fertility expressed as a productivity index (PI), soil density, and the planning horizon (years), as well as a function defining the relationship between changes of the soil productivity and the amount of potential erosion (Pierce et al. 1984). Arguably, the last method is most objective, whereas assessment taking into account the rate of soil formation, itself problematic, greatly underestimates SLT values.

Study Area

The Right-Bank Steppe of Ukraine is a plain rising from sea level northwards to 240 m above sea level and bordered to the north by the Dnieper and Podolsk uplands. The dominant soil parent material is loess which blankets interfluves, marine terraces, and ancient river terraces. The climate is continental with warm summers; drought is common. The main soils are *Ordinary chernozem* and *Southern chernozem*. *Ordinary chernozem* have a thick humus horizon (up to 1 m) with 6–8% humus in the plough layer, big reserves of nutrients (especially phosphorus and potassium), high base-saturation and near-neutral reaction, and a heavy texture that promotes a water-stable soil structure. With adequate rainfall, the potential productivity of these soils is almost unlimited. *Southern chenozem* have a thinner humus horizon (50–60 cm) with a lesser humus content, overlying a well-developed calcic horizon; commonly, these soils are saline. The Right-bank Steppe suffers from soil erosion driven by intense summer rains, sloping relief, and a large proportion of arable (more than 80%) and irrational land use; about half of the arable land is eroded, compared with one-third of the country a whole.

Field and Laboratory Methods

Definition of SLT, taking account of admissible reduction of productivity through soil erosion over a given period of time, provides a quantitative assessment of soil fertility by means of a productivity index (IP). To identify criteria, two profiles of *Southern chernozem* and four profiles of *Ordinary chernozem* were compared in 10-cm layers to a depth of 120 cm (Table 9.1). Dry bulk density was measured on undisturbed core samples (State Standard Ukraine 1998); particle-size distribution, $pH_{soil\ solution}$, content of organic matter, and available phosphorus and potassium were measured on disturbed samples (State Standard Ukraine 2002, 2004b, 2007a, b).

Table 9.1 Location and description of investigated soils

No.	Soil species	Coordinates of control soil profile		MIP	V
1.	*Ordinary chernozem,* not eroded (BSOne)	N47°53′08,5″	E031°48′10,6″	0.7554	0.0009
2.	*Ordinary chernozem,* eroded (BSOe)	N47°53′06,1″	E031°48′26,0″	0.6885	0.0009
3.	*Southern chernozem,* eroded (BSSne)	N46°55′20,5″	E031°40′56,2″	0.7753	0.0008
4.	*Southern chernozem,* (BSSe)	N46°54′35,4″	E031°40′04,4″	0.4143	0.0006
5.	*Ordinary chernozem,* not eroded (BSOne-2)	N47°53′28,8″	E031°49′11,3″	0.7857	0.0009
6.	*Ordinary chernozem,* eroded (BSOe-2)	N47°53′03,1″	E031°49′17,0″	0.6509	0.0007

Modified IP Model

Change of *IP* over a given time is at the heart of definition of SLT. The original *IP* model, developed for *Mollisols* in the Midwest of the USA with respect to maize and soya bean cropping, concerned only so-called *irreplaceable* soil attributes (Pierce et al. 1983, 1984; Gantzer and McCarty 1987). The parameters were as follows: water-retention capacity (A_i), bulk density (C_i), $\mathrm{pH}_{\mathrm{soil\ solution}}$ (D_i), and a root weighting factor (WF) recording the proportion of root mass per unit mass of soil. Each parameter was normalized as a fraction of layer-by-layer values. The basic form is:

$$\mathrm{IP} = \sum_{i=1}^{n} (A_i \times C_i \times D_i \times \mathrm{WF}), \tag{9.1}$$

where *IP* is the productivity index, with $i = 1, 2, \ldots, n$ representing the different soil layers.

The so-called *irreplaceable* soil properties in Eq. (9.1) are actually regulated by fertilizers, irrigation, and tillage; consequently, various authors following the proposed structure of the index have inserted additional indicators of soil fertility—*e.g.* content of water-soluble salts (Larson and Stewart 1990; Mulengeraa and Payton 1999), adsorbed sodium (Doll and Wollenhaupt 1985), and water permeability (Burly et al. 1989).

For our own modified productivity index, we have taken account of extensive Ukrainian research on soil potential (*bonitat*) ratings. Medvedev and Plisko (2006), drawing on a wealth of earlier work on the relationship between soil quality and crop yields, propose 9 indicators of soil fertility: thickness of the root layer, content of humus, $\mathrm{pH}_{\mathrm{soil\ solution}}$, clay content, bulk density, content of mobile phosphorus and available potassium, depth to the gley horizon (impeded drainage), and specific resistance of the soil (the work needed to implement cultivation procedures). In the droughty steppe, a very important factor of fertility is the range of available moisture (RAM)—the ability of the soil to provide crops with water. We would also modify the thickness of the root layer, accounted for in Eq. (9.1) through the WF index. Specific resistance and depth to a gley layer have no direct bearing on productivity; soil texture is homogenous and unchanging over a wide area (clay content lies within the range of 50–60%) so it makes no difference to the productivity index, either according to the degree of erosion or different subtypes of *chernozem*, and may be excluded from the list of indicators defining the productivity index; also RAM, which is defined by soil texture, does not change in our soils, neither between horizons nor on degree of erosion. In calculating the productivity index, we have also chosen to modify the interpretation of several indicators; the soil profile data show dispersion in respect of nutrients and maintenance of humus, especially in the plough layer, probably connected with different applications of manure and fertilizer.

We argue that it is not the multiple of the values of the individual indicators that integrates their influence on productivity but, rather, a more complicated averaging procedure—the geometric mean of these parameters. It is important to emphasize that the averaging procedure takes place without indicator WF. Thus, for a soil thickness of one metre, the modified soil productivity index (MIP) is calculated using the formula:

$$\mathrm{MIP} = \sum_{i=1}^{10} (h_i \times ph_i \times \gamma_i \times \rho_i \times \kappa_i)^{0,2} \times \mathrm{MWF}_i, \quad (9.2)$$

where i number of soil layer (i = 1, 2, 3, …, 10); h_i, ph_i, γ_i, ρ_i, κ_i are, respectively, the impact on the value of the modified soil productivity index (MIP) in the ith layer of soil humus content, pH, bulk density, and the content of mobile phosphorus and mobile potassium, and MWF_i is the modified root weighting factor. Each of these parameters is a dimensionless quantity normalized to unit parts (0–1); the higher the value, the more suitable for crop growth.

Considering the evaluation of each component of the right-hand side of Eq. (9.2):

Estimation of the influence of humus as a source of nutrients and its beneficial influence on many physical, chemical, and biological properties of *chernozem* is described in many papers (Medvedev and Plisko 2006; Nosko 2006). However, humus-driven productivity is most apparent in soils of medium or low humus content; when the humus content is more than about 3.5%, greater humus content is not reflected in greater crop yields. And so

$$h_i = \begin{cases} h/3.5, & \text{if} \quad h \le 3.5\% \\ 1, & \text{if} \quad h > 3.5\% \end{cases}, \quad (9.3)$$

where h humus content, %.

The effect of $pH_{soil\ solution}$ on the fertility of *chernozem* is also recorded in several publications (Gubareva 1991; Medvedev and Derevyanko 1991; Medvedev and Plisko 2006). Analysis of published data in the range of pH that has been investigated (6.0–8.5) yielded the following relationship, which has been used to calculate the modified productivity index:

$$\mathrm{pH}_i = -0.067 \times (\mathrm{pH})^2 + 0.875 \times (\mathrm{pH}) - 1.863, \tag{9.4}$$

where $pH = pH_{soil\ solution}$.

The effect of soil bulk density on the fertility of *chernozem* has been reported by Pierce et al. (1984), Medvedev et al. (2004), and Medvedev and Plisko (2006). From these data, we derived the following quadratic equation specifying the influence of bulk density on soil productivity:

$$\gamma_i = -5.414 \times \left(Y^2\right) + 12.959 \times (Y) - 6.806, \tag{9.5}$$

where Y bulk density, g/cm^3.

The role of available forms of phosphorus and potassium in the calculation of the modified productivity index (MPI) was determined by the following equations:

$$\rho_i = \begin{cases} \frac{P_2O_5}{45}, & \text{if} \quad P_2O_5 \leq 45 \\ 1, & \text{if} \quad P_2O_5 > 45 \end{cases} \tag{9.6}$$

and

$$\kappa_i = \begin{cases} \frac{K_2O}{300}, & \text{if} \quad K_2O \leq 300 \\ 1, & \text{if} \quad K_2O > 300 \end{cases} \tag{9.7}$$

where P_2O_5 and K_2O are contents of available phosphorus and potassium, mg/kg.

Equations (9.6) and (9.7) integrate data from Nosko (1990, 2006), Nosko et al. (1996), and Medvedev and Plisko (2006) and State Standard Ukraine (2004a), all showing that there is no further increase in crop yields where the content of mobile P_2O_5 exceeds 45 mg/kg and the content of available K_2O is greater than 300 mg/kg.

Finally, to indicate the share of roots in each layer, compilation of published data (in particular, Stankov 1964, and Pierce et al. 1984) showed that for the main arable region, this parameter can be calculated by:

$$\mathrm{WF}_i = 0.5 \times e^{-0.05 \cdot \eta}, \tag{9.8}$$

where η depth of soil layer, cm.

Calculation of SLT

In general terms, SLT reflects the tolerated loss of soil fertility over a given period of time, say 50 or 100 years:

$$G = \theta \times \frac{\Delta P}{T} \quad (9.9)$$

where G SLT, t/h/year; θ dimensionless coefficient; ΔP tolerable change of fertility, dimensionless size; and T—planning horizon, years. If a change of fertility may be expressed in terms of transformation of the modified productivity index (MPI), then admissible change of soil fertility (ΔMPI) will be measured as a difference between an initial value of the productivity index (MPI_{IN}) and its value at time T (MPI_T):

$$\Delta P = \Delta MPI = MPI_{IN} - MPI_T. \quad (9.10)$$

Recalculation of changes in fertility in terms of millimetres of soil loss is accomplished using the index V, determined by dividing the amount of change in value MPI in the soil profile at a certain depth; it shows the rate of change of soil fertility resulting from soil erosion. Pierce et al. (1984) consider that V should be calculated for the top 500-mm layer. However, analysis of natural and economic conditions in the Right Bank Steppe and calculation using an adequate mathematical model of soil erosion (Svetlichny et al. 2004) show that even with the current land use, soil erosion will not exceed 15–25 t/ha/year. So, even with worsening erosion in future, not more than 250 mm of soil will be washed away over the planning period of 100 years. Therefore, for indicator V, it is necessary to count changes of MPI over a maximum of the topmost 30 cm. In the study of the distribution of MPI on all six profiles, calculations showed that the value of the indicator V varies between 0.0006 and 0.0009 (Table 9.1).

Thus, the change of soil fertility for a particular period of time in millimetres per layer is equal to:

$$\Delta P = \frac{(MPI_{IN} - MPI_T)}{V}$$

Given the above and turning to the definition of SLT with the recalculation of the soil loss tolerance in tons per hectare per year, we get the following equation:

$$G = \frac{[10 \times (MPI_{IN} - MPI_T) \times \gamma]}{(V \times T)}, \quad (11)$$

where G SLT, t/ha/year; γ bulk density of topsoil, g/cm^3; V the rate of change in erosion process of the MPI 300-mm top layer, mm^{-1}; and T the planned time period, years (50 or 100). The remaining symbols are the same.

Results and Discussion

From Eq. (11), the degree of soil loss tolerance depends on the change in the modified productivity index (MPI) and an indicator of its rate of change through erosion of the topsoil (*V*). In turn, various values of *V* depend on features of distribution of MPI in the uppermost part of the soil profile. Figure 9.1 depicts the distribution of MPI in a metre layer of eroded and not-eroded *Ordinary chernozem* and *Southern chernozem*. The analysis shows that not-eroded *Ordinary chernozem* (BSOne) has a rather bigger MPI value than eroded profile (BSOe) in practically every layer. This is because the not-eroded soil contains more humus and mobile P

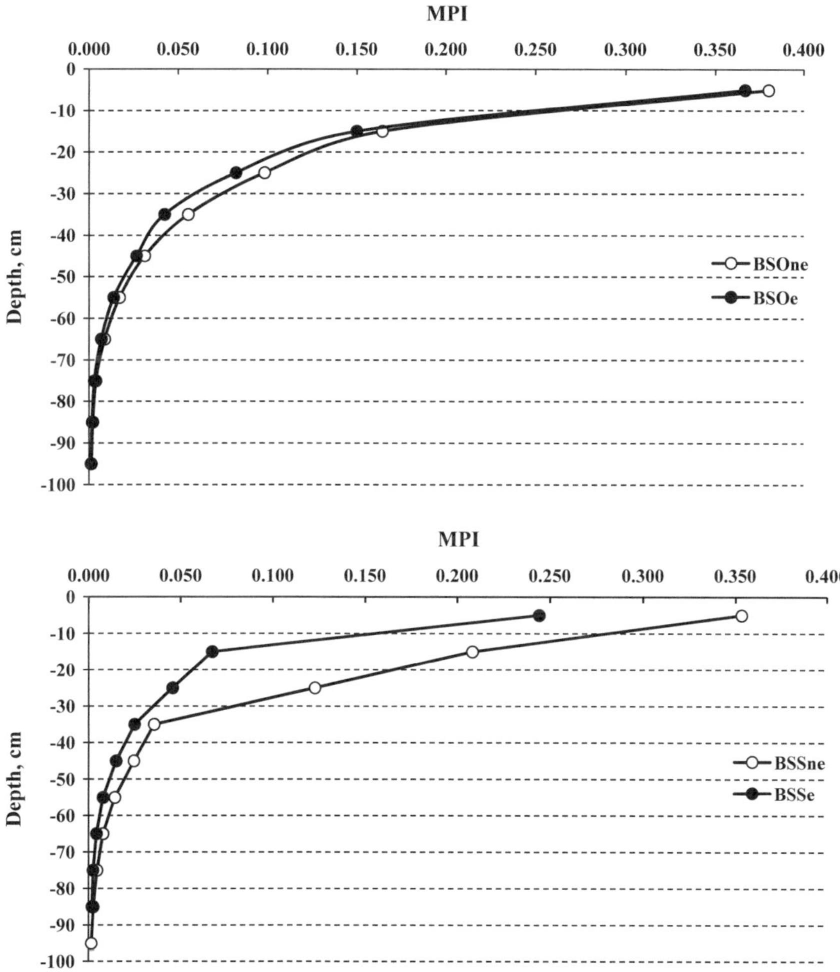

Fig. 9.1 Modified productivity index (MPI) distributions for 100 cm soil depth

Table 9.2 Attributes of *Ordinary chernozem* profiles standardized to 0–1

Layer (cm)	Humus content		Mobile phosphorus content		Mobile potassium content		Bulk density		pH	
	BSOne	BSOe	BSOne	BSOe	BSOne	BSOe	BSOne	BSOe	BSOne	BSOe
0–10	1.000	1.000	1.000	0.838	1.000	1.000	0.937	0.937	0.942	0.942
10–20	1.000	1.000	0.280	0.216	0.872	0.610	0.693	0.843	0.964	0.928
20–30	1.000	1.000	0.209	0.152	0.883	0.539	0.857	0.902	0.964	0.842
30–40	1.000	1.000	0.147	0.081	0.915	0.462	0.857	0.911	0.928	0.821
40–50	1.000	1.000	0.166	0.134	0.819	0.384	0.693	0.812	0.799	0.799
50–60	0.903	0.880	0.137	0.117	0.736	0.323	0.592	0.592	0.821	0.750
60–70	0.851	0.471	0.094	0.118	0.750	0.300	0.441	0.441	0.695	0.750
70–80	0.697	0.429	0.125	0.109	0.659	0.328	0.100	0.100	0.666	0.695
80–90	0.589	0.406	0.104	0.099	0.691	0.362	0.100	0.100	0.666	0.695
90–100	0.571	0.331	0.123	0.143	0.619	0.384	0.100	0.100	0.666	0.666

and K in practically all layers. Other parameters of fertility considered in the calculation of MPI (bulk density and pH) hardly change under the influence of erosion (Table 9.2). Such situation has an essential influence on the total value of MPI. Its value on not-eroded soil is equal 0.7554, whereas eroded soil only 0.6885 (Table 9.1).

Change in MPI in the upper 300-mm layer occurs almost simultaneously in not-eroded and eroded soils (Fig. 9.1); the value of V for these soils is the same at 0.0009. The second profiles of *Ordinary chernozem* (BSOne-2 and BSOe-2) give some different results: the value MPI for not-eroded soil (0.7857) is, again, more than that for eroded soil (0.6509), and for the same reason—there is higher maintenance of humus and available nutrients. However, an indicator of rate of change in MPI in the erosion process of top 300-mm layer is significantly different: in the first case of V is 0.0009 and in the second is only 0.0007.

The values of MPI for *Southern chernozem* are radically different from those in *Ordinary chernozem* (Fig. 9.1): MPI of not-eroded soils in the topmost 30-cm layer is much greater than values for eroded soils. This is because, in eroded *Southern chernozem*, the maintenance of humus and nutrients (especially phosphorus) decreases substantially; also, the eroded soil has high bulk density (Table 9.3). Therefore, the modified efficiency index on not-eroded soil is 0.7753 but on eroded soil, only 0.4143.

The rate of change in MPI in the top 300-mm layer for eroded soils is less than in not-eroded soil—approximately 0.20 units (0.23 for not-eroded soils)—resulting in lower values of the V indicator—0.0006 on eroded soil compared with 0.0008 for the not-eroded analogue.

The above features of the modified productivity index and the index rate of change through erosion of the MPI 300-mm top layer of soil (V) are reflected in the

Table 9.3 Attributes of *Southern chernozem* profiles standardized to 0–1

Layer cm	Humus content		Mobile phosphorus content		Mobile potassium content		Bulk density		pH	
	BSOne	BSOe	BSOne	BSOe	BSOne	BSOe	BSOne	BSOe	BSOne	BSOe
0–10	1.000	0.829	1.000	0.347	1.000	1.000	0.619	0.337	0.993	1.000
10–20	1.000	0.643	1.000	0.287	1.000	1.000	0.535	0.100	0.993	0.996
20–30	1.000	0.600	1.000	0.103	0.926	0.553	0.504	0.100	0.988	0.964
30–40	0.866	0.511	0.057	0.077	0.654	0.516	0.373	0.100	0.981	0.973
40–50	0.606	0.446	0.220	0.080	0.565	0.540	0.301	0.100	0.981	0.942
50–60	0.586	0.434	0.267	0.083	0.521	0.265	0.224	0.100	0.964	0.953
60–70	0.586	0.226	0.287	0.097	0.480	0.258	0.143	0.100	0.973	0.973
70–80	0.457	0.091	0.310	0.190	0.487	0.241	0.143	0.100	0.964	0.953
80–90	0.400	0.091	0.377	0.303	0.441	0.485	0.100	0.100	0.973	0.953
90–100	0.329	0.051	0.420	0.357	0.465	0.474	0.100	0.100	0.964	0.942

Table 9.4 Soil loss tolerance, t/ha/year, depending on acceptable decrease of productivity and planning period: 50/100 years

No.	Soil species	1%	2%	3%	4%	5%
1.	*Ordinary chernozem*, not eroded (BSOne)	2.1/1.1	4.2/2.1	6.3/3.2	8.4/4.2	10.5/5.3
2.	*Ordinary chernozem*, eroded (BSOe)	2.0/1.0	4.0/2.0	6.0/3.0	8.0/4.0	9.9/5.0
3.	*Southern chernozem*, eroded (BSSne)	2.7/1.3	5.3/2.7	8.0/4.0	10.7/5.3	13.3/6.7
4.	*Southern chernozem*, eroded (BSSe)	1.9/0.9	3.8/1.9	5.7/2.8	7.5/3.8	9.4/4.7
5.	*Ordinary chernozem*, not eroded (BSOne-2)	2.3/1.1	4.5/2.3	6.8/3.4	9.1/4.5	11.3/5.7
6.	*Ordinary chernozem*, eroded (BSOe-2)	2.5/1.2	4.9/2.5	7.4/3.7	9.8/4.9	12.3/6.1

calculations of SLT according to Eq. (9.11) at five levels of decreased soil productivity (1, 2, 3, 4, 5%) and two planning periods (50 and 100 years) (Table 9.4). As seen, SLT depends mainly on the value of the modified productivity index; the higher the value, the higher the SLT obtained so, at all levels of decrease in efficiency, values for not-eroded *chernozem* will be higher than for eroded soils. However, if there is a big difference in V values and little in MPI values, as in a case with the BSOne-2 and BSOe-2 profiles, then SLT on eroded soils can be larger, than on not-eroded.

The received SLT values for the conditions of the Right-Bank Steppe of Ukraine may be recommended for using erosion control in this region. They are of the same order as SLT values computed for analogous *chernozem* soils in Northeast China (Duan et al. 2012) and the USA (Pierce et al. 1984).

Conclusions

1. The fertility of *chernozem* of the Right-Bank Steppe of Ukraine is assessed by means of the modified productivity index (MPI), which is defined as the sum of values of productivity of each 10-cm layer for the topmost metre depth of the soil. The value for the productivity of each layer is defined by the geometric average of the following parameters: humus content, content of available phosphorus and potassium, bulk density, and pH, each normalized from 0 to 1. The modified root weighting factor (MWF) is not averaged.
2. The final version of mathematical model of definition of SLT depends on change in MPI during the planning period (T), the bulk density of the top layer of the soil, and an index of the rate of reduction of MPI through soil erosion.
3. For *chernozem* soils of the Right-Bank Steppe of Ukraine, SLT depends on the value of the modified productivity index (MPI) and the parameter that determines the speed of change in MPI in the top layers of the soil (V). SLT values are generally greater for not-eroded *chernozem,* compared with eroded analogues. However, where there is only a small difference in fertility between eroded and not-eroded soils, then the index of speed of change in MPI in the erosion process (V) will be significant.

References

Alexander EB (1988) Rates of soil formation: implications for soil-loss tolerance. Soil Sci 145 (1):37–45

Burly JB, Thomsen CH, Kenkel N (1989) Productivity equation for reclaiming surface mines. Environ Manag 13(1):631–638

Chornyy SG (1999) Estimation of soil loss tolerance for Ukrainian steppe soils. Ukr Geogr J 4:18–22 (Ukrainian)

Doll EC, Wollenhaupt NC (1985) Use of soil parameters in the evaluation of reclamation success in North Dakota. Bridging the gap between science, regulation, and the surface mining operation. In: ASSMR second annual meeting, Denver, pp 91–94

Duan X, Xie Y, Liu B et al (2012) Soil loss tolerance in the black soil region of Northeast China. J Geogr Sci 22(4):737–751

Gantzer CJ, McCarty TR (1987) Predicting corn yield on a claypan soil using a productivity index. Trans Am Soc Agric Eng 30:1347–1352

Gubareva DH (1991) Soil physico-chemical factors. In: Medvedev VV (ed) Soil-ecological conditions of cultivation of agricultural crops. Urogay, Kiev, pp 36–49 (Russian)

Larson WE, Stewart BA (1990) Thresholds for soil removal for maintaining cropland productivity. In: Proceedings of the soil quality standards symposium, 82nd annual meeting of the Soil Science Society of America, San Antonio TE, pp 6–14

Medvedev VV, Derevyanko RG (1991) Optimal ecological models of soil. In: Medvedev VV (ed) Soil-ecological conditions of cultivation of agricultural crops. Urogay, Kiev, pp 59–73 (Russian)

Medvedev VV, Lindina TE, Laktionova TN (2004) The bulk density of the soil (genetic, environmental and agronomic aspects). 13th Publishing House, Kharkiv (Russian)

Medvedev VV, Plisko IV (2006) Valuation and qualitative assessment of the arable lands of Ukraine. 13th Publishing House, Kharkiv (Russian)
Medvedev VV, Laktionova TN, Dontsova LV (2011) Water properties of Ukraine soils and the water availability of crops. Apostrophe, Kharkiv (Russian)
Mulengeraa MK, Payton RW (1999) Modification of the productivity index model. Soil Tillage Res 52:11–19
Nosko BS (1990) Phosphate regime of soil and fertilizer efficiency. Urogay, Kyiv (Ukrainian)
Nosko BS (2006) Anthropogenic evolution of chernozems. 13th Publishing House, Kharkiv (Ukrainian)
Nosko BS, Lisovoy NV, Stolyar VM (1996) Potassium in the soil of Ukraine and the efficiency of potassium fertilizer. IPA UAAN, Kharkiv (Russian)
Pierce FJ, Larson WE, Dowdy RH, Graham WAP (1983) Productivity of soils: assessing long-term changes due to erosion. J Soil Water Conserv 38:39–44
Pierce FJ, Larson WE, Dowdy WE (1984) Soil loss tolerance: maintenance of long-term soil productivity. J Soil Water Conserv 39(2):136–138
State Standard Ukraine (1998) Soil quality. Determination of bulk density on dry weight. ISO 11272:1998 (Ukrainian)
State Standard Ukraine (2002) Soils. Determination of mobile compounds of phosphorus and potassium by modified Machigin method. 4114: 2002 (Ukrainian)
State Standard Ukraine (2004a) Soil quality. Indicators of soil fertility. SS U 4362: 2004 (Ukrainian)
State Standard Ukraine (2004b) Soil quality. Method for determination of organic matter. SSU 4289:2004 (Ukrainian)
State Standard Ukraine (2007a) Soil quality. Determination of granulometric composition analysis by pipette method, after Kachinskiy. 4730:2007 (Ukrainian)
State Standard Ukraine (2007b) Soil quality. Determination of pH. ISO 10390:2007 (Ukrainian)
State Standard of Ukraine (2009) Soil quality. Soil erosion. Terms and definitions of basic concepts. SSU 7118:2009 (Ukrainian)
Stankov NZ (1964) The root system of crops. Kolos, Moscow (Russian)
Svetlichniy AA, Chornyy SG (2007) Basic of soil erosion science. University Books, Sumy (Ukrainian)
Svetlichniy AA, Chornyy SG, Shvebs GI (2004) Science of soil erosion: theoretical and applied aspects. University Books, Sumy (Russian)
US Dept Agriculture (2014) National soil survey handbook. Natural Resources Conservation Service. http://soils.usda.gov/technical/handbook. Accessed 10 Sept 2015
Wischmeier WH, Smith DD (1978) Predicting rainfall erosion losses: a guide to conservation planning. United States Dept Agriculture, Agriculture Handbook 537, Washington DC

Chapter 10
Assessment of Problems of Soil Contamination Using Environmental Indicators

Yevhen Varlamov and Oksana Palaguta

Abstract A system is proposed for assessing the condition of the natural environment in Ukraine. This is based on a study of the ecological indicators that are already in use in particular regions and in the country as a whole. Consideration of environmental performance indicators (EPIs) is used to assess the state of the natural environment in Eastern Europe, Caucasus, and Central Asia.

Keywords Environmental performance · Environmental performance indicator · Environment · System monitoring

Introduction

The agro-industrial complex comprises more than half of Ukraine's productive assets, generates about half of the gross domestic product, and, in turn, impacts the environment—not least by pollution of soil and water. In Eastern Europe, Caucasus, and Central Asia (EECCA), the environmental performance indicator (EPI) is the main tool for assessing the status of the environment (UNECE 2007a). Well-chosen indicators based on enough time series data can show key trends and help describe causes and effects, as well as enable us to monitor the implementation of environmental policy and to evaluate its effectiveness. Therefore, we propose a system of EPIs to assess the state of the environment (hereafter the EPI system). It will facilitate the tracking of trends in status of areas under surveillance and assess the impact of environmental management both across the Ukraine and in particular administrative districts. Furthermore, it will promote the development of the information base of the state environmental monitoring system which should be a basis for policy development, management decisions concerned environmental safety, and public information.

Y. Varlamov (✉) · O. Palaguta
Ukrainian Scientific and Research Institute on Ecological Problems,
6 Bakulina St, 61166 Kharkiv, Ukraine
e-mail: varlamov.niiep@gmail.com

D. Dent and Y. Dmytruk (eds.), *Soil Science Working for a Living*,
DOI 10.1007/978-3-319-45417-7_10

The EPI system should have the following applications:

- A tool for estimating the state of the environment and the effectiveness of environmental policy;
- Providing information on environmental problems for managers and policy makers;
- Determination of the main factors impacting the environment;
- Information for planning economic development while minimizing the negative impact on the environment;
- Monitoring the effectiveness of implementation of environmental protection measures;
- Improving the public awareness of environmental problems according to Ukraine's obligations under international conventions and agreements in the field of environment protection.

Proposals

The EPI system is based on the recommendations of the UNECE Working Group on Monitoring and Assessment of Environment set out in the *Guidelines for EECCA* (UNECE 2007b) and experience of the application of environmental indicators in Ukraine (Ministry of Ecology and Natural Resources 2011; UNECE 2007c). It should cover the main areas of environmental monitoring and the economic zones of industries that have a direct impact on the environment; and it should allow progressive improvements in the components of the system based on the practical experience of its application (Varlamov and Palaguta 2013). Tab. 10.1 lists 39 proposed EPIs under nine headings (directions).

The ***Agriculture*** dimension includes the following EPIs:

- *Adding mineral and organic fertilizers* The amount of mineral and organic fertilizers applied per unit of arable land.
- *Pesticide* The amount of pesticide applied per unit of land area, which increases the risk of harmful effects.

Land and soil includes the following EPIs:

- *Withdrawal of land from production* Index of land withdrawal for transport infrastructure, landfill, waste dumps, tailings, and waste rock dumps. Includes direct urban and industrial development.
- *Areas affected by soil erosion* Total land area and the share of agricultural land affected by wind and water erosion.
- *Structure of agricultural land* Area of agricultural land, as well as the distribution of agricultural land under arable, hayfields, pasture, perennial crops, and fallow.
- *Irrigated and drained lands* Total area of reclaimed land, including irrigated and drained lands.

Table 10.1 Indicators of the state of the environment in Ukraine

Direction (*i*)	EPIs (*j*)
Air pollution	Emissions of pollutants into the air
	Air quality in urban areas
	Radioactive contamination of the atmosphere
	Use of ozone-depleting substances
Climate change	Air temperature
	Precipitation
	Greenhouse gas emissions
Water resources	Renewable freshwater resources
	Freshwater intake
	Household water consumption *per caput*
	Loss of water
	Reuse and recycling of freshwater
	Quality of drinking water
	Biochemical oxygen consumption and concentration of ammonia in river water
	Biogenic substances in freshwater
	Biogenic substances in coastal waters
	Contaminated wastewater
Biodiversity and forests	Natural areas under special protection
	Forests and woodland
	Protected species
	Trends in the number and distribution of selected species
Land and soil	Withdrawal of land from production
	Areas affected by soil erosion
	Structure of agricultural land
	Irrigated and drained land
Energy	Total energy consumption
	Gross domestic use of energy
	Energy capacity
	Energy consumption from renewable sources
Agriculture	Use of mineral and organic fertilizers
	Use of pesticides
Transport	Passenger traffic
	Freight traffic
	Motor vehicles by type of fuel
	Average age of the motor vehicles
Waste management	Waste
	Processing and recycling of waste
	Final waste disposal
	Transboundary movements of hazardous wastes

A methodology has been developed for integrated assessment of the environment in particular regions, which should facilitate the identification of key areas for action to improve the regional environment. This boils down to selecting the most informative environmental performance (EP) factors forming a part of each EPI, valuation of indicators, and their linear conversion with weighting factors. In general, the EPI for individual administrative units is formed by linear combination of the individual, weighted EPs in two steps:

First, standardized assessment of each EPI (A_i^j) should be carried out on the status of environment directions. To obtain the normalized evaluation of each EPI (A_i^j), we need to:

(1) Normalize all the EPs that are part of the EPI. Before continuing with the conversion of EPs into a single EPI, each EP should be normalized so that each will be measured on an N-point (dimensionless) scale. For this purpose, each EPI is calculated by the formula:

$$\left|\xi\widetilde{A}_i^j\right| = \frac{\left|\xi A_i^j\right|_{\text{3BiT.piк}}}{\left|\xi A_i^j\right|_{\max}} \times 10, \qquad (10.1)$$

where

i	means the number of directions, according to Table 10.1;
j	the number of EPIs the direction, according to Table 10.1;
$\left\|\xi\widetilde{A}_i^j\right\| - \xi$	EP, which is required to determine the j—of the EPI for the i—the direction;
$\left\|\xi A_i^j\right\|_{\text{3BiT.piк}}$	the value of ξ − EP for the year to determine the j—of the EPI for the i—the direction;
$\left\|\xi A_i^j\right\|_{\max}$	the maximum value of ξ − EP, which is selected from the list of years for which the calculated j—the EPI for i—the direction;
10	the maximum value of N-point (dimensionless) scale.

(2) Determination of the weighting coefficients ($w\,\xi b_i^j$) for the EP is carried out by an expert. Values of weighting coefficients for EPI are between 0 and 10 ($0 < \xi b_i^j < 10$).

(3) In evaluating the EPI as a whole, all the EPs (ξA_i^j) that are needed for a specific EPI (A_i^j) are calculated by the formula:

$$\left|A_i^j\right| = \frac{\sum_{\xi} b_{\xi}\left|\xi\widetilde{A}_i^j\right|}{\xi}, \qquad (10.2)$$

Table 10.2 Calibration according to the rating assessment for each dimension

Class	Class status	Ecological status
F_1	$0 \leq F_1 = 2$	Normal
F_2	$2 < F_2 \leq 4$	Minor deviation from the normal state
F_3	$4 < F_3 \leq 6$	Significant violation of state
F_4	$6 < F_4 \leq 8$	Dangerous violation of state
F_5	$8 < F_5 = 10$	Critical condition

where

$|A_i^j| - j$ The EPI for i—the direction;
$\sum_\xi b_\xi|\xi\tilde{A}_i^j|$ The sum of all ξ − EPs that are needed to determine the j—of the EPI for the i—the direction;
ξ The number of ξ − EPs that are part of the j—to the EPI.

The resulting EPI value is within the range 0–10. The greater the value of the EPI, the more critical the situation in the corresponding dimension.

In the second phase, a cumulative assessment of the environment along the dimensions (A_i) in the region is carried out. To carry out an integrated assessment of the environment in the region by dimensions, we need to do the linear conversion used in the first phase of the EPI dimension detection by the formula:

$$|\tilde{A}_i| = \frac{\sum_j \tilde{A}_i^j}{j}, \qquad (10.3)$$

where

$|\tilde{A}_i|$ an integrated assessment of the environment for i—the dimension;
$\sum_j \tilde{A}_i^j$ the sum of all j − EPI for i—the dimension;
i the number of j − EPs, which are part of i—the dimension.

Thus, the integrated assessment of environment will be defined in the range from 0 to 10.

To plan for evaluation and action to protect the environment, dimension by dimension, we recommend a 10-point scale assessment for each dimension, as well grouping the calculated EPI values into classes (Table 10.2) which will represent the status of each of the dimensions based on the level of deviation from a standard value.

For example, a class F_1 deviation might be found in territory virtually unaffected by human activities, e.g. a nature reserve. Class F_5 deviation is the characteristic of areas with extreme man-made pressure on the system in the dimension of the EPI. Thus, the 10-point scale gives the opportunity to compare different regions (administrative areas) with one another.

Conclusions

- Objective information is needed for the development and implementation of environmental protection measures, as well as for the effective management of natural resources. In Ukraine, information on environmental condition, the influence of various factors on it, and the use of natural resources are all based on environmental monitoring.
- Currently, environment assessment in Ukraine is by no means comprehensive. The EPI system will enable a comprehensive evaluation of the environment and the effects of using natural resources at the national level and within particular administrative districts.
- The practical value of the EPI system is to improve efficiency and environmental performance at different levels (national, regional, departmental, etc.), as well to raise public awareness about the environmental situation.

References

Ministry of Ecology and Natural Resources (2011) National report on the State of Environment in Ukraine in 2011. Ministry of Ecology and Natural Resources of Ukraine. http://www.menr.gov.ua/docs/activity-dopovidi/NacDopovid2011.pdf

UNECE (2007a) Environmental indicators and environmental assessment reports: Eastern Europe, Caucasus and Central Asia. UNECE Working Group on Monitoring and Assessment of Environment, New York/Geneva

UNECE (2007b) Environmental monitoring: guidelines for the application of environmental indicators in Eastern Europe, Caucasus and Central Asia. http://www.unece.org/fileadmin/DAM/env/europe/monitoring/Belgrade/CRP1.Indicators.Ru.MK.pdf

UNECE (2007c) Environmental performance reviews: Ukraine, second review. UNECE Working Group on Monitoring and Assessment of Environment, New York/Geneva

Varlamov N, Palaguta OA (2013) System of environmental indicators for assessing the state of the environment in Ukraine. Scientific statements Belgorod State University, Science 7, 160, Release 24. http://unid.bsu.edu.ru/unid/res/ved/detail.php?IBLOCK_ID=106&SECTION_ID=571&ELEMENT_ID=251068

Chapter 11
Mathematical Tools to Assess Soil Contamination by Deposition of Technogenic Emissions

Olexandr Popov and Andrij Yatsyshyn

Abstract Mathematical models are developed of soil contamination by deposition of impurities emitted into the surface layer of the atmosphere by industry. Modelling enables the definition of the average level of contaminants arriving at the soil surface for particular periods (day, month, year, etc.) and forecasting zones of long-term loading on the soil. This information will support effective managerial decisions to provide the necessary level of ecological safety in zones of potential hazard.

Keywords Soil · Contamination · Mathematical models · Ecological safety

Introduction

Nuclear power is a fundamental sector of Ukraine's economy, providing up to half of the country's electricity. Currently, the four nuclear power plants (NPPs) within the National Nuclear Energy Generating Company *(Energoatom)* house 15 operational reactors with an installed capacity of 13,835 MW. The *Energy Strategy of Ukraine up to 2030* projects a total share of electricity from nuclear power of 52% depending not only on financial and economic factors but, also, on safe operation of the enterprise.

Development of nuclear power should be based on the studies of its environmental impact during accident-free operation, taking into account the whole complex of factors relating to impact on public health and, also, on the environment. Like any big industrial complex, nuclear power carries environmental risk, in this case thermal, radiation, chemical, and biologic pollution which can activate dangerous geodynamic processes that cause a real threat to public health and the environment (Popov 2014a).

O. Popov (✉) · A. Yatsyshyn
Institute of Environmental Geochemistry of the National Academy of Sciences, 34a Palladina Prospect, Kyiv-142 03680, Ukraine
e-mail: sasha_popov1982@mail.ru

D. Dent and Y. Dmytruk (eds.), *Soil Science Working for a Living*,
DOI 10.1007/978-3-319-45417-7_11

Problem Description

To control the radiation and chemical exposure at Ukrainian NPPs, observation zones in areas influenced by nuclear plants operate systems of ecological and radiation monitoring. The various operational systems of radiation monitoring at Ukrainian NPPs generally meet the needs of existing plants and current requirements. But as to the assessment of non-radiation impacts on environment, there is no effective system in place (Lysychenko and Popov 2014). In this regard, Energoatom contracted the Institute of Environmental Geochemistry of the National Academy of Sciences of Ukraine to develop technical proposals to deliver information and an expert system for evaluation of the impact of nuclear power on the environment (EcoIES). The system should ensure the collection, processing, analysis, storage, and preservation of information for comprehensive environmental assessment of the environment and anthropogenic impact on the population in the observation zone of NPPs. The present authors are undertaking this task.

Barbashev et al. (2014) and Popov (2014b) have elaborated a system that includes various modules, one of which is mathematical modelling and prediction of anthropogenic loadings and environmental risk assessment for the NPP protective zones. Here, we describe mathematical tools to evaluate soil contamination due to chemical emissions of NPPs that are deposited on the soil surface from the surface layer of the atmosphere.

Problem Solving

The main factor of soil contamination by non-radioactive substances is deposition from the atmosphere of emissions from auxiliary facilities: boilers, diesel generators, transport, water and sanitation works, repair and building workshops, etc. Monitoring of chemical contamination within the observation zones is carried out by staff of accredited laboratories of the ecological and environmental protection departments, following the approved *Schedule of sampling and performing of chemical analysis of soil, sludge, and sediment*. Samples are collected in accordance with *Regulation control on soils and sediments within the NPP in the sanitary protection zone and observation zone* and analyses carried out by the appropriate laboratory (Energoatom 2013). For example, in a separate division of Khmelnytska NPP, the following samples are taken every 6 months (KhNPP 2013):

- Soil near slurry reservoir—2 sample points,
- Soil near the site of composting—2 sample points,
- Protective zone of cooling pond—2 sample points,
- Protective zone of the roads from Khmelnytska NPP to Komarivka and to Bilotyn—2 sample points,
- Mouth of Horyn River—2 sample points, and
- Production site—5 sample points.

This low frequency and number of sampling points for the analysis of soil in the 30-km observation zone of KhNPP does not reveal the real distribution of soil chemical contamination and constrains the effectiveness of ecological safety management.

Mathematical modelling enables us to define the areas of anthropogenic impact on the soil and the environmental risks and losses, which should be the basis for the effective management of environmental safety.

Since the main source of soil chemical contamination in the NPP observation zone is the deposition of impurities from the atmosphere, the model must include the main factors influencing the spread of contaminants in the air and their deposition on the ground. In view of the complexity and variety of processes and factors responsible for atmospheric pollution, there are many models of various types (Popov 2010; Popov and Yatsyshyn 2013).

As we know, meteorological conditions change over time; wind changes direction and speed, and also, the state of the atmosphere changes. To determine the average pollution over time, the random nature of these factors must be considered, so we should use a stochastic model of contamination dispersal. Here, we construct a stochastic, mathematical model of soil contamination by deposition of heavy impurities from the surface layer of the atmosphere. It will determine the distribution of mid-level concentrations of pollutants at the soil surface over the long term, taking account of the emissions of impurities in the atmosphere from various sources within the NPP.

Development of a Mathematical Model of Soil Contamination

Considering the spread of contamination from the emission sources over a specified time period T (say a month, a season, or a year) depending on the meteorological conditions for the place where this emission source is located, let there be n typical meteorological situations, the duration of each of which is t_i. Ignoring the transition processes, we may assume that the spread of pollution for the period T is continuous in time and depends on alternating, different types of meteorological situation. The distribution of pollution in the region may be calculated according to the direction of movement of the air mass and the period of time that the wind blows from any particular direction; so we can estimate the spread of contamination that corresponds to one direction of air movement and, then, estimate the average pollution values that correspond to a particular season. After the calculations for each meteorological situation, the average pollution concentrations for the period $T = \sum_{i=1}^{n} t_i$ will be:

$$C_{\text{av}} = \sum_{i=1}^{n} C_i t_i \Big/ \sum_{i=1}^{n} t_i = C_1 t_1 / T + \cdots + C_n t_n / T = C_1 p_1 + \cdots + C_n p_n = \sum_{i=1}^{n} C_i p_i \quad (11.1)$$

where C_i—value of the pollutant concentration which can be used for a particular meteorological situation with specific duration t_i; p_i—frequency or probability of i—meteorological situation.

Equation (11.1) is a statistical model of air pollution proposed by Marchuk (1982). To arrive at the average concentration of pollution in a particular area for a certain period of time, we must know the characteristics of the meteorological regimes during the study period. Meteorological mode is determined by wind direction, speed of air masses, and rainfall.

Wind Speed and Direction

The metrological service measures average daily wind speed and direction. Average daily wind direction corresponds to one of the 8 main directions: W, SW, S, SE, E, NE, N, and NW. Typically, the period of time for which the average concentration of pollution is determined is a month, or a year. Table 11.1 and Fig. 11.1 show an example of data on the frequency of wind directions and lulls for a region.

In Fig. 11.1, $P_{\mathrm{E}}, P_{\mathrm{N-E}} \cdots P_{\mathrm{S-E}}, P_{\mathrm{L}}$ depict the repeatability as the percentage of east, northeast, and other directions, including the lull, so that:

$$P_E + P_{\mathrm{N-E}} + P_{\mathrm{N}} + P_{\mathrm{N-W}} + P_{\mathrm{W}} + P_{\mathrm{S-W}} + P_{\mathrm{S}} + P_{\mathrm{S-E}} + P_{\mathrm{L}} = 100\%$$

Typically, in problems where probability is used, one can operate with a relative value of probability but not with percentages, so the percentage probability is divided by 100. All the following calculations will be performed in terms of relative probability.

Table 11.1 The probability distribution of the wind directions

Wind direction	E	NE	N	NW	W	SW	S	SE	Lull
Probability (%)	P_{E}	$P_{\mathrm{N-E}}$	P_{N}	$P_{\mathrm{N-W}}$	P_{W}	$P_{\mathrm{S-W}}$	P_{S}	$P_{\mathrm{S-E}}$	P_{L}

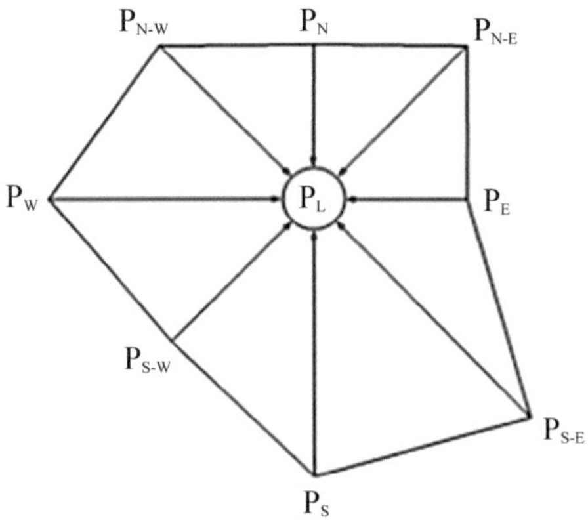

Fig. 11.1 Example of the wind rose for a particular period of time

Table 11.2 Probability distribution of wind speed

Wind speed (m/s)	Lull, 0–1	1–2	2–3	3–4	…	$(k-1)-k$
Probability	$p_1 = P_L$	p_2	p_3	p_4	…	p_k

Wind speed and repeatability of wind speeds, by gradations, are presented in Table 11.2.

For these data: $\sum_{i=1}^{k} p_i = 1$. We also require information about average u_{av} and maximum u_{max} wind speed for the study period.

Taking the wind rose into account, the formula for the average pollution concentrations (11.1) will be:

$$C_{av} = C_E P_E + C_{N-E} P_{N-E} + C_N P_N + C_{N-W} P_{N-W} + C_W P_W + C_{S-W} P_{S-W} + C_S P_S + C_{S-E} P_{S-E} + C_L P_L \quad (11.2)$$

where $C_E, C_{N-E}, \ldots, C_{S-E}$—stationary pollution distributions that meet the east, northeast, southeast wind at a certain average speed of the air masses of this area; C_L—distribution of contamination taking lulls into account.

For accurate information on pollution, it is not enough to consider only the probability of wind directions that correspond to the eight main vectors; it is also necessary to determine the probability of other wind directions. To resolve this issue, we can apply the method of interpolating probabilities with intermediate wind direction with a certain step $\Delta\varphi$ (Popov 2010). Figure 11.2 is an exemplar wind rose with intermediate wind directions, each of which is turned at an angle relative to the previous $\Delta\varphi$.

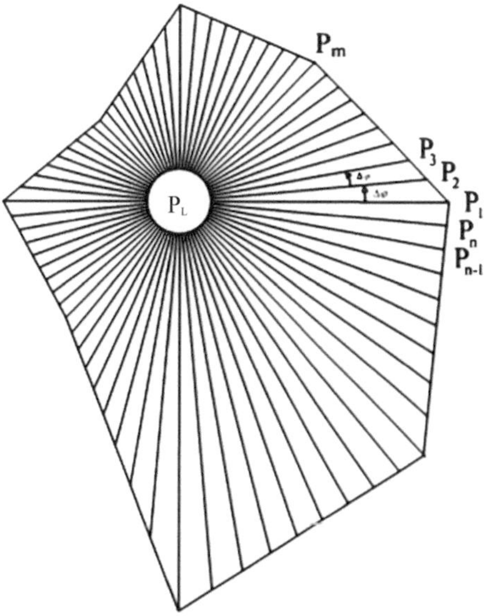

Fig. 11.2 Wind rose with intermediate wind directions

For interpolation, let us divide the whole wind rose on n direction with the step $\Delta\varphi$. For an eight-corner-rhomb wind rose, the most convenient step for interpolation is $\Delta\varphi = 5^\circ$ or $\Delta\varphi = \pi / 36$ radian. Then, the number of destinations of contamination spread will be $n = 360^\circ/5^\circ = 72$. Interpolation of intermediate values of wind direction probability at such a gradation may be carried out geometrically (Popov 2010). For different periods, it will appear as follows:

(1) $1 \le n \le 10,\ \overline{\mathrm{P}}_{1-10} = \mathrm{P}_{\mathrm{E}} \sin\alpha_1 / \sin\left((n-1)\frac{\pi}{36} + \alpha_1\right),$
where $\alpha_1 = \left|arctg\ \left(P_{\mathrm{N-E}} / \left(P_{\mathrm{N-E}} - P_{\mathrm{E}}\sqrt{2}\right)\right)\right|$

(2) $11 \le n \le 19,\ \overline{P}_{11-19} = P_{\mathrm{N-E}} \sin\alpha_2 / \sin\left((n-10)\frac{\pi}{36} + \alpha_2\right),$
where $\alpha_2 = \left|arctg\ \left(P_{\mathrm{N}}\sqrt{2} / P_{\mathrm{N-E}} - 1\right)\right| + \frac{\pi}{4}$

(3) $20 \le n \le 28,\ \overline{P}_{20-28} = P_{\mathrm{N}} \sin\alpha_3 / \sin\left((n-19)\frac{\pi}{36} + \alpha_3\right),$
where $\alpha_3 = \left|arctg\ \left(P_{\mathrm{N-W}} / \left(P_{\mathrm{N}}\sqrt{2} - P_{\mathrm{N-W}}\right)\right)\right|$

(4) $29 \le n \le 37,\ \overline{P}_{29-37} = P_{\mathrm{N-W}} \sin\alpha_4 / \sin\left((n-28)\frac{\pi}{36} + \alpha_4\right),$
where $\alpha_4 = \left|arctg\ \left(P_{\mathrm{W}}\sqrt{2} / P_{\mathrm{N-W}} - 1\right)\right| + \frac{\pi}{4}$

(5) $38 \le n \le 46,\ \overline{P}_{38-46} = P_{\mathrm{W}} \sin\alpha_5 / \sin\left((n-37)\frac{\pi}{36} + \alpha_5\right),$
where $\alpha_5 = \left|arctg\ \left(P_{\mathrm{S-W}} / \left(P_{\mathrm{W}}\sqrt{2} - P_{\mathrm{S-W}}\right)\right)\right|$

(6) $47 \le n \le 55,\ \overline{P}_{47-55} = P_{\mathrm{S-W}} \sin\alpha_6 / \sin\left((n-46)\frac{\pi}{36} + \alpha_6\right),$
where $\alpha_6 = \left|arctg\ \left(P_{\mathrm{S}}\sqrt{2} / P_{\mathrm{S-W}} - 1\right)\right| + \frac{\pi}{4}$

(7) $56 \le n \le 64,\ \overline{P}_{56-64} = P_{\mathrm{S}} \sin\alpha_7 / \sin\left((n-55)\frac{\pi}{36} + \alpha_7\right),$
where $\alpha_7 = \left|arctg\ \left(P_{\mathrm{S-E}} / \left(P_{\mathrm{S}}\sqrt{2} - P_{\mathrm{S-E}}\right)\right)\right|$

(8) $65 \le n \le 72,\ \overline{P}_{65-72} = P_{\mathrm{S-E}} \sin\alpha_8 / \sin\left((n-64)\frac{\pi}{36} + \alpha_8\right),$
where $\alpha_8 = \left|arctg\ \left(P_{\mathrm{E}}\sqrt{2} / P_{\mathrm{S-E}} - 1\right)\right| + \frac{\pi}{4}$

This derives the value $\overline{P}_i$ of non-probabilistic content as far as $\sum_{i=1}^{72} \overline{P}_i > 1$.

To get the final division of probabilities of directions, we need to get $\overline{P}_i$ and then divide it by $\sum_{i=1}^{72} \overline{P}_i$.

In this way, we will arrive at the necessary probabilities: $P_i = \overline{P}_i / \sum_{i=1}^{72} \overline{P}_i$

In as much as the spread of pollution in the atmosphere is affected by wind speed, we need to take account of all wind speeds observed during the time T, not just model averages, and we should calculate the relative frequency of wind speeds by geographic directions. For example, to obtain the relative frequency of wind speeds from an easterly direction, we must multiply the probability of this area $P_{\mathrm{E}} = P_1$ by probabilities of corresponding wind speeds p_{1-k} (Table 11.2). In the same way, we get the relative frequency of wind speeds in other directions. Table 11.3 shows the calculation procedure.

Atmospheric Stratification

Stratification of the atmosphere has a significant effect on the shape of emissions plume, and this cannot be neglected when we try to determine the concentration of

Table 11.3 Determination of repeatability of wind speeds in different directions of the wind rose

Wind speed (m/s)	Wind direction				
	E, 0°	5°	…	355°	Total
Lull, 0–1	$p_1 = P_L$				$p_1 = P_L$
1–2	$P_1 \cdot p_2$	$P_2 \cdot p_2$	…	$P_{72} \cdot p_2$	p_2
2–3	$P_1 \cdot p_3$	$P_2 \cdot p_3$	…	$P_{72} \cdot p_3$	p_3
…	…	…	…	…	…
$(k-1)-k$	$P_1 \cdot p_k$	$P_2 \cdot p_k$	…	$P_{72} \cdot p_k$	p_k
Total					$\sum_{i=1}^{k} p_i = 1$

pollution. There are six basic conditions of the atmosphere, characterized by the temperature gradient, i.e. the distribution of temperature with height (Popov 2009a). Meteorological services do not provide information about the probability distribution of atmospheric condition but, since stratification depends to a large degree on wind speed, we may consider the following dependences, which were obtained experimentally (Popov 2009b). Figure 11.3 depicts the probability distributions of atmospheric stratification according to wind speed. From these relationships, we can create a table of probability distributions of the stratification of atmosphere depending on the wind speed (Table 11.4). Extrapolation of data to the left was performed to determine the number of atmosphere stratification distributions at a wind speed of 1.5 m/s. Using values for wind speed taken at mid-partial intervals, we arrive at the probability of the *i* class of atmosphere stability due to the *j* wind speed—p_{ji}.

Thus, the basic parameters for statistical modelling of pollutants in the atmosphere are defined. Using the above results, a statistical model can be written in general terms. To build statistical models of the average level of pollutants deposited from the atmosphere over a long period of time, we may consider the

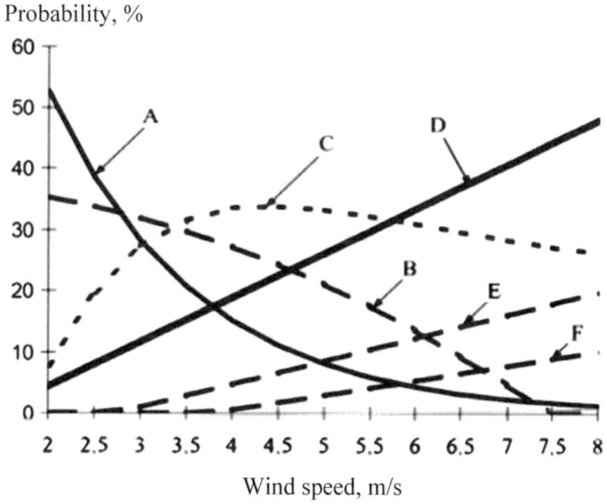

Fig. 11.3 Probability distribution of atmospheric stratification depending on the wind speed

Table 11.4 Probability distribution of the atmosphere, depending on the wind speed

State of the atmosphere	Wind speed (m/s)					
	1.5	2.5	…	j	…	k
1. A	0.64	0.38	…	p_{j1}	…	p_{k1}
2. B	0.35	0.35	…	p_{j2}	…	p_{k2}
…	0	0.19	…	…	…	…
i.	0.01	0.08	…	p_{ji}	…	p_{ki}
…	0	0	…	…	…	…
6. F	0	0	…	p_{j6}	…	p_{k6}
Total	1	1	1	1	1	1

distribution of emissions in a certain direction and wind speed; let it be an easterly wind with speed $u_1 = 1.5\ \mathrm{m/s}$. Then, taking into account the probabilistic division of the atmosphere, the model may be written as follows:

$$C_{\mathrm{av}} = P_{11}\sum_{i=1}^{6} p_{1i}C\big(Q,\ u_{1i},\ a_i,\ b_i,\ c_i,\ H_{ef_1},\ f,\ x,\ y,\ z\big) \tag{11.3}$$

where P_{11}—probability of an east wind at speed $u_1 = 1.5\ \mathrm{m/s}$ (Table 11.3); p_{1i}—probability of the respective state of the atmosphere at wind speed $u_1 = 1.5\ \mathrm{m/s}$ (Table 11.4); C—model through which the concentration of the pollutants is calculated; Q—the quantity of emissions (g/s); u_{1i}—wind speed at the height of the mouth of the delivery pipe when the wind speed at the height of the emissions plume $u_1 = 1.5\ \mathrm{m/s}$; a_i, b_i, c_i—parameters that define the state of the atmosphere; H_{ef_1}—effective height to which the plume of emissions lifts at the wind speed at an altitude of the wind vane $u_1 = 1.5\ \mathrm{m/s}$; f—coefficient that determines the change with the time of concentration of pollution due to the interaction with the environment (washout of sediments, chemical transformation, absorption by underlying surface, etc.), (1/h).

Further, it is necessary to add up (11.3) at different wind speeds. We will have:

$$C_{\mathrm{av}} = \sum_{j=1}^{k} P_{1j}\sum_{i=1}^{6} p_{ji}C\left(Q,\ u_{ji},\ a_i,\ b_i,\ c_i,\ H_{ef_j},\ f,\ x,\ y,\ z\right) \tag{11.4}$$

Expression (11.4) can be defined as a statistical model of pollution dispersion for the wind direction (easterly).

Finally, to get a general formula for the statistical model, add up (11.4) for all directions of the wind, not neglecting the lull:

$$C_{\mathrm{av}} = \sum_{m=1}^{n}\sum_{j=1}^{k} P_{mj}\sum_{i=1}^{6} p_{ji}C\left(Q,\ u_{ji},\ a_i,\ b_i,\ c_i,\ H_{ef_j},\ f,\ x_m(x,\ y,\ \varphi_m),\ y_m(x,\ y,\ \varphi_m),\ z\right) + C_{\mathrm{L}}P_{\mathrm{L}} \tag{11.5}$$

where H_{ef_j}—effective height of the emissions plume during j speed, [m]; u_{ji}—the wind speed at the effective height of the emissions plume H_{ef_j} at the wind speed at the air vane u_j [m/sec] for the i state of the atmosphere; φ_m—the angle, which gives the wind speed effect; $x_m(x, y, \varphi_m) = x\cos\varphi_m + y\sin\varphi_m$, $y_m(x, y, \varphi_m) = -x\sin\varphi_m + y\cos\varphi_m$—formulae of transition to another coordinate system associated with the rotation direction of propagation of pollutants at the angle φ_m in comparison with the easterly direction; C_{L}—model for determining the distribution of pollutants during lulls; and P_{L}—probability of lull.

Following Popov (2010), the concentration of pollution in a lull can be estimated by the following mathematical model:

$$C_{\mathrm{L}} = C\,(H_{ef} - L) \tag{11.6}$$

where $C\,(H_{ef} - L)$—model as a function of difference including wind $(H_{ef} - L)$, L—the height of the lull layer, (m).

In this way, a mathematical model was derived to determine the spatial distribution of emissions (from stationary sources) in the surface layer of the atmosphere (SLA) of an area of interest, taking into account the dynamics of meteorological factors over a long period of time. The basis of the model (11.5) is a function of many variables C, which is a deterministic model that can determine the concentration of pollutants during fixed weather conditions. That model may be, for example, the well-known model that is used by IAEA or any other K model (Popov 2010).

Where soil contamination is due to the deposition of heavy impurities from the atmosphere, features of distribution are primarily determined by the deposition rate of the impurities, determined by their density and size. Aerosols occur in wide range of sizes—from thousandth of a micron to hundreds of microns. The rate of gravitational deposition w_g (sm/s) of spherical particles can be found by Stokes's formula:

$$w_g = 1.3 \cdot 10^{-2} \rho_n r_n^2 \tag{11.7}$$

where ρ_n—the density of dust particles, (g/sm^3); r_n—the radius of dust particles [microns].

In this case, the deterministic model for the formula (11.5) can be a model obtained by Berliand while solving turbulent diffusion equation for the heavy impurity (Popov 2010):

$$\begin{aligned} C\,(x, y, 0) = {} & QH_{ef}^{n+\omega(1+n)} u_1^{1.5+\omega} \Big/ \left(2\,(1+n)\,\Gamma\,(1+\omega)\sqrt{\pi k_0 x}(k_1 x)^{1+\omega}\right) \\ & \times \exp\left(-u_1 H_{ef}^{1+n} \,/\, ((1+n)^2 k_1 x) - y^2 \,/\, (4k_0 x)\right) \end{aligned} \tag{11.8}$$

where $\omega = w_g / (k_1(1+n))$; u_1—the wind speed at the height of air vane; k_1—vertical turbulent diffusion coefficient at the height of air vane; n—dimensionless parameter that depends on the stability of the atmosphere; Γ—gamma function. Parameter k_0 is defined by the formula $k_0 = k_z(h) / u(h)$, where the vertical turbulent diffusion coefficient depends on the height h; h—the highest level of SLA.

The distribution of impurities from the SLA has significantly affected their interaction with environment. The most significant impact is leaching of impurities, chemical transformation, and reaction with the underlying surface (Popov 2010):

$$C_{\alpha\neq 0} = C_{\alpha=0} \cdot \exp\left[-\alpha x / u_{H_{ef}}\right] \tag{11.9}$$

where $C_{\alpha=0}$—concentration distribution without interaction with the environment and $u_{H_{ef}}$—wind speed at the height emission sources. Coefficient α is an integrated parameter that includes chemical transformation (oxidation, neutralization) of output impurities from precipitation, absorption of underlying surface, dry deposition of heavy impurities, etc. The formulae for such calculation are given in detail by Popov (2010).

The final step in the creation of mathematical model to estimate soil pollution is to determine the amount of heavy impurities that fall on the soil surface. The average quantity of impurities deposited on the underlying surface with the square S during emission period t is defined by the formula:

$$J_{av}(x, y, t) = t(w_g + \beta k_z) / h \iint_S C_{av}(x, y)\, dx\, dy \tag{11.10}$$

where β—parameter that characterizes the interaction of impurities with the underlying surface.

Thus, in the EcoIES system, we have derived a suite of mathematical models to assess soil contamination due to deposition of heavy impurities from the SLA. Their application should improve the efficiency of decision-making for environmental security in the NPP observation zone.

Conclusions

The information content of chemical monitoring of soils in the Ukrainian NPP observation zone falls far short of requirements. Because sampling is carried out at only a few points in the NPP zone of influence, it does not reveal the real distribution of levels of chemical contamination, which constrains the effectiveness of management of ecological safety of NPP.

To solve this problem, modern mathematical tools have been applied to assess soil contamination due to chemical emissions from NPP that end up as deposition of heavy impurities from the surface layer of the atmosphere. The mathematical model can determine the average amount of impurities that falls on the soil surface in the NPP observation zone for any specified period of time. This will enable the determination of the actual area of long-term anthropogenic impact on the soil so as to assess environmental risks and losses, which may be a basis for effective management decisions that ensure the appropriate level of environmental safety in the NPP observation zone.

References

Barbashev SV, Lysychenko GV, Popov OO (2014) Expanding the functionality of radiological monitoring of the environment in NPP areas for effective management decision making. Nucl Energy Environ 2(4):12–18 (Ukrainian)

Energoatom (2013) Assessment of non-radiological factors of SD NPP SI NAEC Energoatom on the environment 2012. Report SI NAEC Energoatom (Ukrainian)

KhNPP (2013) Assessment of non-radiological factors of SD Khmelnytska NPP SI NAEC Energoatom on the environment, 2012. Report SD KhNPP (Ukrainian)

Lysychenko GV, Popov OO (2014) The current state of the information system of environmental monitoring in zones of NPP Ukraine. Scientific Papers of the Pukhov Institute of Modelling Problems in Power, NAS of Ukraine, 71 (Ukrainian)

Marchuk GI (1982) Mathematical modeling of environmental problems. Nauka, Moscow (Russian)

Popov OO (2009a) Mathematical model of the distribution of technogenic pollution from energy enterprises. Scientific Papers of the Pukhov Institute of Modelling Problems in Power, NAS of Ukraine, vol 51, pp 73–84 (Ukrainian)

Popov OO (2009b) Stochastic model of surface air pollution from energy enterprises (an example of TPP). Scientific Papers of the Pukhov Institute of Modelling Problems in Power, NAS of Ukraine, vol 53, pp 10–21 (Ukrainian)

Popov OO (2010) Mathematical and computer modelling of man-made pressures on the city atmosphere from stationar, point-sources of pollution. PhD thesis in Engineering Science: 01.05.02 Pukhov Institute of Modeling Problems in Power, NAS of Ukraine (Ukrainian)

Popov OO, Yatsyshyn AV (2013) Mathematical model of distribution of pollution from the effects of non-radiation hazardous facilities on the environment. Model Inf Technol 69:11–21 (Ukrainian)

Popov OO (2014a) NPP impact on environmental security of adjacent areas. Scientific Papers of the Pukhov Institute of Modelling Problems in Power, NAS of Ukraine, vol 70, pp 11–20 (Ukrainian)

Popov OO (2014b) The concept of information and expert systems for environmental impact assessment of NPP on the environment. In: Materials of the XXXIII annual scientific technical conference-modelling, 15–16 Jan 2014: Theses. Pukhov Institute of Modelling Problems in Power, NAS of Ukraine, 5–6 (Ukrainian)

Part III
Soil Fertility, Degradation and Improvement

Chapter 12
Theoretical Problems of Improving Agrochemical Terminology

Anatoly Khristenko

Abstract Various terms are in common use to characterize the provision of nutrients by the soil: assimilable, labile, available, mobile, exchangeable, active, and others. But there is no unambiguous interpretation of any of these terms. Since all the chemical methods of assessing nutrients in the soil use the same principle, it is not correct to say that, according to a specific method, an element is *more available* than according to another method. Moreover, weak correlation between the data obtained by different methods indicates random or systematic errors. The lack of clear definitions leads to misunderstanding and makes it hard to develop accurate soil diagnostic methods, and hard to solve the nutrient-use efficiency problem. Therefore, further theoretical research is needed to improve agrochemical terms.

Keywords Soil diagnostics · Theory · Agrochemistry terms · Availability of elements

Introduction

In most countries, one of the main ways to safeguard food security is to increase the use of fertilizers. This brings its own problems in the shape of increasing anthropogenic load on the environment, increasing cost of the obtained production and, often, reducing its quality. Precise information about the trophic state of soils enables the optimization of crop nutrition; furthermore, precise assessment of soil fertility improves the efficiency of fertilizer use and minimizes the negative impact of agrochemicals on the environment.

However, there is an ill-considered issue with the usual methods of soil quality assessment, in particular those using acidic extractants such as Kirsanov (pH 1.0),

A. Khristenko (✉)
Agrochemistry Department, Sokolovskyi Institute for Soil Science and Agrochemistry Research, 4 Tchaikovsky St, Kharkiv 61024, Ukraine
e-mail: khristenko.an@mail.ru

D. Dent and Y. Dmytruk (eds.), *Soil Science Working for a Living*,
DOI 10.1007/978-3-319-45417-7_12

Brau-Kurtz 2 (pH 1.0), Mehlich 1 (pH-1.2), Arrhenius (pH 2. 0) Chirikov (pH 2.5), Van Lierop/Kelowna (pH 2.7), Truog (pH 3.0), Egner–Riehm (pH 3.6), Egner–Riehm–Domingo (pH 4.2). Each of these chemical methods may provide an objective measure of certain attributes of certain soils—generally those soils on which the methods and the interpretation of their results were developed. But soil attributes such as pH, clay content, and phosphate minerals such as apatite significantly affect the results of such assessments.

Consideration of this issue shows that objective information about the state of the soil cannot be obtained without better theory. For instance, it is believed that acid solutions imitate the action of acid excretions of plant roots: in reality, the process of absorption of nutrients from soil by roots is far more complicated. Theoretical constraints often lead to methodological problems. The error in determining the nutrient content of arable soils may be as much as 100–300%, creating an illusion of natural heterogeneity of soils in relation to nutrients. However, soils contain various amounts of thermodynamically stable compounds of phosphorus and potassium (apatite, felspar, etc.) and the phosphorus or potassium in these minerals is not directly accessible to plants—nevertheless, these compounds are partially extracted by acid or alkaline solutions in the process of chemical analysis.

Previously, the author has considered several theoretical aspects of the trophic status of soils (Khristenko 2014), taking the view that the phosphate and potash system in soils is an open, many-faceted thermodynamic system in which spontaneous, energetically advantageous decay processes are balanced by synthesis processes (Trofimov 1997). It has been shown that the dynamic equilibrium level of hard-worked soils is on the border between *low* and *medium* provision of phosphorus and potassium—but the absence of clear definitions leads to misunderstanding of the phenomenon and complicates the task of developing precise methods of soil diagnostics and practical recommendations. Various terms are used to describe soil fertility in regard to the supply of nutrients: assimilable, available, mobile (denoting the capacity factor); availability (the intensity factor); exchangeable, active, and others—but there is no unambiguous interpretation of any of these terms. For all these reasons, we need theoretical research to improve agrochemical terms.

Methods

Methods based on various principles were used in this research: chemical methods (acid, alkaline and salt); exchange chromatography; radioactive indicators; biological methods (field, vegetation, and laboratory experiments with plants); and statistical analysis of the bank of data (based on DBMS Access 98). The information bank contains analytical data on about 2000 standard soils of Ukraine and other CIS countries drawn from our own data and the literature.

Results and Discussion

In the first half of twentieth century, two terms were introduced to characterize plant nutrient availability—*accessible* and *available*. In English, the term *available* was usually used to denote nutrients that are accessible to plants. Later, authors in the CIS countries began to use the term *mobile,* which is understood as an element that can easily be moved in the soil or as the concentration of, say, P_2O_5 and K_2O in the soil solution (Vorobyova et al. 1995). Such an understanding has given rise to the well-known definition by Pryanishnikov (1963):

> Digestible P_2O_5 is not a specific chemical compound and does not actually occur in the soil.

Perhaps, *available* is not the best term to use in relation to nutrient elements in the soil because only the plant can answer the question: “Is this compound or element available to plants, or not?” Soil analysis does not determine *available* phosphorus or potassium but merely some part of total phosphorus or potassium. The nutrient will become *available* when it is determined by the activity of plants and this is how indices of a soil’s capacity to provide nutrients are developed. In any case, the concept *mobile* has no meaning. Practically, all the phosphorus, potassium, and microelements are in the solid phase of the soil. These nutrient elements still need to be found and taken up by plants—but many people forget that the methods for the determination of the so-called *available* forms of most nutrients are not direct but indirect.

Comparing data obtained by *hard* (acid and alkaline) methods and *soft* methods (salt) is a logical contradiction:

- It is considered that, first, the capacity factor (mobile, exchangeable, available, etc.) is determined. Then the intensity factor (degree of mobility) is determined. But, in different soils, there may be no correlation between the indices of capacity and intensity.
- If we accept this logic then, despite the fact that the entire available phosphorus (potassium, microelements, etc.) is available, some phosphorus is considerably more available.

Research reveals that differences in estimates of soil fertility (obtained by different methods) were caused by either random or systematic error. For example, it was found that the increasing natural P_2O_5 content defined by the Chirikov method in the plough layer of soils of Ukraine (from very low to very high) is an illusion.

The apparent increase is caused by partial dissolution of apatite minerals in acetic acid solution: in Fig. 12.1, soils are placed in order of increasing P_2O_5 content by the Chirikov method (0.5M CH_3COOH at pH 2.5) but, in reality, the phosphorus in these minerals becomes available only through mineral weathering. By contrast, the determination of phosphorus by the Olsen method (0.5M $NaHCO_3$ at pH 8.5) shows a phosphate level that corresponds to the boundary between the low and medium provision. Empirical data from field tests confirm that this last diagnosis is accurate; on all kinds of arable soils with a natural supply of N, P, and

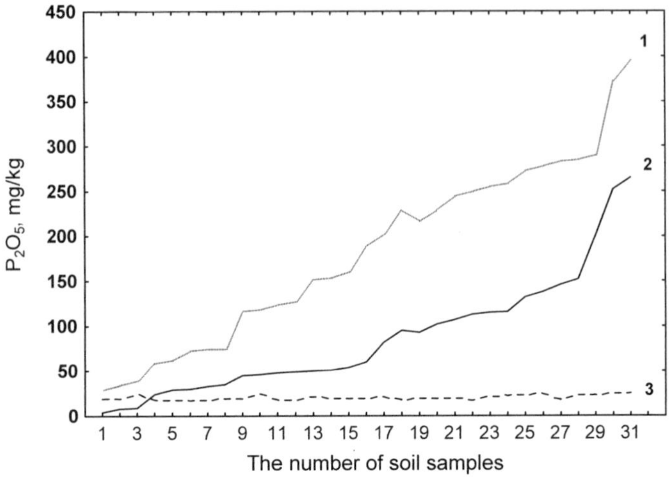

Fig. 12.1 Estimates of the natural P_2O_5 content in the plough layer: *1* apatite (Ca-P fraction by the method of Chang and Jackson; *2* P_2O_5 by Chirikov method); *3* P_2O_5 by Olsen's method

K, fertilizer-efficiency is high—and the effectiveness of phosphate fertilizers on chernozem is the highest in Ukraine.

Now, *capacity* is a value that indicates how much of something is in place—but not that what it is possible to extract from there. What can be got out of the soil is described by the concentration of nutrient ions—but the term *intensity factor* is not used in Russian or Ukrainian languages (Janiszewski 1996). Indeed, Sokolov (1968) was right in stating:

> The use in agrochemistry of terms from physical chemistry or thermodynamics can lead to misconceptions about soil phosphates.

We may take it that the principle of action of all chemical methods is identical, so to assert that, according to data of one particular method, phosphorus or potassium is more *available* than according to another method is, simply, not correct (those who are able to distinguish degrees of mobility may disagree). Determining the value of any index using different methods, we must come to the same result. The discrepancy between the estimates (low, medium, high) testifies not to the features of some method or other, but to an elementary error (random or systematic).

To avoid such misunderstandings in the names of all national standards developed by us, we used only one term—*available compounds* of phosphorus or potassium. In the following editions of normative documents, the term *mobile* will be replaced by the term *available*. In that case, if the methods are properly selected for the particular soil, the correlation coefficient (r) between the data of these methods is usually higher than 0.8 and this is observed regardless of terminology (*available, availability, active fraction*) (Fig. 12.2).

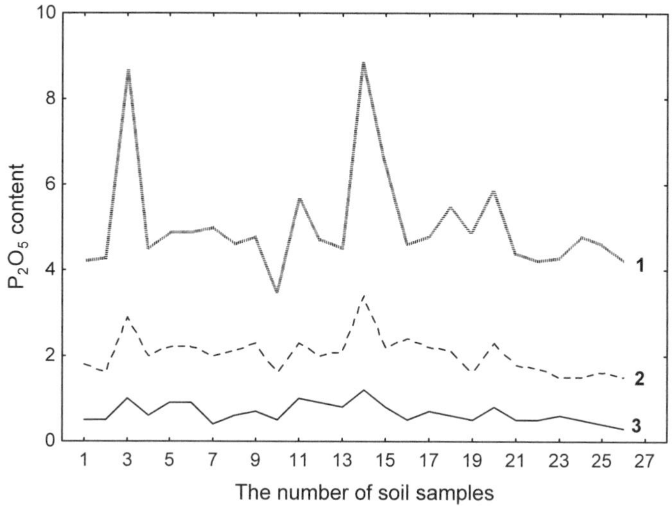

Fig. 12.2 Content of P_2O_5 in the plough layer of the unfertilized or little-fertilized soils: *1* by the method of Chang and Jackson, fraction Al-P, mgP_2O_5/100 g; *2* by the method of Olsen, mgP_2O_5/100 g; *3* by the method of Karpinski–Zamyatina, mgP_2O_5/kg

Faulty theory is responsible for the fact that the salt methods of State Standards of Ukraine by Dashevskiy (DSTU 7603) and Karpinski–Zamyatina (DSTU 4727) remain practically unused, and there is a widespread view that the parameter *intensity factor* is accessory and inadequate.

Faulty theory leads to wrong decisions: e.g. according to Prokoshev and Deriugina (2000), the most appropriate approach to determine available potassium is to use a block of three methods: Pchelkina's, Kirsanov's, and the method based on $CaCl_2$ soil extract. It is suggested that the first method should be use once every 7 years; the second, once every 5 years; the third, annually! At first glance, this proposal might seem plausible—but it is erroneous in theory and will not work in practice. First, a block of three analyses is an expensive measure and, secondly, it is unnecessary. For example, it is possible to treat a number of soil samples with 2M HCl solution (Pchelkin's method) and, then, with a solution of 0.2M HCl (Kirsanov's method). The correlation coefficient between the data of these methods should be close to one (Fig. 12.3)—we should not expect anything else from solutions of hydrochloric acid of different concentration (for easier visualization, the data of the Pchelkin method are shown in mg/10 g soil and those after Kirsanov in mg/kg soil). We have not performed enough analyses using $CaCl_2$ solution, therefore data obtained using another salt extractant (K_2SO_4 solution, method of Dashevsky) are presented.

By the use of three methods together, the following outcome is possible: according to data of one method, a soil is characterized as poorly provided with potassium; according to another, as modestly provided; from data of the third, as highly provided. Now, on the basis of which method should we calculate doses of

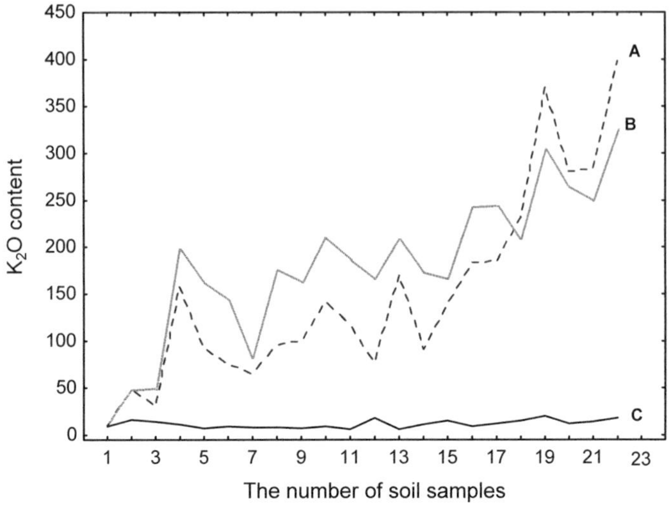

Fig. 12.3 Natural K_2O content in the plough layer of soils with comparable particle size distribution: *A* by Pchelkin's method, mg/10 g; *B* by Kirsanov's method, mg/kg; *C* by Dashevsky's method, mg/kg

fertilizers? Fractionation of potassium compounds is a highly specialized study which may bring to light additional information but a direct link between the potential potassium reserves and available compounds of this element is missing. Therefore, the division of soil potassium (and phosphorus) by distant, near, and direct *reserves* is not really correct—and often leads to the inadequate estimation of a soil's fertility. The theory also shows that attempts to increase the amount of nutrients in the soil without using proper fertilizers (by liming, bringing of acids, phosphorus mobilizing microflora, surfactants, and others) are doomed to failure—they cannot dynamically change that which does not exist.

Thus, there is a problem, the importance of which is underestimated. And a paradox: to date, there is no clear interpretation of the basic term of *agrochemistry,* that is the term characterizing the availability of nutrients to plants. Finding out what exactly is an *available* form of nutrient is not pointless theorizing. Without solving this problem, it is difficult to estimate the real soil fertility, improve the methodology to determine soil fertility, and optimize plant nutrition.

Conclusions

- At present, there is no unambiguous interpretation of many agrochemical terms, including those that characterize the availability of soil nutrients to plants.
- The principle of action of all chemical methods is identical. Therefore, it is not correct to assert that, by one method, an element is more "available" than by

another. The weak correlation between the data obtained by different methods shows random or systematic error.

- The lack of clear definitions leads to misunderstanding of the phenomenon. Moreover, it is difficult to develop reliable soil diagnostic methods and to assess nutrient-use efficiency. We need more theoretical research aimed at improving agrochemical terms.

References

Janiszewski PF (1996) Chemical assessment of phosphate status in soil. Agrochemistry 4:95–116 (Russian)

Khristenko AA (2014) Theoretical aspects of the trophic soil state and regularities of its evolution. Agrochem Soil Sci Special issue 1:172–179 (Russian)

Prokoshev VV, Derjugin IP (2000) Potassium and potassium fertilizers. Ledum, Moscow (Russian)

Pryanishnikov DN (1963) On the question of the use of chemicals in agriculture. Publisher of Agricultural Literature, Moscow, pp 217–242 (Russian)

Sokolov AV (1968) Determination of assimilable phosphates stock in soil, their composition and availability. Pochvovedenie 8:5–16 (Russian)

Trofimov SY (1997) Functioning of soils in biogeocenoses: approaches to description and analysis. Pochvovedenie 6:770–778 (Russian)

Vorobyova LA, Gorobetc AV, Rudakov TA (1995) Methods of assessing phosphorus availability in soils. Pochvovedenie 8:963–968 (Russian)

Chapter 13
Criteria and Parameters for Forecasting the Direction of Irrigated Soil Evolution

Sviatoslav Baliuk, Alexander Nosonenko, Marina Zakharova, Elena Drozd, Ludmila Vorotyntseva and Yuri Afanasyev

Abstract Three main pathways of soil evolution under irrigation are distinguished, depending on the initial soil conditions, the quality of irrigation water and the farming system. Degradation processes that can occur under adverse irrigation conditions are characterized, and a system of criteria and parameters is proposed for assessing the situation, the extent and nature of degradation of irrigated land.

Keywords Irrigated soils · Degradation evolution · Criteria · Parameters

Introduction

The growing season of the Steppe and much of the Forest-Steppe of Ukraine is characterized by abundant heat and sunshine but a significant water deficit. However, the fertile soils, mainly chernozem, offer potential to allay the impact of drought through irrigation. Global warming is expected to reduce the availability and quality of water resources, especially in the southern parts of the country, so this is another good reason to safeguard irrigation capacity.

Under irrigation, the evolution of soils undergoes an acceleration of soil processes, changes away from the natural equilibrium, and several quantitatively and qualitatively new properties emerge. In terms of their agronomic value and impact on crop yields, the effects of these changes can be positive or negative (Baliuk 1996; Baliuk et al. 2009). Parameterization, that is the establishment of specific quantitative characteristics of soil degradation processes that can develop under irrigation, provides an opportunity to optimize the use of agricultural technologies and land improvement measures aimed at improving soil fertility and irrigation efficiency.

S. Baliuk · A. Nosonenko · M. Zakharova · E. Drozd (✉) · L. Vorotyntseva · Y. Afanasyev
Sokolovskyi Institute for Soil Science and Agrochemistry Research, 4 Tchaikovsky St, Kharkiv 61024, Ukraine
e-mail: oroshenie@ukr.net

D. Dent and Y. Dmytruk (eds.), *Soil Science Working for a Living*,
DOI 10.1007/978-3-319-45417-7_13

Materials and Methods

Research has been carried out on processes, criteria and indicators of land degradation under irrigation; organizational, economic, agronomic and land improvement measures to boost fertility; and the conditions, methods and parameters of these measures on degraded irrigated land. Field experiments, monitoring, surveys of ecological and land improvement status (ELIS) of land for industrial purposes, and modelling have been carried out on irrigation systems in the country around Kharkiv (Forest-Steppe), Donetsk and Lugansk (Northern Steppe), Zaporizhzhya and Odessa (Southern Steppe), and Kherson (Dry Steppe). Research included the following:

- Analysis and synthesis of information, expert opinions;
- Standard stationary experiments with statistical processing of results (Dospechov 1985);
- Application of the *key analogue* method to determine ELIS;
- Ecological and land survey of irrigated lands (State Water Management Committee of Ukraine 2004);
- Environmental and ameliorative monitoring (State Water Management Committee of Ukraine 2002a, b);
- Certified and temporarily admitted methods for chemical analysis of water, soil and plants (National Standards of Ukraine 2004, 2006); and
- Statistics and correlation analysis using standard software packages (Dmitriev 2009).

Results and Discussion

Directions of Irrigated Soil Evolution

Irrigation increases the speed of fundamental soil processes and agro-ecosystem evolution. From the outset, it changes the character, direction, rate and spread of many soil processes—in particular the accumulation of soluble sodium and amorphous silica. Trends of gleying and the transformation of secondary minerals are not yet measurable, but we believe that they will become apparent in future. Processes such as the development of alkalinity occur in the first stage of irrigation but will probably be slow. The results of these changes depend on the quantity and quality of irrigation water, climatic and geological conditions, buffer properties of soils, irrigation techniques and the total farming system. Among several pathways of evolution of soil under irrigation, three may be highlighted (Baliuk et al. 2013):

With the use of Class1 water,[1] a sound crop rotation with perennial legumes and comprehensive fertilization, soil fertility is improved through increased available water and nutrients, a positive humus balance and increased biodiversity. In short, a highly productive agro-ecosystem is formed with defined parameters of soil properties. Depending on the crops grown, the productivity of irrigated land is between 2–4 and 7–8 times higher than unirrigated land. Even after cessation of irrigation, productivity remains 10–20% higher than similar unirrigated land for about 10 years.

- Under crop rotations without perennial legumes and under sub-optimal management, there are several negative trends, particularly compaction and a loss of soil organic matter, but productivity is still higher than for similar unirrigated soils.
- Use of Class2 and, especially, Class3 water leads to salinity, alkalinity, impaired physical properties, depletion of nutrient macro- and microelements, and pollution. Under such conditions, even the use of complex land improvement measures can only limit the degradation but cannot eliminate them entirely.

Degradation Processes on Irrigated Soils

Soil degradation may be described by the following:

- *Degree of soil degradation* a comparative level of soil degradation manifest at a particular time;
- *Soil degradation rate* the rate of change in the degree of soil degradation;
- *Type of soil degradation* a group of processes and deterioration of soil quality with the same range of mechanisms and results; and
- *Number of connections* the number of types of soil degradation diagnosed simultaneously in the same soil (Khitrov 1998).

According to the degree of deviation from optimum soil fertility, we may distinguish degrees of degradation:

- *Not degraded* soils in which productivity matches their natural fertility (up to 5% deviation of properties and productivity from optimal, which is within possible experimental error);
- *Slightly degraded* some deterioration of properties and performance caused by irrigation and reduced productivity that does not exceed 20%;
- *Moderately degraded* reduced productivity within 20–50%; and
- *Strongly degraded* productivity reduced by more than 50%.

[1]Class1—suitable for irrigation, Class2—limited suitability, Class3—not suitable according to State Standards of Ukraine (1994) and State Water Management Committee of Ukraine (1999).

The main types of land degradation under irrigation:

- *Hydrogeological* rise of groundwater and, consequently, flooding and waterlogging;
- *Mechanical* reduction of the thickness of topsoil due to wind, water and irrigation erosion, and mechanical removal of soil horizons and sedimentation;
- *Physical* dispersion of the topsoil, loss of favourable soil structure, reduced permeability, crusting, compaction and slaking of topsoil and subsoil and the resulting deterioration of water, air and temperature regimes;
- *Chemical* salinity, alkalinity, reducing the amount of calcium through dissolution and leaching resulting in violation of the carbonate geochemical barrier that prevents the downward migration organic compounds and entrained elements. Chemical degradation is manifest in the reduction of humus content and quality, increasing its mobility, depletion of nutrients in layers exploited by roots, and contamination by toxins such as heavy metals, fluoride and radionuclides; and
- *Biological* reducing the number, diversity and function of soil microbial ecology and other fauna, soil contamination by pathogenic microorganisms.

The degree and the rate of soil degradation under irrigation are determined by the original properties of the soil, the quality of irrigation water, reclamation status of irrigated lands, irrigation equipment and technology, and the farming system.

Rising Groundwater

The main degradation process on irrigated lands in Ukraine is rising groundwater and related flooding, waterlogging and compaction caused by overlapping changes in water supply and drainage, regulation of river flow, seepage from canals and other conduits, and siltation of rivers and other natural water bodies. Excess application of irrigation water may cause flooding and waterlogging, not only in naturally wet landscapes with meadow-chernozem, meadow and marsh soils but also in previously well-drained soils due to water-regulating effects. And human influences may be exacerbated by natural factors such as impeded drainage and above-average rainfall.

Salinity and Alkalinity

Salinity may be manifest as *initial*, natural accumulation of salts in the soil resulting from the evaporation of subsoil water and the salinity of soil parent materials; or *secondary* accumulation of salts in the soil caused by changes in water regime, often under irrigation. Secondary salinization is usually caused by the upward movement of soluble salts from deeper layers and subsoil waters or inflow with mineralized irrigation water. According to the distribution of salts in the soil, salinity is distinguished as follows: *deep*—caused by mineralized waters lying at some depths or salt-bearing sediments; or *surface*—characterized by the accumulation of salts in the upper soil layers. At the time of detection, salinity may be categorized as follows: *modern* (present day); *residual* (salts remaining from a previous hydromorphic stage); and *relict* (remaining from older, palaeogeographic conditions (Rode 1975).

Irrigation may induce alkalinity by introducing water of hydrogen carbonate-sodium composition where the soil is weakly buffered against alkalinity; where the groundwater is sodic and lies closer than 3 m from the surface; or under in anaerobic conditions where, in the presence of organic matter and in the absence of oxygen, sulphides generated by sulphate-reducing bacteria may be regenerated as soda:

$$Na_2SO_4 + 2C \rightarrow 2CO_2 + Na_2S$$
$$Na_2S + CO_2 + H_2O \rightarrow Na_2CO_3 + H_2S.$$

The negative impact of alkalinity is felt by plants and microorganisms and, also, through the dispersal of soil colloids which results in severe deterioration of soil physical properties (Romashchenko et al. 2003). This is most prominent in the Dry and Southern Steppe but not widespread in Ukraine because of strong soil buffering. Irrigation with fresh irrigation waters leads to an increase in sodium from 0.6–1.0 to 1.5–2.0% of exchangeable cations, and from 3 to 10% with the use of saline irrigation waters. Adsorption of sodium is rapid in the first 2–3 years of irrigation but slows over 3–5 years to achieve a quasi-stationary state; over a long period of irrigation, alkalinity moves deeper down the soil profile. Increase in adsorbed sodium increases the electro-kinetic potential from 20 to 39–46 mV, hydrophilicity, soil dispersal and swelling. At the same time, the infiltration rate and drainage rate are reduced (State Water Management Committee of Ukraine 2002a, b, 2004).

Agro-Physical Degradation

Under freshwater irrigation, transformation of soil minerals is observed over periods greater than 15–20 years. Shrink-swell smectite and hydromica–montmorillonite minerals are diminished by 10–30%: at the same time, the content of amorphous silica is increased 1.5–3 times. The increased hydrophilic colloids may impair agro-physical properties, but negative transformation and migration may be mitigated by ameliorants containing calcium that lessen the charge of colloids, their coagulation and the intensity of alkaline hydrolysis. The magnitude of agro-physical degradation—compaction, slaking and crusting—is not due to the lack of appropriate supervision services; according to experts, these processes are developed mainly in areas where saline soils are common.

Pollution by Heavy Metals (HM)

The HM content of irrigation waters of Ukraine is only slightly dependent on natural zonal features and is quite independent of mechanisms for the formation of complex heavy metal content (Baliuk et al. 2009). Along with the increased aridity across the transition from forest-steppe to steppe, there is a tendency for a slight increase in HM content in irrigation water sources. Lack of clear natural zonal differentiation of HM can be explained by anthropogenic factors; irrigation waters exhibit the least change of the chemical composition in ecologically clean regions—although there may be slightly increased concentrations of Pb and Co in smaller reservoirs of slow run-off in closed and small river basins.

Water bodies affected by local and regional pollution with Pb, Cd and Co are of limited suitability or unsuitable for irrigation. Otherwise, the HM content of irrigated topsoils is within the local background and not more than double average values, which indicates homogeneity of geochemical landscapes and uniformity of biogeochemical factors of HM migration. In general, significant changes of gross HM content in the plough layer were not found; the mean gross HM content in irrigated and rain-fed soils differs by ±1–5%. In cases of local and regional pollution and soils under long-term irrigation by poor quality water, the content of Fe, Mn, Pb and Ni is increased by 5–20%. In general, soil is a natural buffer controlling the transfer of chemical substances in the atmosphere and hydrosphere by accumulating potentially toxic compounds.

Although there is a variation in the content of reactive HM in topsoil, under irrigation with good quality water, the average content of available HM forms is significantly lower than the maximum permissible concentration and differs little from the baseline values and unirrigated analogues. In irrigated soils, average concentrations of most HMs tend to increase 1.1–1.5 times due to the additional inputs of water, change in ion-salt composition of soil solution and transfer of reserves of HM to mobile forms:

1. Because of greater soil water content, HM content increases after the first 4 years of irrigation. This is associated with changes in ion-salt composition of soil solution that increase the HM availability.
2. Concentration of HM stabilizes after 8–12 years of irrigation, coinciding with the stabilization of ion-salt composition of soil solution and the formation of a new dynamic equilibrium.
3. After 16–30 years, the content of some metals decreases—perhaps due to increased leaching and biological removal from topsoil and an increase in $pH_{soil\ solution}$ from 6.6 to 7.4.

Long-term use of manure and fertilizer under irrigation tends to increase the concentration of available Mn, Pb, Cu, Co, Cr, Cd and Zn (in the range 1.1–1.8 times) but without creating a risk of contamination of plants or significantly improving their micronutrient supply. So there appears to be no danger of HM accumulation in soils or plants when good quality irrigation water is applied. However, where there is local or regional pollution, and where limited-use or unsuitable water is applied, the changes in content of metals in soils are more serious, even the average concentrations of Zn, Pb, Cd, Cr, Co and Cu may be significantly higher than background content (up to 50 times), and the maximum concentration is higher than the maximum permissible concentration.

Industrial and vehicle emissions of HM are an important source of available HM forms in irrigated soils. Compared with the gross HM content, there is a significant increase in the availability of metals and their migration ability, increased danger of high amounts of HM in plants and migration of elements within the soil. Moreover, irrigation water and soils are contaminated by elements that can be accumulated in agricultural products in quantities equal to or above the maximum permissible concentration. This underscores the need for control over the HM content in all components of the *irrigation water–irrigated soil–plants* system.

Changes in the Status and Functioning of Microbial Ecosystems

25-years' irrigation of *Typical chernozem* with Class1 water increases the population of all groups of microorganisms in the soil. Trends in the structure of microbial biomass of irrigated soil have also been detected in increasing parameters of oligotrophicity and mineralization (Baliuk et al. 2009).

Moderate irrigation (rates of 300–1500 m^3/ha) with good water restructures the microbial ecology of *Chernozem*, with a tendency to reduce the fungal component and increase the denitrifiers. However, prolonged intensive irrigation of *Typical chernozem* under vegetable crops led to biological degradation with the decrease in the number of major groups of organisms, increasing mineralization and accumulation of phytotoxic substances in the soil.

Irrigation with mineralized waters causes deep and persistent degradation with significant effects on water and air, salt and nutrient regimes of soil, its physical, chemical and agrochemical properties, and changes in the habitat of microbial communities. Microbial communities respond with structural and functional changes; 12–13 years' irrigation has led to a significant reduction in microorganisms and inhibition of major ecological and trophic groups—notably by oligotrophy, indicating a deterioration of its trophic regime.

Parameterization of Soil Changes Evaluation Under the Influence of Irrigation

For a comprehensive evaluation of changes in the land and soil that can be influenced by irrigation, land improvement and land degradation, we have developed a set of criteria and parameters (Table 13.1). Evaluation of irrigated lands using this system allows optimization of their use and application of agricultural technologies by determining the conditions, methods and parameters for the application of measures to conserve and enhance their fertility. This will significantly increase crop yields and efficiency of irrigated agriculture in Ukraine as a whole.

Table 13.1 Integrated evaluation of irrigated soils according to the degree of the degradation

Indices	Soil without degradation	Degree of degradation		
		Slight	Moderate	Strong
Salinity 0–50 cm				
Toxic salts, $_{e}Cl^{-}$ meq/100 g soil	<0.3	0.3–1.5	1.5–3.5	>3.5
Ca:Na in water extract	>2.5	2.5–1.0	1.0–0.5	<0.5
Sodicity 0–30 cm				
Na^{+} + K^{+}% sum of cations, clay soils	<3	3–6	6–10	>10
Na^{+} + K^{+}% sum of cations, sandy soils	<5	5–8	8–12	>12
$aNa/\sqrt{a}Ca$	<1	1–3	3–7	>7
Kachinksy factor of dispersivity,%	<10	10–20	20–30	>30
Alkalinity 0–30 cm				
$_{p}H_{water}$	<7.8	7.8–8.5	8.5–9.0	>9.0
$HCO_3^{-} - Ca^{2+}$ meq/100 g soil	<0.5	0.5–1.0	1.0–2.0	>2.0
CO_3^{2-} meq/100 g of soil	<0.1	0.1–0.3	0.3–0.9	>0.9
$_{p}H - _{p}Na$	<4.0	4.0–5.0	5.0–5.5	>5.5
Humus status 0–50 cm				
Decrease of humus content, % from initial	0	0–10	10–20	>20
Agro-physical state 0–30 cm				
% air-dry aggregates 0.25–10 mm	>70	60–70	40–60	<40
% water-stable aggregates >0.25 mm	>45	35–45	25–35	<25
Equilibrium density of composition g/cm^3, clay soils	<1.3	1.3–1.4	1.4–1.6	>1.6
Equilibrium density of composition, g/cm^3, sandy soils	<1.3	1.3–1.5	1.5–1.7	1.7
Pollution 0–100 cm				
HM content in Zn equivalents mg/kg of soil	<25	25–50	50–100	>100
Water-soluble F mg/kg of soil	<6	6–10	10–20	>20

Conclusions

- Irrigation significantly influences the character, direction, speed and spread of many soil processes.
- Three main pathways of the soil evolution under irrigation are distinguished as follows:
  - With the use Class1 water and the best agronomic practices, irrigation enhances the natural fertility of the land; the productivity of agro-ecosystems is increased two- to eightfold. Even after cessation of irrigation, a positive after-effect of irrigation lasts for 10 years or more.
  - Even with the application of Class1 water, poor agronomic practices and inadequate resource provision cause land degradation—loss of humus, compaction, even soil erosion; but productivity under irrigation is higher than similar unirrigated soils.
  - Use of Class2 or Class3 water (suitable for limited use or unsuitable for irrigation) causes land degradation—flooding, salinity, alkalinity, transformation of clay minerals, deterioration of biophysical and hydrological properties, nutrient depletion, soil contamination or pollution, and inhibition of microbial ecosystems. Under such conditions, even the use of complex land improvement measures can only limit these processes but cannot eliminate them.
- A set of criteria and parameters is proposed for the comprehensive evaluation of quantitative and qualitative changes in land and soils that can be influenced by irrigation, land improvement and degradation.

References

Baliuk SA (1996) Irrigated chernozem of forest-steppe and Northern steppe of Ukraine: assessment, protection and improvement of fertility. Thesis, Dr Agricultural Science, Kharkiv (Russian)

Baliuk SA, Romashchenko MI, Stashuk VA (2009) Scientific basis of conservation and sustainable use of irrigated lands in Ukraine. Agricultural Science Publishers, Kyiv (Ukrainian)

Baliuk SA, Romashchenko MI, Stashuk VA (2013) Complex of measures against degradation on irrigated lands of Ukraine. Agricultural Science Publishers, Kyiv (Ukrainian)

Dmitriev EA (2009) Mathematical statistics in soil science. Lybrokom, Moscow (Russian)

Dospechov BA (1985) Fieldwork methods (with the basics of statistical processing of the research results). Agropromizdat, Moscow (Russian)

Khitrov NB (1998) Soil and soil cover degradation: concepts and approaches for obtaining estimates. In: Anthropogenic degradation of soil and its prevention measures (Abstracts). Moscow, pp 20–26 (Russian)

National Standards of Ukraine (2004, 2006) Collection of National Standards of Ukraine in the fields of soil science, agrochemistry and soil protection, vols 1, 2. Kharkiv (Ukrainian)

Rode AA (ed) (1975) Glossary of soil science. Nauka, Moscow (Russian)

Romashchenko MI, Sushko OO, Savchuk LP, Kulbida M (2003) On some problems of agricultural science in relation to climate change. Scientific Report Information, Kyiv (Ukrainian)
State Standard of Ukraine (1994) Quality of natural water for irrigation. Agronomic criteria DSSU 2730-94. Kyiv (Ukrainian)
State Water Management Committee of Ukraine (1999) Protection of water, soil and plant resources from pollution by heavy metals under irrigation, DND 33-5.5-06-99. Kharkiv (Ukrainian)
State Water Management Committee of Ukraine (2002a) Organization and management of environmental and reclamation monitoring. Part 1, irrigated lands. DND 33-5.5-01-97. Kharkiv (Ukrainian)
State Water Management Committee of Ukraine (2002b) Instructions for saline soil reclamation of irrigated lands of Ukraine DND 33-5.5-11-02. Kharkiv (Ukrainian)
State Water Management Committee of Ukraine (2004) Manual of soil and salt touble-shooting on irrigated lands of Ukraine. Guide 2 to DND 33-5.5-11-02. Kharkiv (Ukrainian)

Chapter 14
Sustainability of Soil Fertility in the Southern Steppe of Ukraine, Depending on Fertilizers and Irrigation

Valentyna Gamajunova

Abstract Farming in the Southern Steppe of Ukraine is mining humus and nutrients. There comes a point when soil degradation is irreversible: sustainability requires complying with the fundamental laws of agriculture—in particular, sound crop rotation and return of nutrients to balance removal by the crops. Short- and long-term field experiments on typical *Kastanozen* and *Chernozem* reveal that provision of adequate nutrients and water gives consistently high crop yields and these factors significantly change the main indicators of soil fertility: humus content, gross and moving content of NPK, and water–physical properties (as well as the content of arsenic and heavy metals). Combined use of organic and mineral fertilizers is the most effective way to stabilize crop yields and soil fertility; organic fertilizers stabilize soil structure which, in turn, enhances the infiltration of rainfall. Combined organic–mineral fertilizer in crop rotation increases the efficiency of water utilization on average by 20–30%, in very dry years by 30–40%.

Keywords *Kastanozem* · *Chernozem* · Humus · Soil nutrients · Fertilizer regime · Sustainability

Introduction

Judged by any indicator of soil fertility, Ukraine is the richest country in the world. But while the soil is the main means of production and must be used, it should also be conserved. In developed countries, legislation spells out the inadmissibility of land degradation that leads to a loss of soil function, productivity and ecosystem services but, in Ukraine, no one takes responsibility for land degradation—indeed, the government has suspended programs to improve soil fertility (Baliuk 2014).

V. Gamajunova (✉)
Mykolayiv National Agrarian University, 9 Paryzka Komuni St,
Mykolayiv 54020, Ukraine
e-mail: gamajunova2301@gmail.com

D. Dent and Y. Dmytruk (eds.), *Soil Science Working for a Living*,
DOI 10.1007/978-3-319-45417-7_14

The productivity and economic efficiency of the farming system depend on the structure of the sown area and crop rotation. A sound rotation, in which each crop is grown following the best predecessor, is the most important element in maintaining soil fertility—increasing the productivity of arable land by 15–20% without any additional costs. Fertilization according to the special needs of each crop in rotation reduces costs and contains weeds, pests and diseases—without resorting to chemical means of protection; soil fertility and soil moisture are conserved; and biological nitrogen is accumulated by legumes.

Research has established that to maintain soil fertility in the Southern Steppe zone of Ukraine, each hectare of crop rotation needs to generate, annually, 7–8 t of organic matter and 80–100 kg of nutrients under rain-fed farming and 12–15 t/ha and 240–260 kg/ha, respectively, under irrigation. Such productivity is not achieved nowadays. In recent years, farming has been mining residual humus and nutrients; the soils are more and more degraded, and there comes a point when degradation processes are irreversible. Arresting soil degradation means observing the fundamental laws of agriculture—in particular, the law of returning nutrients to the soil by the application of mineral and organic fertilizers to balance removal by the crops. Moreover, fertilization is the most effective factor for crop yield and, also, improves the quality of the crops (Dobrovolsky 2001; Filip'yev et al. 2009).

The yield deficit also arises from insufficient rainfall or, rather, unproductive water loss—an increasingly serious issue the context of global warming. As the vast majority of farms have shunned a sound crop rotation, neglecting the legumes and grasses in favor of sunflower cultivation, we have observed wastage of humus and loss of soil structure that inhibits the infiltration of rainfall and accumulation of water reserves in the soil (Krutya and Tarariko 2000).

In itself, optimum fertilization contributes to the efficient utilization of soil moisture by crops. Long-term trials show that water consumption by unfertilized winter wheat is 526 m^3 per tonne of grain, whereas with complete fertilizer, consumptive use of water is 336 m^3/t–36% less than the unfertilized control (Gamajunova 2004). Similarly, optimal supply of nutrients reduces the consumptive use of water to produce a unit of dry matter of corn and sugar beet by 20–25% compared with the soils of low nutrient security. Variable weather may cause annual fluctuations of crop yields within ±40–50%, but these fluctuations are much less when soil fertility is maintained.

Field Experiments

To provide reliable data and determine changes in the main indicators of soil fertility and crop productivity, long-term and short-term field experiments have been undertaken on typical dark brown earth (*Kastanozem*) and black earth (*Chernozem*) on irrigated fields of the Institute of the National Academy of Agrarian Sciences of Ukraine and Mykolayiv National Agrarian University, respectively.

Table 14.1 Humus content in the 0–30-cm layer of dark brown soil under fertilization and irrigation

Variant of experiment	Humus content (%)		Humus loss or gain	
	Initial	After 4th rotation	Absolute (%)	Mean annual (kg/ha)
Without fertilizer	2.26	2.11	−0.15	−277.5
P_2O_5	2.26	2.19	−0.07	−129.5
N	2.26	2.23	−0.03	−55.5
NPK	2.26	2.25	−0.01	−13.5
NPK + 80 t/ha farmyard manure per rotation	2.26	2.35	+0.09	+166.5

Trends of Topsoil Humus Content

On the *Kastanozem* of the Ingulets irrigation system, after four rotations of a 7-field rotation that includes 36% lucerne, the annual loss of humus from the topsoil of unirrigated soil amounted to 70–90 kg/ha—somewhat more under irrigation. Humus content was maintained by the application of complete fertilizer (Table 14.1).

Trends of Crop Yield

Fertilizer application influences soil physical properties, microbiological activity, and the content of micronutrients. In turn, the soil's nutrient status significantly affects both crop yields and quality, both with and without irrigation. Irrigated crops, producing higher yields, require more nutrients. This is illustrated in Table 14.2 which summarizes the results. Again, we see that provision of adequate nutrients and water gives consistently high crop yields and these two factors significantly change the main indicators of soil fertility: humus content (Table 14.1), gross and moving content of NPK, and water–physical properties.

Trends in Arsenic and Heavy Metals

It is known that arsenic and heavy metals are added to the soil with fertilizer; in the case of arsenic in nitrogen fertilizer—the quantity varies from 2.2 to 120 mg/kg (Karpova 1991). Irrigation water can also pollute where the arsenic content exceeds the maximum allowable concentration. Soil samples were selected from 20-cm increments to a depth of 1 m in June 2001. Total arsenic was extracted with 0.2 M HCl and determined by arsenic–molybdenum blue after removal of arsenic in the

Table 14.2 Productivity of crop rotation depending on irrigation and fertilizers

Variant	Productivity of crop rotation c/ha fodder units				Over 30-year period	Increase %		
	I	II	III	IV		From fertilizers	From irrigation	From irrigation and fertilizers
Without irrigation or fertilizer	337.1 / 42.1	360.5 / 45.1	251.4 / 35.9	223.9 / 32.0	1173 / 39.1	–	–	–
Without irrigation + NPK_1	394.7 / 49.3	455.9 / 56.7	265.4 / 37.9	244.7 / 35.0	1359 / 45.3	15.9	–	–
Without irrigation + NPK_2	408.3 / 51.0	489.2 / 61.2	267.6 / 38.2	233.4 / 33.3	1398 / 46.6	19.2	–	–
Irrigation, without fertilizers	428.8 / 53.6	414.9 / 51.9	450.0 / 64.3	369.0 / 52.7	1663 / 55.4	–	41.7	–
Irrigation + NPK_1	506.0 / 63.3	566.8 / 70.9	563.8 / 80.5	501.5 / 71.6	2138 / 71.3	28.7	57.4	182.3
Irrigation + NPK_2	698.1 / 87.3	659.0 / 82.4	599.1 / 85.6	551.7 / 78.8	2508 / 83.6	50.9	79.4	213.8

Above the line—the amount per crop rotation; below the line—in an average year
NPK_1—recommended dose of fertilizers for each rotation culture for rain-fed conditions
NPK_2—the same conditions for irrigation

form of arsine and its subsequent oxidation of iodine solution and measuring the absorbance of the complex relative to a reference solution for wavelength $\lambda = 750$ nm. The experimental data demonstrate that the amount of total and mobile arsenic is increased by systematic application of increasing doses of mineral fertilizers (Table 14.3).

Irrigation without fertilizers had almost no effect on the content of mobile forms of arsenic. However, the total arsenic content of both 0–20 and 0–100-cm irrigated soil was lower than the rain-fed analogue (1.35 and 0.92 as opposed to 1.71 and 1.42 mg/kg, respectively, in Table 14.4). This reduction in the arsenic content in soil under irrigation may be connected with higher crop yields (Table 14.2) and enhanced removal of chemical elements by these crops.

Table 14.3 Arsenic in the 0–20-cm layer of dark brown soil after 30 years of irrigated crop rotation

Variant of experiment	Arsenic (mg/kg)			
	Without irrigation		Irrigated	
	Total	Mobile	Total	Mobile
Fallow (background content)	0.98 ± 0.05	0.03 ± 0.001	0.98 ± 0.05	0.03 ± 0.001
Without fertilizer	1.35 ± 0.06	0.07 ± 0.002	0.92 ± 0.03	0.06 ± 0.002
$N_{90}P_{60}K_{30}$	2.15 ± 0.06	0.05 ± 0.001	2.15 ± 0.09	0.37 ± 0.020
$N_{150}P_{90}K_{60}$	3.10 ± 0.16	0.10 ± 0.005	3.40 ± 0.17	0.54 ± 0.020

Table 14.4 Total arsenic accumulation in soil layers, depending on irrigation and fertilizers

Soil layer (cm)	Total arsenic, mg/kg soil						
	Fallow (background)	Without irrigation			Irrigated		
		Without fertilizer	$N_{90}P_{60}K_{30}$	$N_{150}P_{90}K_{60}$	Without fertilizer	$N_{90}P_{60}K_{30}$	$N_{150}P_{90}K_{60}$
0–20	0.98	1.35	2.15	3.10	0.92	2.15	3.40
20–40	2.05	1.55	2.40	2.50	1.75	2.55	1.75
40–60	1.20	1.40	2.70	2.75	1.45	2.30	2.75
60–80	1.55	1.75	1.75	1.55	1.55	1.55	1.38
80–100	0.92	0.98	1.95	0.98	1.45	1.55	0.92
0–100	1.34	1.41	2.19	1.98	1.42	2.02	2.04
Least significant difference	0.17	0.17	0.27	0.26	0.17	0.23	0.21

Table 14.5 Water-absorbing capacity of soil depending on organic fertilizers (means 2008–2010)

Variant experiment	Infiltration (mm/h)	% compared with control
Without fertilizers	11.69	100.0
Manure 30 t/ha	13.60	116.0
Green fertilizer (peas)	15.30	130.9
Green fertilizer (rape)	14.73	126.0

Trends in Available Water

Combined use of organic and mineral fertilizers is the most effective way to stabilize crop yields and increase soil fertility. Organic fertilizers contribute to increased yields in dry years by providing reactive humic substances that stabilize soil structure which, in turn, enhances infiltration of rainfall. Combined organic–mineral fertilizer in crop rotation increases the efficiency of water utilization on average by 20–30%, in very dry years by 30–40%.

Nowadays, farmyard manure is hardly available, so, as an alternative, we investigated cereal straw and green manure. Straw was incorporated in the topsoil immediately after harvesting winter wheat, stubble-seeded oats/pea mixture for the green mass, and maize. Table 14.5 shows the beneficial effect on water infiltration and water retention by the soil.

Over 3 years of growing maize, compared with the unfertilized control, prior application of 30 t/ha farmyard manure increased the soil water by 25.6%; straw incorporation at 6 t/ha increased soil water by 34.6%. The residual effect under the following winter wheat crop was an increase of 16.3% from farmyard manure and 22.8% from straw incorporation. These data indicate that maintenance of soil organic matter is essential to maintain the soil's infiltration capacity and water retention capacity. In itself, the water-holding capacity of organic matter is 5–10 times higher than the mineral fraction of the soil, but the greater benefit comes from the maintenance of a stable crumb structure. Incorporating straw, cornstalks, and green manure is more effective than adding farmyard manure.

Postharvest crop residues distributed on the fields accelerate the infiltration of water and arrest runoff. Earlier research (Gamajunova 2006) showed that incorporation of straw into the soil increases the accumulation of moisture in the topsoil by 15–20% compared with no incorporation of straw.

Combined Effects

Many studies have shown that improving the soil nutrient regime and availability of moisture by applying fertilizer increases crop yields. Table 14.6 shows the example of *soriz* (grain sorghum) grown in different crop sequences in crop rotation where,

Table 14.6 Effect of fertilizers and predecessors on the yield of soriz, t/ha

Field crop rotation (factor A)	Background supply (factor B)								
	Without fertilizer			Straw + $N_{60}P_{40}$			$N_{60}P_{40}$		
	2004	2005	2006	2004	2005	2006	2004	2005	2006
Peas–winter barley–soriz	4.26	3.18	3.30	5.29	4.41	4.50	5.42	4.55	4.61
Sunflower–winter barley–soriz	4.05	2.98	3.11	5.00	4.09	4.16	5.14	4.22	4.28
Maize–winter barley–soriz	4.06	3.01	3.13	5.17	4.23	4.30	5.27	4.33	4.40
Winter barley–maize–soriz	3.95	2.88	3.02	4.92	3.93	4.12	5.12	4.16	4.24
Winter wheat–sunflower–soriz	3.81	2.70	2.82	4.82	3.92	4.10	5.09	4.11	4.20
Average for factor B	4.03	2.75	3.08	4.84	4.12	4.24	5.21	4.27	4.35
Least significant difference $_{05}$, t/ha	*2004 p.*			*2005 p.*			*2006 p.*		
Factor A	0.071			0.051			0.12		
Factor B	0.084			0.098			0.19		
Combination of factors A and B	0.130			0.270			0.21		

after harvesting cereals, the straw is incorporated into the soil as organic fertilizer (Table 14.6).

The average grain yield of *soriz* with all predecessors without fertilizers is 3.35 t/ha. With the combined use of a link of crop rotation and organic fertilizer (such as ploughing in chopped straw after harvesting) and application of recommended dose of fertilizer ($N_{60}P_{40}$), the yield is raised to 4.46 t/ha, an increase of 1.11 t/ha or 33% compared with the unfertilized control. Application of $N_{60}P_{40}$ without straw raised the grain yield to 4.61 t/ha, an increase over the control of 1.26 t/ha or 36.7%. The lesser yield associated with straw application is associated with the uptake of nutrients, especially nitrogen, by microorganisms that decompose the organic matter. However, the temporarily fixed nutrients subsequently become available to plants through the mineralization of the straw and microbial organic matter. The highest crop yields were recorded the relatively wet year of in 2004 but the benefit of fertilizer is seen even more in dry years, which indicates more efficient use of water by fertilized plants.

And it should be noted that combined organic and mineral fertilizers also significantly improve the quality indicators of farm produce.

References

Baliuk SA (2014) Recommendations of the Ukrainian Society of Soil Scientists and Agro-Chemists 2010–2014: Urgent Tasks for the Future. In: Agricultural Chemistry and Soil Science—special issue on the IX Congress of the Ukrainian Society of Soil Scientists and Agro-chemists, Plenary lectures: protection of soil—the basis of sustainable development of Ukraine, Kharkiv, vol 1, pp 3–17 (Ukrainian)

Dobrovolsky GV (2001) Saving soil fertility is the important environmental problem of the XXI century. In: Soils and fertility at the turn of the century. II Congress of Belorussian Soil Scientists' Society. Theoretical and practical soil science problems, Minsk, vol 1, pp 74–75 (Russian)

Filip'yev ID, VVGamayunova VV, Baliuk SA (2009) Systems of crop fertilization. In: SA irrigated land of Ukraine. Agricultural Science Publishers, Kyiv, pp 279–299 (Ukrainian)

Gamajunova VV (2004) Current status, problems and prospects of fertilizer usage in irrigated agriculture of the Southern Zone of Ukraine. In: Gamajunova VV, Filip'yev ID, Sidyakina AV (eds) Soil science, agricultural chemistry, agriculture, forestry, no. 1. Kharkiv National University Bulletin, Kharkiv, pp 181–186

Gamajunova VV (2006) Phytosanitary monitoring fertility of irrigated soils. Training guide for agrochemical survey of agricultural land. Kherson (Ukrainian)

Gamajunova VV (2014) Change of soil fertility in the Southern Steppe of Ukraine under the influence of fertilizers and approaches to their effective use in modern agriculture. In: Agricultural Chemistry and Soil Science—special issue on the IX Congress of the Ukrainian Society of Soil Scientists and Agro-chemists, Plenary Lectures. Kharkiv, vol 1, pp 38–47 (Ukrainian)

Karpova EA (1991) Arsenic in soils and plants. Chem Selsk 4:30–34 (Ukrainian)

Krutya VM, Tarariko OG (2000) Agriculture in low moisture (scientific and practical conclusions). Agricultural Science Publishers, Kyiv (Ukrainian)

Chapter 15
Evolution of Potassium Reserves in Chernozem Under Different Fertilizer Strategies, and Indicators of Potency

Yevheniia Hladkikh

Abstract The after-effect of potassium fertilizers on *Typical chernozem* is investigated in a long-term field experiment carried out since 1969 in the Forest-Steppe zone of Ukraine. The modified soil is characterized by elevated contents of mobile potassium and transformation of exchangeable and non-exchangeable forms. The evolution of the soil's potash reserve under long-term application of different doses of fertilizer enhances the productivity of the crop rotation: an increase of 17–25% in variants with fertilizers applied to the reserve (25 years of after-effect) and 79–80% in options with systematic application of fertilizer. Systematic monitoring of mobile potassium allows accurate recommendations on fertilizer application, especially for demanding crops that respond to potassium fertilizer.

Keywords Chernozem · Potassium fertilizer · Available potassium · Potassium reserve · Anthropogenically modified soil evolution · After-effect of fertilizer application · Crop rotation

Introduction

High and stable levels of soil fertility and productivity require systematic, scientifically reasoned application of agrochemical resources. Adding mineral fertilizers is the most effective way to create optimal plant nutrient status. Potassium is one of the big three nutrients; the amount of total potassium in soil nearly always exceeds that of phosphorus and nitrogen (Gorbunov 1963). Many experiments confirm the role of mineral fertilizers in maintaining potassium reserves; Nosko and Babynin (1995) and Nosko (1998, 1999) showed that application of potassium fertilizers significantly increased the content of available potassium in soil; under different

Y. Hladkikh (✉)
Department of Agrochemistry, Sokolovskyi Institute for Soil Science and Agrochemistry Research, 4 Tchaikovsky St, Kharkiv 61024, Ukraine
e-mail: ye.hladkikh@ukr.net

D. Dent and Y. Dmytruk (eds.), *Soil Science Working for a Living*,
DOI 10.1007/978-3-319-45417-7_15

terms and doses of fertilizer application, the content of available potassium is correlated with its supply from fertilizer (correlation coefficient 0.75).

There is a belief that the *Chernozem's* reserves of potassium are inexhaustible and, indeed, chernozem is outstanding in its stock of total potassium—amounting to 2–2.5% by mass. Even so, most crops respond positively to potassium fertilizers because, without fertilizer, cropping inevitably depletes potassium reserves—whether mobile, easily soluble, exchangeable, or non-exchangeable forms. Their depletion reflects a close interrelation between various processes of evolution of soil potassium reserves after tillage of virgin soil. Sokolov (1968) gives an example of potassium deficiency on *Chernozem* at the Mironovskaya Experimental Station where many years of traditional farming removed 65 tonne/ha of potassium. Again, Mamontov (1972) established that, compared with virgin soil, the 0–20 cm soil layer in the Western Steppe of Ukraine lost 15–16% of its total potassium over more than a century of arable; water-soluble and exchangeable potassium were also depleted, indicating a strong transformation processes in soil potassium reserves.

A wealth of research on systematic application of potassium fertilizers is summarized by Sobachkin (1978), Prokoshev (1985) and Kulakovskaia (1990); in all cases, there was an increase in the efficiency of potassium fertilizers from rotation to rotation. The same regularity is observed across a wide range of investigations (Dubets and Dudas 1981; Szewczuk et al. 2008, 2009; Safora 2010; Bernardi et al. 2011). Analysis of results from short-term trials on the influence of mineral fertilizers on the potassium status in soil and on crop-use efficiency shows, in most cases, only the direct fertilizer effect. With such a research philosophy, it is hard to ascertain the residual accumulation of potassium in soil and K-delivery to plants. We need to look in more detail at the direct action of fertilizer, its after-effects, and the dynamics of transformation of potassium in the soil after high doses of fertilizers.

Materials and Methods

Experimental Site

As far back as 1969, a long-term field experiment was begun on *Typical chernozem* at the ISSAR Grakivske Experimental Station in Kharkiv Region to investigate the transformation and after-effect of different kinds, rates, terms, and conditions of mineral fertilizer application on the soil's potassium regime. During the period 1969–1983, three high-dose applications of mineral fertilizers (200, 400, and 600 kg/ha) were made to create four levels of nitrogen, phosphate, potash, and nitrogen–phosphorus–potassium agrochemical backgrounds (natural, medium, heightened, and high). The experimental field was laid out in 360 variants with different doses, types of fertilizers, and frequency of their application.

For our investigation, we selected variants of high potassium and nitrogen–phosphorus–potassium agrochemical backgrounds. Prior to soil sampling, six

Table 15.1 Scheme of soil sampling and levels of fertilization on variants of the six crop rotations

Variant	Agrochemical background	Total amount of soil-assimilated nutrients from fertilizers (Kg/ha)		
		N	P_2O_5	K_2O
1	Virgin fallow	0	0	0
2	Control, without fertilizer	0	0	0
3	Manure, 140t/ha (Background)	560	280	700
4	Background + K_{1800} (into reserve, after-effect since 1983)	560	280	2500
5	Background + $N_{1800}P_{1800}K_{1800}$ (into reserve, after-effect since 1983)	2360	2080	2500
6	Background + $N_{1800}P_{1800}K_{1800}$ (into reserve) + 1N1P1K (single rate[a] of systematic fertilization during crop rotations)	3590	3280	3490
7	Background + 2N2P2K (double rate of systematic fertilization during crop rotations)	2990	2680	2860

[a]For cereals, a single dose of NPK is 60 kg/ha; for the beet and maize, 90 kg/ha

rotations of a six-field crop rotation (vetch and oats for green forage, winter wheat, sugar beet, barley, maize silage, and winter wheat) had been completed. Sampling of the 0–20 cm layer before application of fertilizers showed pH 5.5, humus content 3.9–4.5%, total N 0.22%, total P 0.12%, and total K 2.05%.

Experimental Design and Agricultural Practices

Soil samples were collected from seven agrochemical backgrounds with different levels of fertilization (Table 15.1) and analyzed to determine various dynamic forms of potassium: available (in extract of 0.5 M CH_3COOH); easily soluble (in extract of 0.03% $MgSO_4$); exchangeable (in extract of M CH_3COONH_4), and hydrolysable (in extract of 2 M HCl). Weakly linked and non-exchangeable forms of potassium were estimated by computation.

Results and Discussion

Soil Available Potassium

After ploughing up virgin *Typical chernozem*, cropping in a six-field rotation without fertilizer depleted available potassium in both 0–20 cm and 20–40 cm layers by one third (Fig. 15.1).

Transfer of potassium from fertilizers to the soil reserve increased the content of available potassium from 79 mg/kg (in the control) to 93–100 mg/kg (variants 4

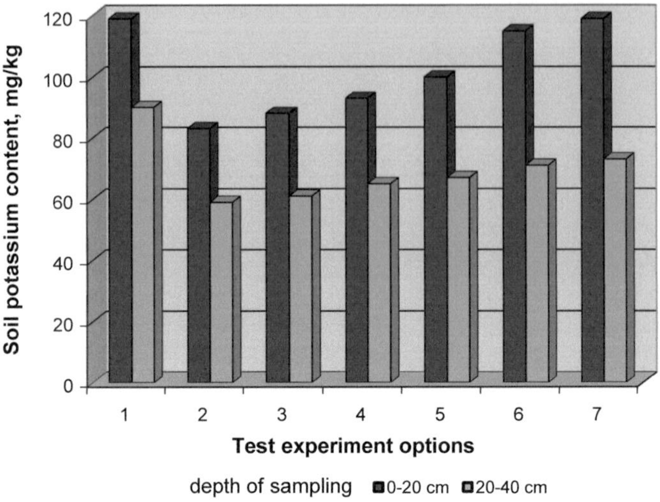

Fig. 15.1 Evolution of available potassium contents in soil at different levels of fertilization

and 5) in spite of a negative potassium balance (−748 and −1828 kg/ha, respectively) during the 25-year period of after-effects.

For the experimental variant with highest rate of potassium fertilizer (variant 4), assessment of the dynamics of available potassium during the 36 years (from 1969 to 2008) of six complete crop rotations (Fig. 15.2) testifies to evolution of the

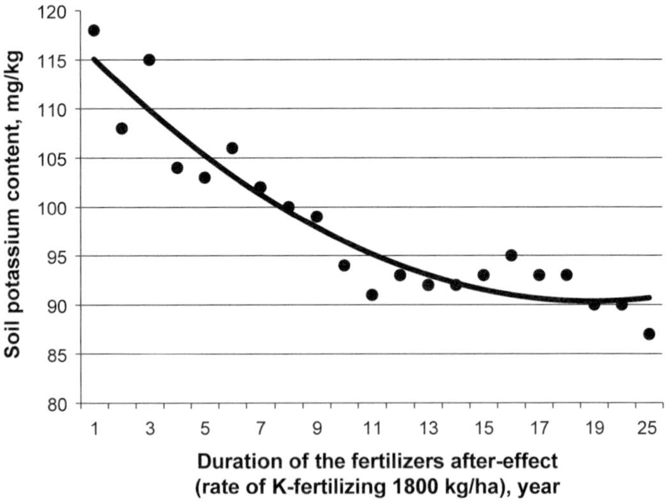

Fig. 15.2 Variant 4: Forecast mobile potassium contents in 0.5 N CH_3COOH extract from 0 to 20 cm

potassium reserves and establishment of a new equilibrium level for the different forms of potassium in the soil. In this variant, the amount of available potassium declined with increasing duration of the after-effect period. This is may be explained by its transformation to a less-soluble form but, after completion of the fourth crop rotation (15 years of after-effect), available potassium stabilized despite the continued negative potassium balance.

Systematic application of mineral fertilizers including potassium (variants 6 and 7) promoted a rise in contents of available potassium (by 27–37% in comparison with the control) and replenishment of the reserves to the level in virgin land. Thus, by the end of the sixth crop rotation, systematic application of fertilizers had maintained elevated levels of available potassium, even under a negative K balance (−314 and −1338 kg/ha).

Transformation of Potassium Reserves in the Soil

The main factor in shifting the equilibrium potassium regime after ploughing up virgin land is the application of potassium fertilizers. Continuous application of mineral fertilizers leads to transformation of the mineral component of the soil. Barshad and Kishk (1970) and, also, Graf and Reichenbach (1972) observed clay mineral transformations in the active fraction of soil, particularly in the plough layer; montmorillonite and vermiculite clays with shrink-swell crystal lattices undergo transformation into fixed-lattice chlorite. These changes are accompanied by a transformation of reserve potassium—so that potassium fertilizers promote increased availability of soil potassium resources.

In experimental variants that impose different amounts and timing of fertilizer, the most important indicator of transformation of reserve potassium is the interrelation between various forms of potassium (easily soluble, exchangeable, non-exchangeable, weakly linked, and hydrolyzed). Without application of fertilizers, there was no essential change in content of easily soluble, exchangeable, or weakly linked potassium forms in the plough layer and the subsoil compared with virgin *Typical chernozem*—although there was a tendency of these indicator values to decrease—whereas the content of hardly soluble (hydrolyzed and non-exchangeable) potassium forms is changing more dynamically—increasing by 14.3 and 16.0%, respectively (Table 15.2). This indicates a change of soil potassium reserves that proves beneficial in terms of the absorption of soluble and exchangeable potassium.

In experimental variants in which potassium fertilizer was added to the reserve (K_{1800} and $N_{1800}P_{1800}K_{1800}$), there were slight but significant changes in the structure of the reserves in the plough and subsoil layers. The content of easily soluble potassium is hardly different from the control, while exchangeable K increases from 169 to 182 mgK_2O/kg; a significant increase is also registered in weakly linked and non-exchangeable forms of potassium.

Table 15.2 Dynamics of potassium fractional composition (0–20 cm) under different levels of fertilization

Experiment variant	K_2O contents, mg/kg				
	Easily soluble	Exchangeable	Non-exchangeable	Weakly linked	Hydrolyzed
1	30	169	1613	139	1444
2	29	165	1842	136	1677
3	30	169	1740	139	1571
4	28	182	1932	173	1730
5	26	179	1980	154	1801
6	35	305	2101	269	1796
7	39	357	2125	318	1768
LSD (5%)	6	11	52	16	41

The greatest changes in potassium reserves were registered after systematic and reserve application of potassium fertilizers in-between crop rotations (variants 6 and 7). Here, after six crop rotations, exchangeable K exceeds the control index by 140–192 mgK_2O/kg; easily soluble and weakly linked potassium has increased by 16–30 and 52–55%, respectively. A considerable portion of potassium is assimilated into hydrolyzed and non-exchangeable forms—their contents increase by over 300 and 500 mg/kg, respectively. These changes are found throughout the upper metre of the soil.

Potassium Balance in Soil and Crop Productivity

The generalized data on potassium balance in the soil after six crop rotations indicate that removal of potassium from the soil (by harvesting crops) over and above the inputs from fertilizer gradually diminishes to a definite level compared with the control variant (of the order of 3016 kg/ha). This definite level is typical of experimental variants with joint $N_{1800}P_{1800}K_{1800}$ application into the reserve and with single rate of systematic applications of NPK fertilizer (314 kg/ha) where the least-deficit balance is arrived at Table 15.3.

In the variant with 25 years of after-effect of potassium fertilizer application (K_{1800}) into the reserve, the average annual increased removal of potassium was 20.7 kgK_2O/ha. At the same time, this variant produced a significant yield increase of crop-link rotation which, indirectly, indicates the establishment of an equilibrium level of potassium regime that supported residual reserves of all fractions of potassium. In all variants with the introduction of nitrogen–phosphorus fertilizer, potassium removal is greatly increased as a consequence of high increases of harvest. Nosko and Babynin (1995) also demonstrated that, in *Typical chernozem,* residual reserves of potassium not only raise the productivity of the crop rotation (sugar beet in particular), but also improve crop quality (in this case, the sugar content).

Table 15.3 Balance of potassium in soil after six rotations on six test-fields and yield-efficiency of a crop rotation link

Experimental variants	K-fertilizer application	K-removal (kg/ha)	Balance (kg/ha)		Yield-efficiency of a crop rotation link (vetch–oats/maize/winter wheat)	
			After 36 years	Annual average	Forage units (t/ha)	Increment to control variant (%)
1	100	3116	−3016	−83.7	10.6	–
2	700	3010	−2310	−64.2	11.0	3.4
3	2500	3248	−748	−20.7	12.5	17.4
4	2500	4328	−1828	−50.7	13.3	25.2
5	3490	3804	−314	−8.7	191.9	80.3
6	2680	4018	−1338	−37.1	191.0	79.5

Conclusions

- Addition of high doses of potassium fertilizer into the soil reserves and systematic fertilization are reflected in the composition of the soil potassium reserves—which reflects transformation of all forms of potassium.
- Increasing the after-effect period (from 13 to 25 years) reduces the effect of fertilizer to the level of the control by the transformation of soluble forms of potassium to less-soluble forms.
- In unfertilized experimental variants over six rotations, the K-stock in the soil evolved under a negative potassium balance. However, the percentage of easily soluble, exchangeable, and weakly linked potassium forms still remains stable, showing a tendency to regain the original level typical of virgin soil.
- Systematic monitoring of mobile potassium allows accurate recommendations on fertilizer application, especially for demanding crops that respond to potassium fertilizer.

References

Barshad I, Kishk FM (1970) Factors affecting potassium-fixation and cation exchange capacity of soil vermiculite clays. Clays Clay Miner 18:127–137

Bernardi AC, Gimenez LM, Machado PL (2011) Variable-rate application of potassium fertilization for soybean in Brazil. Electron Int Fertil Corresp Q Corresp Int Potash Inst 27:14–18

Dubets S, Dudas M (1981) Potassium status of a dark brown chernozem soil after sixty-six years of cropping under irrigation. Can J Soil Sci 61:409–415

Gorbunov NI (1963) High-dispersion minerals and methods of studying them. Academy of Science USSR, Moscow (Russian)

Graf V, Reichenbach H (1972) Factors of mica transformation. In: Potassium in soil, proceedings of the 9th colloquium of the International Potash Institute, Landshut, Federal Republic of Germany. International Potash Institute, Berne, pp 33–42

Kulakovskaia TN (1990) Optimization of agrochemical soil plant nutrition systems. Agropromizdat, Moscow (Russian)

Mamontov VT (1972) Influence of long-term agricultural use and fertilization on contents of potassium in deep chernozem soil of West Ukrainian Forest-Steppe. J Agrochem Soil Sci 22:49–55 (Russian)

Nosko BS (1998) Efficiency of potassium fertilizer on soil with different levels of fertility. In: AE Johnston (ed) Essential role of potassium in diverse cropping systems. Proceedings of the 16th World Congress, International Potash Institute, Montpellier, 20–26 August. International Potash Institute, Berne, pp 27–31

Nosko BS (1999) Modifications of potassium status in chernozem after ploughing perennial fallow land. Pochvovdenie 12:1474–1480 (Russian)

Nosko BS, Babynin VI (1995) Potassium regime of Typical chernozem and sugar beet harvest after potassium fertilizer application. J Agrochem 11:15–25 (Russian)

Prokoshev VV (1985) Actual issues of agrochemicals of potassium fertilizer. J Agrochem 4:32–40 (Russian)

Safora A (2010) Influence of different potassium fertilizer sources on sunflower production. Soil solutions for a changing world. In: Proceedings of the 19th World Congress of Soil Science, pp 16–18, Brisbane

Sobachkin AA (1978) Studying of efficiency potassium fertilizers in crop rotation of saturation intensive crops (field crop rotation). 100–120 in Research results in prolonged experiments with fertilizers in the zones of the country 5, Moscow (Russian)

Sokolov AV (1968) Geographic laws of fertilizer efficiency. Znanie, Moscow (Russian)

Szewczuk A, Komosa A, Gudarowska E (2008) Effect of soil potassium levels and different potassium fertilizer forms on yield and storability of Golden Delicious. Acta Sci Pol Hortorum Cultus 7(2):53–59

Szewczuk A, Komosa A, Gudarowska E (2009) Effect of different soil potassium soil levels and forms of potassium fertilizers on micro-elemental nutrient status of apple trees in early fruition period. Acta Sci Pol Hortorum Cultus 10(1):83–94

Chapter 16
Ecological Reclamation of Acid Soils

Yuriy Tsapko, Karina Desyatnik and Al'bina Ogorodnya

Abstract Compared with traditional liming practice, applying alternative sources of lime in bands with organic manure reduces leaching from coarse-textured *Sod-podzolic* soil: lime by almost six times, soluble organics and nitrate by 1.8 and 2.9 times, respectively. In *Podzolised chernozem*, the same amelioration technology increases the population of earthworms and microorganisms, thereby activating self-renewing and regulating processes. Perennial grasses, legumes and Sudan grass are effective phyto-ameliorants; their beneficial effects manifest through resilient soil aggregates and the accumulation of organic matter.

Keywords Acid soil · Localized reclamation · Alternative lime sources · Leaching · Phyto-amelioration

Introduction

The beginning of the third millennium has witnessed the beginning of a new phase of development with the realization that nature is not limitless and needs care and protection. For millions of years, soils accumulated rich reserves of biogenic elements and energy; now that we understand the decline of soil fertility caused by mismanagement and pollution, society is seeking ways to encourage sustainable, environmentally oriented use of our common wealth. One example is purposeful activity towards resource conservation and environmental safety in agriculture.

In Ukraine, in recent years, liming to ameliorate soil acidity has been sadly neglected. In landscapes that are dominated by acid soils, there has been a sharp deterioration of soil physical properties, depletion in calcium and magnesium, destruction of buffer mechanisms and weakening of biological sustainability and biodiversity. Lack of calcium in the soil retards root development and plant growth,

Y. Tsapko · K. Desyatnik (✉) · A. Ogorodnya
Sokolovskyi Institute for Soil Science and Agrochemistry Research,
4 Tchaikovsky St, Kharkiv 61024, Ukraine
e-mail: karina.desyatnik@i.ua

D. Dent and Y. Dmytruk (eds.), *Soil Science Working for a Living*,
DOI 10.1007/978-3-319-45417-7_16

leading to lower crop yields; in animals, it causes diseases such as rickets, weak cardiac activity and haemophilia—so lack of calcium in the soil feeds through the food chain to endanger public health. And soil acidification drives toxic, mobile aluminium into surface waters; reservoirs adjacent to acidic soils become acidified and hazards arise because aquatic ecosystems support food chains involving almost all wild animals (Deriy and Ilyuha 2000). Mercury, for instance, occurs in only meagre concentrations in soil but becomes mobile in acid conditions and enters drainage water as monomethyl mercury. This mercury accumulates in the tissues of invertebrates and fish and becomes more concentrated along the trophic chain so that consumption of fish may be poisonous. Furthermore, acidified water dissolves heavy metals from water pipes; drinking this water is bad for us (Ivashura and Orekhov 2004).

Considering the widespread acidity in Ukrainian soils (Baliuk et al. 2012), it is obvious that measures are needed to improve their fertility and agro-environmental status (Mazur and Barvynsky 1993; Martin and Pitblado 1984). The issue is not only to optimize soil fertility but, also, to conserve the resource and enhance the environmental functions of soils. It is well known that liming improves the fertility and agro-ecological properties of acid soils but the cost has become prohibitive. An innovative approach is to make use of waste products as ameliorants. However, care should be taken to avoid harmful consequences—so we need an ecological and functional approach. The Sokolovskyi Institute for Soil Science and Agrochemistry Research has developed a resource-saving and environmentally safe technology of chemical reclamation (Truskavetsky 2003; Tsapko 2003) and this article highlights the impact on soils of lime ameliorants from various sources. We also review prospects for environmentally benign phyto-amelioration.

Local land reclamation technology is a cost-effective way to maintain and improve fertility. Its key mechanism is the fixing of calcium in the soil by application of ameliorants in discrete bands along with organic fertilizers (Truskavetsky 2003; Truskavetsky et al. 2003). Under these conditions, calcium and organic matter combine as calcium humates that resist leaching (Tsapko 2006). Moreover, localised reclamation involves fewer passes of farm machinery that degrade the soil. Each pass of machinery and, especially, each inversion of the topsoil shifts the acid–base balance and nutrient regime—causing significant damage to soil ecology, especially to earthworms and micro-fauna.

Materials and Methods

The study was carried out on two soil types: coarse loamy *Sod-podzolic* and fine loamy *Podzolised chernozem*. In the sandy loam *Sod-podzolic* soil, we studied the intensity of leaching of nutrients under various reclamation technologies: the traditional application of fertilizers and calcium ameliorants to the soil surface followed by ploughing; maintenance application, similar to the above but using only one quarter to one third the amount of lime and fertilizer; and local reclamation technology, ploughing fertilizers and ameliorants to a depth of 20–25 cm and reducing the amount by 5–8 times compared with traditional practice.

In a small-area field experiment on fine loamy *Podzolised chernozem,* we determined the effects of different calcium ameliorants on numbers of micro- and meso-fauna in the plough layer. We also investigated the effects of lucerne, soybean, mustard, Sudan grass, lupin and sainfoin on soil condition.

The selection of soil samples and determination of physical and chemical parameters were made using certified, conventional methods. Accounting of soil invertebrates was performed using Gilyarova's method with hand excavation and breaking up of soil samples (Byzova et al. 1987).

Results and Discussion

Figure 16.1 contrasts the intensity leaching of nutrients from coarse loamy *Sod-podzolic* soil under maize with different ameliorative technologies. Leaching of calcium (lime) is reduced by local cultivation compared with traditional practice: sixfold in respect of lime (from 215.5 to 35.3 kg/ha), by 1.8 times in respect of water-soluble organic matter (from 21.7 to 12.1 kg/ha), and by 2.9 times in respect of nitrates (from 43.0 to 14.9 kg/ha). Deployment of organic fertilizers in bands in the subsoil was the most effective in terms of reducing unproductive losses of

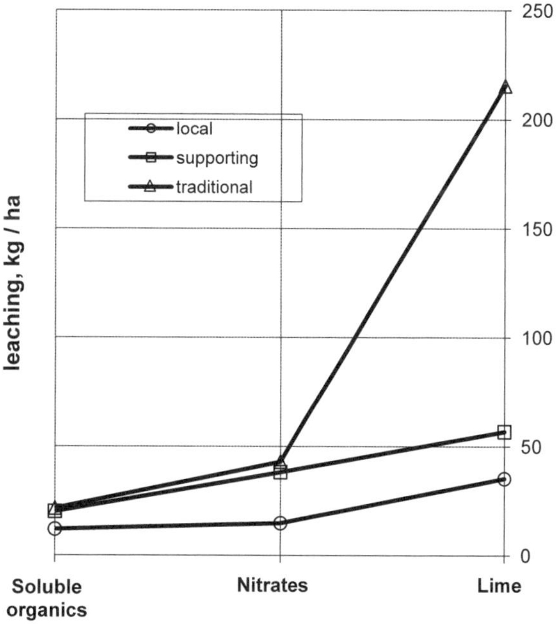

Fig. 16.1 Seasonal leaching of materials from the root zone of sod-podzolic soil under different reclamation practices

Table 16.1 Change in numbers of micro- and meso-fauna in the surface (0–20 cm) layer of *Podzolised chernozem* influenced by lime ameliorants (individuals/m^2)

Variant	Micro-fauna (*Micro-arthropoda*)						Meso-fauna (*Lumbricidae*)		
	Collembola			*Oribatida*					
	1[a]	2	3	1	2	3	1	2	3
Control	213	320	160	80	160	80	9.0	14.6	16.0
Slaked lime	–	160	80	346	480	240	16.3	33.6	21.3
Dolomite	293	320	213	186	240	186	8.7	30.0	12.3
Cement dust	320	373	240	160	186	80	17.3	10.3	12.0
Red sludge	80	160	80	133	160	–	20.6	6.0	8.0
$HIP_{0.5}$	58.4	77.9	38.9	72.8	38.9	38.9	1.1	4.0	2.1

[a]1—2012; 2—2013; 3—2014

nutrients. Thus, local reclamation technology improves the quality of subsoil and surface water.

In *Podzolised chernozem*, liming encourages optimal conditions for crop growth and for the soil organisms that make humus. Given that self-regulation and self-renewal of soil fertility depend on soil organisms, we counted their numbers according to the kind of calcium ameliorant applied. Liming increases the number of *Lumbricidae* through changes in soil pH—so the numbers of all kinds of earthworms were increased by all ameliorants—except red sludge in the first and second year after application when their numbers declined, indicating a toxic effect. Similar changes are observed in soil micro-fauna (Table 16.1).

In *Podzolised chernozem,* liming creates a favourable habitat for earthworms, which are intolerant of acidity (Holhoeva 2004). Earthworms consume plant debris, mineral soil particles and microorganisms, and void casts enriched with lime and humus, which combine as stable complexes including calcium humates. The casts are, in themselves, water-stable soil aggregates. And they stimulate biological activity; their micro-flora generate enzymes, antibiotics, amino acids, vitamins and other biologically active substances that destroy pathogens (Bityutsky et al. 2005). Thus, the increase in the number of earthworms in a limed soil activates self-renewing, regulative soil processes and improves its agro-ecological condition.

Calciferous glands in the earthworm gut neutralize acids formed during the decomposition of organic matter—so the casts have a neutral reaction; paradoxically, even in the absence of liming, they can neutralize soils of low pH. For example, in Kharkiv region, acid soils have not been limed since the 1980s yet the area affected by acidity has declined. We guess that neutralization of soil acidity has been accomplished by biological factors and, above all, the increase in the earthworm population under reduced chemical loads (mineral fertilizers, herbicides, fungicides, etc.). Worms mix the soil to a depth of two metres, reaching the underlying loess that contains 25–30% calcium carbonate so a substantial amount of $CaCO_3$ enters the upper layers of soil in worm casts. This natural soil reclamation contributes to self-renewal and self-regulation of soil fertility.

It has been known for a long time that liming contributes to the formation of soil aggregates, coagulates colloids and improves soil tilth and aeration—but there was no scientific explanation. In our view, the combination of lime with organic matter in the form of calcium humates may be explanation enough. Most likely, the benefits accrue indirectly due to improvement in the habitat for earthworms. And earthworms improve soil agro-ecology by forming water-stable structure, improving aeration and water permeability, enriching the soil in humus and nitrogen, inhibiting pathogenic micro-flora and revitalizing beneficial micro-flora.

It needs to be said that progress in agriculture is due not only to the use of fertilizers, ameliorants, pesticides and other chemicals, but also the potential of plants to create soil fertility. Phyto-amelioration is achieved by introducing into the crop rotation acid-tolerant crops that not only withstand the acid reaction but, also, transport calcium from the lower to the upper layers of soil. This is accomplished by perennial grasses and legumes such as lucerne, sainfoin, lupin and soya. Investigation of the structural aggregates of *Podzolised chernozem* showed that the growth of various phyto-ameliorants significantly affects the ratio of differently sized structural units; the highest degree of structure formation was observed under perennial grasses and lupin. The aggregate index, or *structuring factor*, describes the qualitative and quantitative composition of structural aggregate according to the formula:

$$K = A/B$$

where K is the structuring factor; A—the amount of macro-aggregates (0.25–10 mm)%; B—the amount of aggregates <0.25 and clods greater than 10 mm, %.

The structuring factor in upper 20 cm of the soil after 2 years phyto-amelioration was 8.3 under lucerne, 7.3 under lupin and 7.2 under sainfoin. This effect was due to the ability of the root system to ramify deep into the soil and lift calcium from the lower horizons, as well as the adding organic matter, which serves as a natural glue for soil aggregates (Sokolovskyi 1971; Williams 1949).

Phyto-ameliorants also increase the water resistance of soil aggregates: the greatest water resistance in the 0–20 cm layer after 1 year phyto-amelioration was achieved by sainfoin—0.9 compared with 0.6 under the control; under perennial grasses, the maximum value was reached in 20–40 cm layer coinciding with the greatest root mass. As well as perennial grasses, the annual Sudan grass (*Sorghum Sudanese*) is unrivalled in adding organic matter to degraded soils through its root system.

Conclusions

1. Compared with traditional liming practice, application of technology for *local cultivation* of acidic soil can reduce nutrient losses from leaching: lime by almost six times, soluble organics and nitrate by 1.8 and 2.9 times, respectively.

2. Increase in the population of soil organisms in *Podzolized chernozem* through liming activates self-renewing and regulating processes, thereby improving the agro-ecological condition of the soil.
3. Perennial grasses, legumes and Sudan grass are effective phyto-ameliorants. Their beneficial effects are complex but are displayed through resilient soil aggregates and the accumulation of organic matter.

References

Baliuk SA, Truskavetsky RS, Tsapko YL (eds) (2012) Chemical reclamation of soils (the concept of innovative development). Miskdruk, Kharkiv (Ukrainian)

Bityutsky NP, Solovieva AN, Lukina EI et al (2005) Influence of earthworms on the modification of the population of micro-organisms and activity of enzyme in soil. Soil Sci 1:82–91 (Russian)

Byzova YB, Gilyarov, Dungen V et al (1987) Quantitative methods in soil zoology. Nauka, Moscow (Russian)

Deriy SI, Ilyuha VO (2000) Fundamentals of ecology. Ukrainian Phytosocial Centre, Kyiv (Ukrainian)

Holhoeva LS (2004) Resistance mechanisms of earthworms (*Lumbricidae*) to destabilizing natural and anthropogenic factors. In: Actual problems of preserving the stability of living systems: collected articles of the International Scientific Ecological Conference (Belgorod, Russia, 27–29 September 2004), Belgorod State University Publishing House, Belgorod (Russian), pp 232–233

Ivashura AA, Orekhov VM (2004) Ecology: theory and practice. INZHEK, Kharkiv (Ukrainian)

Martin JP, Pitblado JR (1984) Les pluies acides. Agriculture 484:226–230 (French)

Mazur GA, Barvynsky AV (1993) Degradation of arable sod—podzolized soils and methods of its prevention. Pochvovedenye 1:62–69 (Russian)

Sokolovskyi AN (1971) Soil structure and its agricultural value. Selected works. SPL. Urozhay, Kiev (Russian)

Truskavetsky RS (2003) The buffer capacity of soils and their main function—PPV Nove Slovo, Kharkiv (Ukrainian)

Truskavetsky RS, Tsapko YL, Khristenko SI et al (2003) Local cultivation—an effective method of soil fertility renewal. Agrochem Soil Sci 64:12–16

Tsapko YL (2003) New approaches to determination of requirements of acid soils in liming. Bulletin of Agricultural Science 6:14–17 (Ukrainian)

Tsapko YL (2006) Infra-red spectra of absorption of podzolised chernozem by different technologies of cultivation. Bull KhAI Kharkiv 6S:116–119

Williams VR (1949) Grassland farming systems. Oblizdat, Voronezh (Russian)

Chapter 17
Composition of Mobile Phosphate Fractions in Soils of the Pre-Carpathians Influenced by Drainage, Lime, and Phosphate Fertilizer

Tatiana Tsvyk

Abstract Specific factors of soil formation in the Pre-Carpathians drive a prevalence of iron and aluminium phosphate fractions and a low incidence of calcium phosphate. Given that present soil phosphate status is not satisfactory for agriculture, optimization can be achieved only through fundamental change of regime. Changes in the space profile of phosphate fractions in *brown podzolic gleys (Albeluvisols)* are examined under different drainage conditions and after application of lime and phosphate. It is confirmed that tile drainage not only improves soil aeration but also significantly influences the fractional composition of phosphates.

Keywords Soil phosphate fractions · Albeluvisols · Drainage · Lime and phosphate

Introduction

Specific soil conditions in *brown podzolic gleys* (*Albeluvisols*) in the Ukrainian Pre-Carpathians are responsible for the prevalence among the phosphate fractions of iron and aluminium phosphates and a low proportion of calcium phosphate. In these soils, there is a clear correlation between the content of individual phosphate fractions, the parent rock, the acid–base status, and the composition of microorganisms that transform phosphorus compounds (Smaga 2003, 2005; Nikorych et al. 1998). The specific conditions that give rise to iron/aluminium phosphates, as opposed to calcium phosphate, are weathering and leaching that result in a loss of bases. Remedial treatment simply according to the soil subtype (e.g. brown earths as a whole) makes no agronomic sense: optimization of the phosphate status can be achieved only through a fundamental change of regime.

T. Tsvyk (✉)
Department of Soil Science, Yuriy Fedkovych Chernivtsi National University,
25 Lesi Ukrainki, Chernivtsi 58000, Ukraine
e-mail: ticvik@gmail.com

D. Dent and Y. Dmytruk (eds.), *Soil Science Working for a Living*,
DOI 10.1007/978-3-319-45417-7_17

Installation of pipe drains results in increased leaching (Smaga 2001) but, at the same time, improved soil aeration should be reflected in a different phosphate regime. Drainage has been undertaken over some 100 thousand ha in the Carpathians (Kanivets and Polupan 1988); the most effective way of improving nutrient and base status of the drained soil has proven to be application of lime and ground rock phosphate twice per complete crop rotation (Nazarenko 1981).

Methods

The investigated soils were typical, medium loamy *brown podzolic gleys* under both undrained and tile-drained pasture:

Profile 1 500 m south-east of the village of Ispas, Vyzhnytsya district, Chernivtsi region in flat, undrained pasture cleared of forest 40 years ago and never systematically limed or cultivated. Vegetation: *Lolium perenne, Festuca valesiaca, Lotus corniculatus,* and *Taraxacum officinale.*

Hd (turf) 0–6 cm Root mat
Hegl (humus–eluvial) 6–22 cm Brownish grey powdered with silica, medium loam, granular structure, loose, iron–manganese nodules, abundant fine roots, sharp boundary
Ehgl (eluvial) 22–40 cm Bluish grey powdered with silica, medium loam, subangular blocky, compact, iron–manganese nodules, gradual boundary
EIgl (transitional) 40–72 cm Mottled and blotched brown, ochre and bluish grey with a little silica powder, medium loam, subangular blocky, compact, iron–manganese nodules, gradual boundary
Igl (illuvial) 72–135 cm Bluish grey and blue, medium loam, blocky to prismatic structure, wet
Pgl (transition to bedrock) >135 cm Brownish yellow, medium loam, massive.

Profile 2 2 km south-east of Ispas village, under pasture with pipe drains installed in 1976 at 12-m spacing and a depth of 0.95–1 m.

Hd (turf) 0–4 cm Root mat
Hegl (humus–eluvial) 4–45 cm Pale grey with abundant ochre mottles and silica powder, medium loam, subangular blocky, iron–manganese nodules, gradual boundary
EIgl (transitional) 45–70 cm Mottled brown and bluish grey with silica powdering, coarse loamy, subangular blocky, iron–manganese nodules, many earthworms, gradual boundary
Igl (illuvial) 70–125 cm Blotchy brown, fine loamy, blocky and prismatic structure, small iron–manganese nodules, gradual boundary
Pgl >125 cm Bluish grey, wet

The aim was to establish the patterns of formation of the phosphate fractions as influenced by drainage and chemical reclamation. Soil samples were collected in duplicate from the middle of each horizon from both drained and undrained pastures. The following treatments were applied: (1) control—without treatment; (2) lime—equivalent to 4 t/ha; (3) ground phosphate—equivalent to 2 t/ha; (4) lime at half the standard dose + ground phosphate at half the standard dose; and (5) lime at half the standard dose + ground phosphate at half the standard dose applied 6 months after liming.

The duration of the interaction of soil with chemical ameliorants was 1 and 2 years. From selected samples, the fractional phosphate composition was determined in triplicate by the method of Chang-Jackson and physicochemical properties by conventional methods (Sokolov 1975).

Results and Discussion

By comparing the soils of the same type but with different intensity of present-day gley processes (with and without drainage), we may assess the contribution of drainage status to the formation of different phosphate fractions (Table 17.1).

Phosphate fractions may be differentiated according to their solubility and potential uptake by plants. The content of the weakly bound fractions, assumed to be accessible to plants, increases with depth in drained pasture as opposed to undrained pasture. However, in both drained and undrained soils, the content of weakly bound phosphates is very low.

In the drained profile, the greater expression of podzolization in the upper part of the profile corresponds with the increase in the content of iron and aluminium phosphate. In the undrained soil, the mid-profile and parent rock show a significant increase in iron phosphate (respectively 52.8 and 68.8 mg/100 g soil). Amelioration by drainage reverses this trend—the concentration of iron phosphates in the lower profile and, especially, in the parent material is less by as much as 60%. Trends of aluminium phosphate are not so clear. Calcium phosphates are present only in small quantities; they are considered potentially available to plants, and the sharp increase in their content in the drained soil, especially in the parent material, may be caused by increased migration of calcium.

The different distribution of phosphate fractions in the drained brown soil compared with the undrained gley podzol reflects a slowdown of gley processes and an increase in brown soil formation. The two variants also showed significant differences after incubation with chemical ameliorants over 1 year (Table 17.2).

Application of lime and ground phosphate, together, enhanced the content of weakly bound phosphates—as did the application of phosphate 6 months after liming. Compared with the original soil, the concentration of weakly bound phosphates has increased tenfold as a result of the reduction of acidity, which slows down the formation of mineral phosphates. The influence of chemical ameliorants in the drained pasture is noteworthy, especially when lime and phosphate were

Table 17.1 Composition of phosphate fractions in *brown podzolic gley* soil under pasture

Horizon	Depth (cm)	Weakly bound phosphates		Al–P		Fe–P		Ca–P		Total fractions
		mg/100 g Soil	%	mg/100 g Soil	%	mg/100 g Soil	%	mg/100 g Soil	%	mg/100 g Soil
Undrained										
Hegl	6–22	1.4	3.1	7.2	15.9	34.2	75.5	2.5	5.5	45.3
Ehgl	22–72	1.6	2.4	7.4	11.2	52.8	80.1	4.1	6.2	65.9
Igl	72–135	1.2	2.3	6.1	11.6	45.2	85.8	5.2	9.9	52.7
Pgl	>135	1.6	2.0	3.8	4.8	68.8	86.5	5.3	6.7	79.5
Drained										
Hegl	4–45	1.7	1.9	9.1	10.6	63.2	74.0	11.5	13.5	85.5
EIgl	45–70	1.8	2.0	7.9	8.3	66.3	70.0	19.0	19.7	95.1
Igl	70–125	1.7	3.1	3.3	6.3	40.9	76.6	7.5	14.0	53.4
Pgl	>125	2.2	3.2	9.9	13.9	42.8	60.6	15.7	22.3	70.6

Table 17.2 Fractional composition of phosphates in *brown podzolic gley* soil treated with chemical ameliorant

Horizon	Treatments	Weakly bound phosphates		AI–P		Fe–P		Ca–P		Total fractions
		mg/100 g Soil	%	mg/100 g Soil	%	mg/100 g Soil	%	mg/100 g Soil	%	mg/100 g Soil
Undrained										
Hegl	Lime – standard rate	18.4	37.7	11.1	21.0	15.6	29.6	7.8	14.7	52.8
	Phosphate – standard rate	15.2	36.1	15.8	37.5	9.1	21.6	2.0	4.8	42.2
	Lime + phosphate, each at half the standard rate	18.8	53.9	8.9	25.7	5.1	14.6	2.0	5.8	34.8
	Lime half standard + phosphate half standard after 6 months	20.4	49.3	9.6	23.2	8.2	19.7	3.3	7.9	41.4
Drained										
Hegl	Lime – standard rate	29.1	52.0	8.9	16.1	11.9	21.3	5.9	10.6	55.9
	Phosphate – standard rate	30.2	51.7	10.3	17.5	9.2	15.8	8.8	15.1	58.5
	Lime + phosphate, each at half the standard rate	21.1	32.7	6.6	10.2	21.6	33.5	15.2	23.6	64.6
	Lime – half standard + phosphate half standard after 6 months	38.6	55.8	6.6	9.5	14.4	20.8	9.5	13.8	69.2

introduced separately. Chemical ameliorant had little impact on the content of aluminium phosphate in the drained pasture. Much of the decline in the content of iron phosphate may be due to the weakening of gley processes and lower mobilization of iron. On the other hand, calcium phosphate accumulates more in the drained soil. Thus, the optimal composition of mineral phosphates in *brown podzolic gley* soil may be achieved by drainage and applying lime (4 t/ha) and ground phosphate (2 t/ha) after 6 months.

Comparison of undrained and drained soils made it possible to establish the dynamics of mobile phosphate fractions (Table 17.3). Phosphates soluble in acid reagent (such as Kirsanov's 0.2 M HCl extract) are considered partially available to plants; those soluble in neutral or near-neutral reagents (such as Schofield's 0.01 M $CaCl_2$ extract and Karpynskiy–Zamyatin's 0.03 M K_2SO_4 extract) are increasingly mobile and available.

One year after liming, there was a marked increase in Kirsanov-extractable phosphorus compared with the control in samples from both undrained and drained soils. After 2 years of incubation, twice the amount of phosphate was extracted with

Table 17.3 Indicators of mobile phosphate in *brown podzolic gley* soil after 1 and 2 years of interaction with chemical ameliorant

Variant of experiment	1 year of incubation			2 years of incubation		
	P_2O_5 in 0.03 M K_2SO_4 extract	P_2O_5 in 0.01 M $CaCl_2$ extract	P_2O_5 in 0.2 M HCl extract	P_2O_5 in 0.03 M K_2SO_4 extract	P_2O_5 in 0.01 M $CaCl_2$ extract	P_2O_5 in 0.2 M HCl extract
	mg\l		mg\kg Soil	mg\l		mg\kg Soil
Undrained						
Control	0.001	0.02	31.0	0.001	0.02	31.0
Lime, standard rate	0.011	0.12	35.0	0.009	0.12	59.0
Phosphate, standard rate	0.38	0.16	38.4	0.052	0.07	80.0
Lime + phosphate, each at half standard	0.43	0.31	46.4	0.15	0.068	74.0
Lime, half standard rate + phosphate, half standard rate after 6 months	0.45	0.32	48.8	0.06	0.035	73.2
Drained						
Control	0.001	0.023	27.0	0.001	0.023	27.0
Lime, standard rate	0.021	0.23	33.0	0.011	0.03	52.0
Phosphate, standard rate	0.28	0.30	36.5	0.03	0.02	74.0
Lime + phosphate, each at half standard	0.29	0.29	48. 6	0.18	0.021	93.6
Lime, half standard rate + phosphate, half standard rate after 6 months	0.38	0.11	48.4	0.38	0.020	68.8

0.2 M HCl solution. In versions 1 and 4 of the experiment, soil from undrained pasture showed an increase in P_2O_5 content of 73.2–80 mg/kg soil, which may be explained by the preferential formation of calcium phosphate under the influence of liming, calcium phosphate being relatively well extracted from the soil by 0.2 M HCl. After 2 years of incubation, soils from drained pastures exhibited increase in all mobile phosphates in all versions of the experiment.

One year after application of ground rock phosphate, the content of weakly bound phosphorus increased in samples from both drained and undrained pastures (by 8 mg/kg soil). After 2 years, there was a further increase, especially drained soil. The introduction of lime and ground phosphate, together, provided much the same content of weakly bound phosphorus as application of phosphate alone after 1 year's incubation; after 2 years of its content was lower than values obtained after application of phosphate alone.

Comparisons of the content of P_2O_5 in 0.03 M K_2SO_4 extract were carried out only between different variants of the experiment, but all showed significant differences in phosphate recovery. After a year's incubation, there was a marked increase from joint application of lime and phosphate in samples from the undrained pasture; after 2 years of incubation, the content decreased equally dramatically. Values remained highest in samples from undrained pasture treated with ground phosphate alone, and after the combined lime and phosphate treatment of samples from the tile-drained soil.

In both drained and undrained soils, phosphate extracted by 0.01 M $CaCl_2$ after 1 year of incubation was lower than the concentration in 0.03 M K_2SO_4 extract. After 2 years of incubation, there was a marked reduction in phosphate concentration in 0.01 M $CaCl_2$ extract, bringing it back to the level of initial values (the control).

Because of the low availability of phosphate in the soil, we observe high efficiency of phosphate fertilizer in terms of crop yields—thanks to the increase of weakly bound forms of phosphate. Depending on its solubility, residual phosphate in the soil ensures a future supply of available phosphates, but different soils exhibit differential release of this residual phosphate to plants. This is due to the degree of phosphate saturation of the soil, which causes the ageing of residual phosphates—their transition to more or less insoluble forms. We observed this effect in the experiment. In the first year after applying chemical ameliorant, there was a significant increase of concentration of phosphates in light-saline extracts (assumed to be available to plants); but after 2 years of incubation, 0.2 M HCl solution extracted a higher amount of phosphates, whereas their concentration in light-saline extract was much reduced.

The procedure of chemical amelioration also has considerable influence on the phosphate regime. Regardless of the genetic characteristics of the soil, addition of ground rock phosphate, alone or in combination with lime, increases the availability of phosphorus. However, the advantages of making the soil chemically compatible before adding phosphate are apparent in the drained soil—akin to a brown soil rather than a gley—where the availability of phosphate in the longer term is improved 1.5–2 times compared with simply adding ground phosphate.

Conclusions

1. Under the influence of drainage, amelioration in *brown podzolic gley soil* increases calcium phosphate with a corresponding decrease in iron phosphate.
2. Both drained and undrained soils respond to chemical amelioration, but a greater concentration of weakly bound (plant-available) phosphate is released in drained soil.
3. Optimum composition of mineral phosphates is produced by application to tile-drained soil of 4 t/ha ground lime, followed, 6 months later, by 2 t/ha ground rock phosphate.
4. The highest values of available phosphate are achieved in the first year after application: after 2 years, the concentration of available phosphate does not exceed the natural state.

References

Kanivets VI, Polupan NI (1988) Brown-podzolic gley soil. In: NI Polupan (ed) Ukraine soils and increasing fertility, pp 282–283, vol 1. Vintage, Kiev (Russian)

Nazarenko II (1981) Gley soils. Nauka, Moscow (Russian)

Nikorych VA, Polchyna SM, Smaga IS et al (1998) Transformation of mineral phosphate forms in brown-podzolic gley soils of the Pre-Carpathians due to changes in microbiological and enzymatic activity. Science (Biolohiya-Chernivtsi) 38:72–82

Smaga IS (2001) Impact of drainage reclamation on the ecological state of brown-podzolic gleys of the Pre-Carpathians. Sci Bull Uzh NU Series: Biol 9:45–47 (Ukrainian)

Smaga IS (2003) Ways to use genetic characteristics and the impact of brown earth and brown-podzolic soils on the form of mineral phosphates. In: Genesis, geography and ecology of soils—Collected works, pp 349–354 . Ivan Franko Lviv National University, Lviv (Ukrainian)

Smaga IS (2005) Indicators of phosphate status and properties of brown-podzolic surface-gley soils. Bull Agric Sci 7:17–20 (Ukrainian)

Sokolov AV (1975) Research methods in agrochemistry. Nauka, Moscow (Russian)

Part IV
Soil Contamination, Monitoring and Remediation

Chapter 18
Toxic Soil Contamination and Its Mitigation in Ukraine

Valeriia Kovach and Georgii Lysychenko

Abstract From the days of the Soviet Union, Ukraine inherited many abandoned stores of pesticide and herbicide, their use-by dates long expired, and dumps of toxic waste. These sites are health hazards and a serious factor for regional environmental instability, polluting soils and water resources. Impetus for dealing with them came with ratification, in 2004, of the Stockholm Convention on Persistent Organic Pollutants (POPs) aimed at reducing and phasing-out particularly toxic compounds, and the National Plan for the Stockholm Convention on POPs (2011). Four thousand five hundred dumps of obsolete pesticides and waste products were recorded in Ukraine in 2010; by the end of 2014, these had been cut to 1033. Details are given of the state of hazardous waste from the organochlorine plant operated in Kalush in western Ukraine that illustrates examples of successful removal and disposal of large amounts of toxic wastes but, still, many unresolved issues.

Keywords Toxic contamination · Obsolete pesticides · Potentially dangerous objects · Waste disposal · Hexachlorobenzene (HCB)

Introduction

The production and release of persistent organic pollutants (POPs) have caused global contamination of the environment and a hazard to wildlife and public health. Gross contamination has been documented at production sites and related landfills for various POPs including hexachlorocyclohexane (HCH), polychlorinated biphenyls (PCBs), or perfluorooctanesulfonic acid (PFOS) (Fernández et al. 2013; Jit et al. 2010; Oliaei et al. 2013; Torres et al. 2013a, b; Vijgen et al. 2011;

V. Kovach (✉) · G. Lysychenko
Institute of Environmental Geochemistry of the National Academy of Sciences, 34a Palladina Prospect, Kyiv 03680, Ukraine
e-mail: valeriiakovach@gmail.com

D. Dent and Y. Dmytruk (eds.), *Soil Science Working for a Living*,
DOI 10.1007/978-3-319-45417-7_18

Wycisk et al. 2013). The UNEP *Toolkit for identification and quantification of releases of dioxins, furans and other unintentional POPs* (UNEP 2013b) highlights the production of organochlorine as a major source of POP-contamination. Moreover, the production of organochlorine chemicals has resulted in the release of unintentional POPs (UNEP 2012, 2013a; Verta et al. 2009; Weber et al. 2008; Weber and Varbelow 2013). The need to protect human health and the environment is addressed by the 2004 Stockholm Convention on Persistent Organic Pollutants, ratified by Ukraine in 2004; Article 6 requires countries to develop strategies for identifying POP-contaminated sites.

A big organochlorine plant operated in Kalush in western Ukraine. The Kalush chemical and metallurgical complex (KChMC) was founded in 1968 near the big Kalush-Golinskaya potash deposit. It embraced 12 factories producing fertilizers as well as chlorinated solvents. Subsequently, solvent production was expanded under the Vinyl Chloride Chemical Company within the Oriana-Galev group that operated until October 2001.

A lot of hexachlorobenzene (HCB) waste was produced and disposed of by the Kalush factory. In 2008, it was discovered that HCB and other pollutants had been released into the soil, groundwater and the Sapogi-Limnytsia river (a tributary of the Dniester), suggesting that the containers holding the HCB waste had corroded and released pollutants into the environment. In view of the imminent threat to the Dniester river basin that supplies drinking water to millions of people, by Presidential Decree and relevant Decree of the Cabinet of Ministers of Ukraine, it was decided to excavate and remediate the toxic waste (President of Ukraine 2010; Cabinet of Ministers of Ukraine 2010). This study summarizes the magnitude of the problem at Kalush, remediation activities at the site from 2010 to 2013 and the current situation. We also draw some conclusions that may apply to similar sites and to POP waste imports to the European Union.

Materials and Methods

Sampling

Sampling and analysis of soil samples, water and excavated HCB waste were carried out in three stages. First, before work on HCB waste disposal, 8 wells were drilled at the disposal site (Polygon) to a depth of 8–12 m; six of these wells were equipped for hydrochemical monitoring. During the excavation, soil samples were taken for analysis of HCB content; excavations were made to 2–2.5 m and 4 samples (each 100 g) were taken from between 0.5 and 1.5 m and 2 samples from the bottom of the pit. The samples from the individual pits were combined and oven-dried at 110 °C. The composite samples, finely ground in a jasper mortar, were used for extraction and other measurements. Water from the observation wells after their pumping and the restoration was collected by a special sampling device designed by Simonov.

In the second stage, every day during excavation of waste from Polygon, composite samples were taken from Big-Bags loaded with excavated wastes for export. One kg of waste was taken from each Big-Bag; these were mixed and a 1-kg composite sample was collected. From selected samples, analysis of the content of HCB and chlorine was carried out at the Institute of Environmental Geochemistry of the NAS of Ukraine. Control samples were analysed by gas chromatography/mass spectrometry at the Scientific Research Institute of Ecological Problems of the Ministry of Ecology of Ukraine.

In the third stage, after removing the HCB waste and backfilling with uncontaminated soil, the Polygon surface soil was analysed. One-hundred-gram samples were collected from four points on each side and one in the centre of a 5×5 m sampling envelope and analysed as above. Also, water samples from observation wells were analysed.

Extraction and Analysis

Screening and monitoring focused on HCB as the indicator material. Other organochlorine impurities were identified previously by the Central Epidemiological Station of Water Transport at Illichivs'k and the laboratory of the Marzeiev Research Institute of Hygiene and Medical Ecology, as well as the UNEP/OCHA mission; these other pollutants had a higher variability that precluded adequate monitoring. Water samples were extracted with hexane by shaking for 10 min in an automated shaker. For the measurement/screening of HCB in wastes and soil, samples were extracted with benzene. Identification and measurements were performed using a Carlo Erba gas chromatography electron capture detector (ECD) equipped with a capillary column (30 m, diameter 0.32 mm, film thickness 0.5 μm) was used with the following instrumental settings chosen for rapid analysis (temperature of the injector 230 °C, isothermal column 230 °C, ECD detector 270 °C). With the selected parameters, the retention time of hexachlorobenzene on the column was 2.88 min. HCB was characterized by retention time and mass, and quantified by external standard calibration. Some screening was also performed by thin layer chromatography after extraction.

Soil samples were randomly screened using an Element 02 X-ray fluorescence (XRF) spectrometer to determine the elemental composition. Measurements of XRF meet the requirements of Category III of precision analysis which is used for mass analysis of geological samples in exploration, reserve calculation and control analyses.

Results and Discussion

Environmental Situation and Threats from Former Mining Activities

Potash deposits were opencast-mined at the Dombrovskii quarry. During the operational period (1967–2000), the extracted volume was 52.5 million m^3. From the northern part of the quarry, 14.6 million tonne of kainite was as dug out from a depth of 63 m; the southern part of the quarry produced 17.3 million tonne of kainite and 20 million tonne of waste rock from a depth of 127 m. The waste dumps and the quarry itself were destroyed by salt karst. The quarry is now flooded; more than 16 million m^3 of brine has accumulated and, near the surface, its salt concentration is about 80 g/l. This brine is seeping SSE towards the river Limnitsia.

According to expert assessment, the largest threats in the area are the old Dombrovskii quarry; the tailings of Mine Waste Dumps No. 1 and No. 2; the salt and sludge dumps at the edges of the quarry; the potash mines at Kalush, Holyn, Khoty and Novo-Holyn; and the HCB waste deposit near Kalush (the Polygon). Lack of rehabilitation, sediment control structures and maintenance, together with the unstable slopes of Mine Waste Dump No. 1, have led to erosion, filling up of drains and uncontrolled run-off (containing brine) into surrounding waterways and into the Dombrovskii quarry. The waste dumps are also a major, if not the major, source of brine polluting the aquifer in the direction of the river Limnitsia (UNEP/OCHA Environment Unit 2010).

In affected areas of Kalush City and the villages Kropyvnyk and Sivka Kalushska, the population is threatened by the unstable ground surface and salinization and pollution of the drinking water. To a very limited extent, the underground mines have been backfilled with solid materials but, for the most part, the mines have been flooded with brine, in some cases with fresh water. Flooding with brine does not necessarily guarantee the stability of the inter-chamber pillars and, as a result of the collapse these pillars, the ground above the mine is subsiding. Several sink holes have appeared, varying in size from a few metres to over 10 m in diameter. While the impact of sink holes is localized, there will be casualties if houses or apartment blocks were affected—and there are some 1300 residential units and 23 industrial facilities in the area of intensive mining. According to the authorities, 4328 people are at risk and in need of resettlement; the local authority has identified an area for resettlement which could house over 6000 people, a school, and two kindergartens (UNEP/OCHA Environment Unit 2010).

Solvent Production, HCB Waste Generation and Landfill

Solvent production in Kalush started in 1973 with an estimated production capacity of 30,000 tonne per year. The solvents included carbon tetrachloride and tetrachloroethene; also, dichloroethene (ethylenedichloride EDC) was manufactured for PVC production. By-products included carbon tetrachloride, hydrogen chloride and hexachlorobutadiene. In the process, some of the tetrachloroethene is condensed to HCBD and, further, to HCB. According to company records, the HCB waste was originally a yellow crystalline substance containing tetrachloroethane (C_2Cl_4) 2–3%, hexachloroethane (C_2Cl_6) 7–11%, hexachlorobutadiene (HCBD, C_4Cl_6) 20–27% and HCB (C_6Cl_6) 53–67% (LLC Oriana-Galev 2001). In short, it consisted mainly of POPs (HCB and HCBD). A recent assessment of the production of carbon tetrachloride has demonstrated that some further condensation takes place generating octachlorostyrene (OCS), chlorinated naphthalenes (PCNs) and polychlorinated biphenyls (PCBs) as well as small amounts of polychlorinated dibenzo-p-dioxins and dibenzofurans (PCDD/Fs) (Zhang et al. 2015).

All the HCB waste from solvent production as well as wastes from the EDC production from the Oriana-Galev operations was dumped in the Polygon site (LLC Oriana-Galev 2001, 2002; Tsygulieva-Kharkov 2001)—this amounted to about 540 tonne/year between 1973 and 2001; a total of about 15,000 tonne. For the first 3 years (1973–1976), there was no dedicated storage site; HCB waste was, reportedly, dumped at several places in the Kalush Area. The major HCB landfill site is 6 km south-west of the city of Kalush; it was assessed by a 2010 UNEP/OCHA mission that confirmed POP pollution; in addition to HCB, pentachlorobenzene was detected in surface water. According to company records, at the date of decommissioning (October 2001) there were 11087.6 tonne of HCB waste (Hazard Class 1) and 265 tonne of solid residues from equipment (Hazard Class 3). Further evidence could now be provided of storage at Mine Waste Dump No. 1 and/or No. 4 which, today, would be classified as uncontrolled dumping (UNEP/OCHA Environment Unit 2010).

Before the remediation started, several site inspections were conducted at the initiative of the Ivano-Frankivsk Regional Administration to assess the quantity and quality of the waste mixtures and to refine the inventory. The HCB waste was packed in 200-litre steel drums and buried at Polygon in trenches excavated to a depth of 4 m, covered with plastic wrap and backfilled with earth. Due to corrosion of the drums (Fig. 18.1), some of the waste had already migrated, and the results of these assessments differed significantly from the volume of originally disposed wastes.

Assessment of Waste Composition and Toxicity Considerations

The composition of the waste at the Polygon and pollution in the surrounding area was investigated in December 2011 by Central Toxicology Laboratory on Water Transport (Odessa). These studies showed that the proportion of HCB waste can reach 90%. Screening of releases from the site revealed:

Fig. 18.1 Documentary pictures of the Polygon, showing the corroded HCB waste containers and the process of destruction

The vapour phase in the samples of soil contained: 1,1-dichloroethane; chloroform; carbon tetrachloride; trichlorethylene; 1,2,3-trichloroethane; tetrachlorethylene; 1-bromo-1,2, 2-trichlorethylene; hexachloroethane; and 1,3-butadiene-1,1,2,3,4,4-hexachloro (HCBD).

Water samples contained: HCB; chloroform; 1,2-dichloroethane; carbon tetrachloride; tetrachlorethylene; hexachloroethane; and 1,3-butadiene-1,1,2,3,4,4-hexachloro (HCBD).

As a Stockholm Convention-listed POP and the major component of the waste, HCB is of greatest concern—it is highly toxic, resistant to degradation, and accumulates biologically in terrestrial and aquatic ecosystems. It is considered a probable human carcinogen, a proven carcinogen to animals, and highly toxic to aquatic organisms. Emergency measures were needed to prevent its harmful effects (Stockholm Convention 2010). Ukraine has developed several standards for HCB including:

- For surface water: 0.001 μg/l (=1 ng/l) (risk category I),
- For soil: 0.03 mg/kg dry matter (=30 ng/g),
- For air: 0.9 $\mu g/m^3$.

These standards are comparable with those elsewhere. Other compounds of concern in the waste are PeCBz and HCBD, both POPs, and both have higher water solubility than HCB. In one of water samples, PeCBz was detected at higher concentration than HCB (UNEP/OCHA Environment Unit 2010).

Risk Posed by the Dumped Waste and the Decision to Remediate

Storage of the drums of waste in a site permeated by salty groundwater led to corrosion of the drums (Fig. 18.1). As a result, HCB and other toxins were released into the soil and groundwater and present a hazard to the Sapogi-Limnytsia river, a tributary of the Dniester. By Presidential Decree and Decree of the Cabinet of Ministers of Ukraine in 2010, it was decided to remove the toxic waste and to close the Polygon site (President of Ukraine 2010; Cabinet of Ministers of Ukraine 2010). The HCB waste was excavated and, together with the contaminated soil, shipped to the UK, France and Poland for destruction by incineration. Polygon was refilled with clean soil.

Analysis of HCB in the control soil samples, collected after the removal of waste from the Polygon site and refilling with clean soil, demonstrates that Polygon is still contaminated with HCB at around 0.5% (5 g/kg)—170 thousand times higher than the Ukraine LAC limit for dry soil (Institute of Environmental Geochemistry of the National Academy of Sciences of Ukraine 2012). Significant levels of HCB were also recorded in the removed topsoil, which, according to the procedure of remediation (Ministry of Ecology and Natural resources of Ukraine 2010), was stored and subsequently loaded into a neutralized yard. The mass fraction of HCB in overburden is 1.9% (19 g/kg), which is 6330 times higher than the LAC for dry soil.

It is clear that the feasibility study for the HCB waste disposal (Ministry of Ecology and Natural Resources of Ukraine 2010) did not take into account of the long-term migration of HCB waste into the upper soil layers through seasonal changes in groundwater levels, capillary flux and evaporation from the soil surface. Some contamination could also have occurred during remediation and levelling of the surface. Compilation of data on HCB contamination and distribution in groundwater (Fig. 18.2) shows that the contamination is mainly south of Polygon. The concentrations were sometimes in the g/l range—far above the normal water solubility of HCB (see, e.g. Well #2 and #3, Fig. 18.2) indicating highly soluble organochlorine co-pollutants (Institute of Environmental Geochemistry of the National Academy of Sciences of Ukraine 2014). HCB concentrations in wells exceed Ukraine limits ten to more than one hundred-fold. Monitoring of HCB in the Sapogov stream, 200 m from the site, reveals HCB values that exceed the safe level three to five-fold.

Fate of Exported Waste

Most of the recovered HCB waste was exported to hazardous-waste incinerators in France and UK suitable for POPs destruction. But 10,000 tonne was exported to a Polish facility where it was stored in the open, only 50 m from the Baltic Sea (documentary at http://www.ihpa.info/actions/2012/12/09/the-carcinogenic-hcb-by-the-baltic-sea/). Moreover, the capacity of the Polish incinerator is relatively small

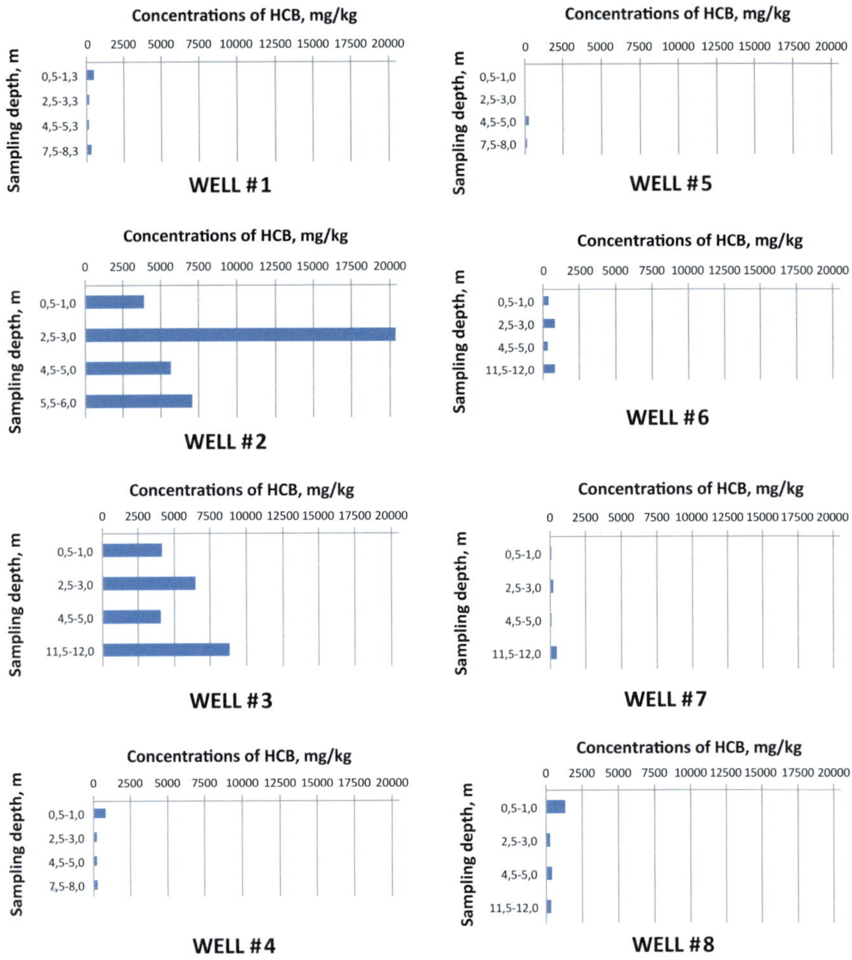

Fig. 18.2 HCB levels in soil at the Polygon wells

and destruction of the waste will take around 2 years. This facility was not suitable to receive and destroy the POP waste.[1] There was a similar case in 2011 in Germany where the ENVIO company imported and processed PCB-containing equipment; the ENVIO staff, factory site and the surrounding area were heavily contaminated with PCB. It is clear that a more rigorous monitoring frame is needed for facilities conducting POP destruction and the local authorities need specific capacity building for the assessment of such sensitive activities. Alternatively, other authorities specialized in such supervision should be responsible.

[1]The Polish prosecutor has opened an investigation.

Conclusions

Ukraine has made a big effort to remediate the worst hazardous POP stockpiles. But despite the excavation and remediation, the Polygon site remains contaminated and requires further assessment and remediation including:

- A detailed assessment of whether all HCB waste has been removed.
- Considering that Polygon is still polluted, further remediation and securing measures. The appropriate international agency might involve foreign experts for further inspection of the site in cooperation with the Institute of Environmental Geochemistry of the National Academy of Sciences of Ukraine. In such an investigation, one outcome should be measures to minimize the migration of HCB, HCBD and other pollutants beyond the site.
- Determination of the HCB and HCBD pollution from Polygon into the surroundings, including the river sediments.
- Whether HCB waste from the factory has been dumped elsewhere in the area.

Feasibility studies for site decommissioning should take account of the long-term migration of POP waste and co-pollutants into the soil, groundwater and the wider environment. For the groundwater in particular, chlorinated compounds with higher water solubility pose a critical risk (see supporting information in Wycisk et al. 2013) and the co-deposition of more-soluble chlorinated organics can trigger a high water solubility of HCB, as observed in the current case in Kalush.

The excavation of the HCB waste in Kalush revealed that steel drums containing corrosive waste can soon become severely corroded and, then, release of the POPs into groundwater and soil. This needs to be considered at other sites where wastes from chlorinated solvent production have been stored or disposed of. The more water-soluble HCBD, which is included in the EU environmental quality norms for fish, might be a good indicator for releases from such storage sites.

According to the State Register of Ukraine, there are more than 24 thousand potentially dangerous objects (PDOs) on its territory, requiring constant attention and supervision. This was a big problem, even in peacetime, and remained unresolved. Today, during hostilities, the situation is even more acute. These PDOs are a potential threat to national security and demand new thinking about their existence and protection.

References

Cabinet of Ministers of Ukraine (2010) Measures related the mitigation of consequences of an ecological situation in the town of Kalush and villages Kropyvnyk and Sivka-Kalush of Kalush district, Ivano-Frankivsk Region. Regulation no 381-p. 02 Mar 2010, Kyiv (Ukrainian)

Central Toxicology Laboratory on Water Transport (2011) Inspection of hexachlorobenzene waste disposal at the Polygon disposal site of toxic wastes of LLC Oriana-Galev in Kalush, Ivano-Frankivsk region, dated 23 May 2011 (Ukrainian)

CH2M Hill International Corp (1991) Feasibility study: hazardous-waste remediation at the Chabarovice site, vol 1. Export trade information report no: PB-92-216027/XAB. Document provided to NTIS by the US Trade and Development Program, Rosslyn

Fernández J, Arjol MA, Cacho C (2013) POP-contaminated sites from HCH production in Sabiñánigo, Spain. Env Sci Pollut Res 20:1937–1950

Heinisch E, Kettru A, Bergheim W et al (2006) Persistent chlorinated hydrocarbons, source-oriented monitoring in aquatic media 4. The chlorobenzenes. Fresenius Environ Bull 15(3):148–169

Heinisch E, Kettrup A, Bergheim W, Wenzel S (2007) Persistent chlorinated hydrocarbons, source-oriented monitoring in aquatic media. 6. Strikingly high contaminated sites. Fresenius Environ Bull 16:1248–1273

IISD Reporting Service (2014) Summary of the tenth meeting of the Stockholm Convention's Persistent Organic Pollutants Review Committee: 27–30 October 2014. Earth Negot Bull 15:214

Institute of Environmental Geochemistry of the National Academy of Sciences of Ukraine (2012) Research and methodological support in the implementation of environmental protection measures of the customer on disposal of hazardous waste of hexachlorobenzene at Polygon in the area of the Dombrowski quarry in Kalush District, Ivano-Frankivsk Region. 3 volumes, State registration no. 0112U007273 (Ukrainian)

Institute of Environmental Geochemistry of the National Academy of Sciences of Ukraine (2014) Research and methodological support for the disposal of hexachlorobenzene hazardous waste at Polygon in the area of the Dombrivskii quarry in Kalush District, Ivano-Frankivsk Region in 2013 (Ukrainian)

Jacob F, Scarberry R, Rosa D (1986) Source assessment of hexachlorobenzene from the organic chemical manufacturing industry. In: Morris CR, Cabral JRP (eds) Hexachlorobenzene: proceedings of an international symposium, Lyon. IARC Scientific Publication 77, OUP, New York, pp 31–37

James P (2009) The supervision of environmental risk: the case of HCB waste or Botany/Randwick? J Env Manag 90:1576–1582

Jit S, Dadhwal M, Kumari H et al (2010) Evaluation of hexachlorocyclohexane contamination from the last Lindane production plant operating in India. Env Sci Pollut Res 18:586–597

Jones K, Barber J, Sweetman A (2005) Hexachlorobenzene-sources, environmental fate and risk characterisation. Euroc Sci Doss. Lancaster University. http://eprints.lancs.ac.uk/id/eprint/2122

LLC Oriana-Galev (2001) Passport of potentially dangerous object. Kalush , Ivano-Frankivsk region. Registration no. 6, 3 Apr 2001 (Ukrainian)

LLC Oriana-Galev (2002) Passport of the waste disposal site "Polygon of solid toxic waste" Kalush, Ivano-Frankivsk region. Registration no. 8, 19 Aug 2002 (Ukrainian)

Ministry of Ecology and Natural Resources of Ukraine (2010) Feasibility study of the project on waste management of hexachlorobenzene located in Dombrivskii quarry and Polygon of solid toxic wastes in the zone of ecological emergency in the town of Kalush and villages Kropyvnyk and Sivka-Kalush of Kalush District, Ivano-Frankivsk Region. National Centre for Hazardous Waste Management, Kyiv (Ukrainian)

Oliaei F, Kriens D, Weber R, Watson A (2013) PFOS and PFC releases and associated pollution from a PFC production plant in Minnesota (USA). Env Sci Pollut Res 20:1977–1992

President of Ukraine (2010) Announcement of a zone of ecological emergency in Kalush and villages Kropyvnyk and Sivka-Kalush of Kalush District, Ivano-Frankivsk Region. Decree of the President of Ukraine no. 145, 10 Feb 2010, Kyiv (Ukrainian)

Stockholm Convention (2010) Additional consideration of new persistent organic pollutants: pentachlorobenzene. 6th POP Review Committee Meeting Geneva 10/2010 UNEP/POPRC/6/INF/21

Torres JPM, Fróes-Asmus CIR, Weber R, Vijgen JMH (2013a) Status of HCH contamination from former pesticide production and formulation in Brazil—a task for Stockholm Convention implementation. Env Sci Pollut Res 20:1951–1957

Torres JPM, Leite C, Krauss T, Weber R (2013b) Landfill mining from a deposit of the chlorine/organochlorine industry as source of dioxin contamination of animal feed and assessment of the responsible processes. Env Sci Pollut Res 20:1958–1965

Tsygulieva-Kharkov O (2001) Persistent organic pollutants: a review of the situation in Ukraine. International POPs elimination project. http://www.ecoaccord.org/pop/ipep/ukr-review.htm. Accessed Sept 2015 (Ukrainian)

UNEP (2010) Additional consideration of new persistent organic pollutants: pentachlorobenzene. 6th POP Review Committee Meeting Geneva 10/2010 UNEP/POPRC/6/INF/21, Geneva

UNEP (2012) Draft risk profile: hexachlorobutadiene. UNEP/POPS/POPRC.8/3, Geneva

UNEP (2013a) Addendum. Risk management evaluation on hexachlorobutadiene. UNEP/POPS/POPRC.9/13/Add.2, Geneva

UNEP (2013b) Toolkit for identification and quantification of releases of dioxins, furans and other unintentional POPs under Article 5 of the Stockholm Convention on Persistent Organic Pollutants. Geneva

UNEP/OCHA Environment Unit (2010) A joint UN—European Commission environmental emergency response mission: technical scoping mission Kalush Area, Ukraine Mar 2010. UNEP/OCHA Geneva

Verta M, Kiviranta H, Salo S et al (2009) A decision framework for possible remediation of contaminated sediments in the River Kymijoki, Finland. Env Sci Pollut Res 16:95–105

Vijgen J, Abhilash PC, Li Y-F et al (2011) HCH as new Stockholm Convention POP—a global perspective on the management of lindane and its waste isomers. Env Sci Pollut Res 18:152–162

Weber R, Varbelow G (2013) The Dioxin/POPs legacy of pesticide production in Hamburg: part 1 securing of the production area. Env Sci Pollut Res 20:1918–1924

Weber R, Gaus C, Tysklind M et al (2008) Dioxin-and POP-contaminated sites-contemporary and future relevance and challenges. Env Sci Pollut Res 15:363–393

Weber R, Watson A, Forter M, Oliaei F et al (2011) Persistent organic pollutants and landfills—a review of past experiences and future challenges. Waste Manag Res 29:107–121

Wycisk P, Stollberg R, Neumann C et al (2013) Integrated methodology for assessing the HCH groundwater pollution at the multi-source contaminated mega-site Bitterfeld/Wolfen. Env Sci Pollut Res 20:1907–1917

Zhang L, Yanga W, Zhanga L, Lib X (2015) Highly chlorinated unintentionally produced persistent organic pollutants generated during the methanol-based production of chlorinated methanes: a case study in China. Chemosphere 22(19):14391–14404

Chapter 19
Heavy Metals in Soils Under the Heel of Heavy Industry

Georgii Lysychenko, Iryna Kuraieva, Anatolij Samchuk, Viacheslav Manichev, Yuliia Voitiuk and Oleksandra Matvienko

Abstract The soils of urban agglomerations dominated by heavy industry exhibit specific geochemical behaviour. New data are presented on total and mobile forms of heavy metals in soils contaminated by steelworks and chemical plants in different parts of Ukraine. The distribution of heavy metals in technogenic soils and the soil geochemical factors influencing the mobility of heavy metals and the mobile forms of metals in soil solution are assessed, and biogeochemical indices of polluted areas are determined. Heavy metal concentrations in soils of polluted areas are tens or hundreds of times higher than background values and maximum allowable concentrations, the soils' buffer capacity is much diminished, and the mobility of heavy metals is increased. Microscopic fungi may be a supplementary criterion for ecological and geochemical studies.

Keywords Heavy metals · Geochemical distribution · Forms of occurrence · Biogeochemical indices

Introduction

Soil is a unique natural resource, the font of terrestrial life, and the original natural indicator of pollution. The degree and the character of soil contamination by heavy metals are of special interest in the field of environmental geochemistry—which embraces the role and significance of natural and technogenic geochemical fields, processes and phenomena in the formation and evolution of life on Earth, and the

---

G. Lysychenko (✉) · V. Manichev
Department of Ecological Safety Problems, Institute of Environmental Geochemistry, Kyiv, Ukraine
e-mail: lysychenko@ukr.net

I. Kuraieva · A. Samchuk · Y. Voitiuk · O. Matvienko
Department of Geochemistry of Anthropogenic Metals and Analytical Chemistry, Semenenko Institute of Geochemistry, Mineralogy and Ore Formation, NAS Ukraine, 34a Palladina Av, Kyiv-142 03680, Ukraine

D. Dent and Y. Dmytruk (eds.), *Soil Science Working for a Living*,
DOI 10.1007/978-3-319-45417-7_19

functioning of the biota, including the influence of anthropogenic factors on ecosystems (Kuraieva et al. 2009). Here we investigate the distribution and geochemical behaviour of heavy metals in soils of Ukraine subjected to the impact of heavy industries. Contamination with heavy metals diminishes the activity of microorganisms and the ecological condition of soils (Kashin 2013). Although there is a considerable Russian and Ukrainian literature (Bondarenko 2011; Dolin 2011; Glazovskaya 1990; Il'in 1997; Mitskevich 1971; Saet et al. 1989; Samchuk et al. 1993, 1998; Vodyanitsky et al. 2012; Zhovinsky 2002), research on the geochemical component of technogenic pollution of soils is increasingly urgent.

Materials and Methods

Study Area

Research was carried out in eastern and northeastern Ukraine on soils under the influence of heavy industries: the steel industry in Mariupol, Alchevsk, and Dniprodzerzhyns'k, and the surrounding areas (the village of Melekine, the Streltsovskaya Steppe, and Dnyprovsko-Orylsky Nature Reserve); and the chemical industry in Shostka and surrounding areas (the Bogdanivsky reserve, and the villages of Obrazhiyivka and Lazarevka).

Field and Laboratory Methods

For each location, one large bulk sample was taken from various soils common in the urban areas. Sampling was stratified in 0–5 and 5–10 cm layers. In the laboratory, the samples were air-dried, crushed with a wooden rolling pin, and screened through a 2 mm steel sieve. Arinushkina's methodology was used to determine the soils' physical and chemical properties (Arinushkina 1970). Subsamples were finely ground in an agate mill and dissolved in concentrated HCl and HNO_3 (3:1) in a microwave system. The total content of heavy metals was assessed by ICP-AES using a PerkinElmer Optima 3200RL instrument. Mobile forms of heavy metals were determined following Pavlotskaya (1974). Statistical analysis was performed using the MS Excel package and maps drawn with the MapInfo program.

Results and Discussion

Soil Chemical and Physical Properties

We have studied both industrially contaminated soils and the soils of surrounding areas (Table 19.1). Soil organic carbon (C_{org}), pH, and the content of adsorbed

Table 19.1 Physicochemical properties of topsoils in contaminated areas and background values

Location of soil samples	C_{org} (mg/kg)	pH	Exchangeable cations * meq/100 g					
			H^+	Ca^{2+}	Mg^{2+}	K^+	Na^+	CEC
Technogenic soils affected by the steel industry								
Mariupol area	4.7	7.7	7.2	2.3	1.2	0.1	1.0	11.8
Alchevsk area	4.5	7.2	7.3	2.4	1.1	0.1	1.1	12
Dniprodzerzhyns'k area	4.6	6.7	10.1	2.3	1.1	0.2	1.2	14.9
Soils of background areas								
Melekine village	6.4	7.4	8.4	38.2	13.0	0.6	0.5	60.7
Shooting Steppe nature reserve	4.8	6.6	5.8	28.2	12	0.3	0.4	46.7
Dnyprovsko-Orilskyy nature reserve	7.8	6.4	10.8	49.3	18.4	0.8	0.6	79.9
Technogenic soils under the influence of the chemical industry								
Chemical factory at Shostka	0.75	5.1	–	2.3	0.9	0.08	0.3	3.58
Zirka	0.96	4.8	–	1.9	0.7	0.1	0.4	3.1
Svema	0.83	4.9	–	1.7	0.6	–	0.3	2.6
Soils of background areas								
Bogdanivsky reserve	0.85	6.3	–	5	2	0.14	0.9	8.04
Obrazhiyivka village	1.42	6.5	–	12.3	0.7	0.1	0.68	13.78
Lazarevka village	1.61	6.4	–	2	1	–	0.57	3.57

cations all decrease in technogenically contaminated soil compared with the background values and, as noted in earlier papers (Zhovinskiy 2002; Kuraieva 2013, 2014), soil physical and chemical properties affect the distribution of heavy metals.

Total Content of Heavy Metals

The content of some heavy metals in samples of soil from the towns of Mariupol and Shostka is presented in Tables 19.2 and 19.3, respectively. Since the contents of metals are specific and depend on the soil parent material and the conditions of soil formation, pollution levels are measured by comparison with control samples that are taken as a background level. The gross concentrations of heavy metals in each of the studied areas are as much as an order of magnitude greater than the background values. Maximum concentrations are found in industrial areas of the cities; contaminants are distributed according to the wind rose.

Table 19.2 Total content of heavy metals in topsoils from steel towns, mg/kg

Element / Sample location	Ni	Cr	Cu	Pb	Co	Ag
Mariupol area, $n = 39$	$\frac{2100}{500}$	$\frac{204}{50}$	$\frac{308}{20}$	$\frac{184}{18}$	$\frac{9}{5}$	n.a.
Alchevsk area, $n = 51$	$\frac{59}{23}$	$\frac{140}{80}$	$\frac{73}{21}$	$\frac{101}{13}$	$\frac{6}{6}$	n.a.
Dniprodzerzhyns'k area, $n = 36$	$\frac{36}{20}$	$\frac{113}{50}$	$\frac{60}{20}$	$\frac{119}{12}$	$\frac{4}{3}$	n.a.

Note Above the line—total concentration of heavy metals in contaminated soils; below the line—background value; *n*—number of samples; n.a.—not analyzed

Table 19.3 Total content of heavy metals in topsoils of the town of Shostka, mg/kg

Element / Sample location	Ni	Cr	Cu	Pb	Co	Ag
Svema, $n = 55$	$\frac{250}{10}$	$\frac{140}{5}$	$\frac{60}{10}$	$\frac{55}{15}$	$\frac{15}{4}$	$\frac{10}{1}$
Zirka, $n = 50$	$\frac{85}{10}$	$\frac{60}{5}$	$\frac{300}{10}$	$\frac{130}{15}$	$\frac{6}{4}$	$\frac{5}{1}$
Shostka plant chemicals, $n = 44$	$\frac{25}{10}$	$\frac{20}{5}$	$\frac{210}{10}$	$\frac{40}{15}$	$\frac{4}{4}$	$\frac{5}{1}$

Note Above the line—total concentration of heavy metals in contaminated soils; below the line—background value; *n*—number of samples

Total Soil Contamination

The geochemical association of heavy metals in the topsoils of Mariupol comprises the following elements: $Pb_{14.4} > Cu_{8.8} > Zn_{5.3} > Cr_{4.3} > Mn_{3.5}$; and for Shostka: $Pb_{53} > Ni_{16} > Cr_9 > Co_5 > Ag_4 > Cu_2$ (the numbers next to the elemental symbol represent concentration ratios).

For the two towns, total contamination index (Z_C) is determined by the method of Saet et al. (1989):

$$Z_C = \sum_{1}^{n} K_C - (n-1), \quad \text{where } K_C = \frac{C_i}{C_\Phi} \tag{19.1}$$

K_c—coefficient of concentration, n—number of accounted anomalous elements; C_i—concentration of the element in the investigated object, C_Φ—background content element.

For Mariupol, the total contamination index (Z_C) was calculated for the following elements: Mn, Ni, Co, V, Cr, Mo, Cu, Pb, Zn, and Sn. For the upper 5-cm layer, Z_C ranges from 3 to 581 with a mean of 38. Almost half of the city suffers dangerously high soil pollution ($Z_C > 32$); high concentrations of heavy metals are found not only around the steel works but also in residential areas. There are two anomalous areas with extremely dangerous levels of contamination ($Z_C > 128$)—one in the town centre and the other in the northwest part of the town. The least contaminated area is the southeastern part of town, upwind of the steel works. For the 5–10 cm soil layer, the contamination index 3–591 with a mean of 43; the 5–10 cm soil layer has a greater loading of heavy metals than the surface layer. Four areas of anomalously high and extremely dangerous level of contamination ($Z_C > 128$) were located: two in the central part of the town (northern Azovstal) and the other two in the west. About two-thirds of the town is characterized by high and dangerous pollution ($Z_C > 32$).

The total soil contamination of the soils in Shostka by Cr, Cu, Pb, Co, and Ag reaches a maximum in the vicinity of the industrial enterprises. The figure is 170 around the facility for production of hydroquinone (Shostka Plant Chemicals) compared with the average of 7; once again, pollution is attenuated downwind of the industrial plants.

Forms of Heavy Metals in Soils

Using Pavlotska's selective extraction, we separated exchangeable (ammonium acetate buffer), reserve and fixed forms of the heavy metals in the soils (Fig. 19.1).

The content of exchangeable forms of heavy metals in the measured soils decreases in the sequence (%): Zn (16.9–18.5) > Pb (5.2–10.6) > Ni (4.5–5.2) > Cu (1.8–3.4) > Cr (0.8–1.4). The share of reserve forms of heavy metals decreases in the sequence: Pb (55.2–57) > Cu (40.7–45.3) > Zn (31.6–39.2) > Ni (23.2–35.4) > Cr (16.9–19.1). Fixed forms of heavy metals decrease in the sequence Cr (79.4–82.3) > Ni (60.1–71.6) > Cu (51.3–57.5) > Zn (40.7–50.3) > Pb (32.4–39.6). In the background areas, the sequence of heavy metals

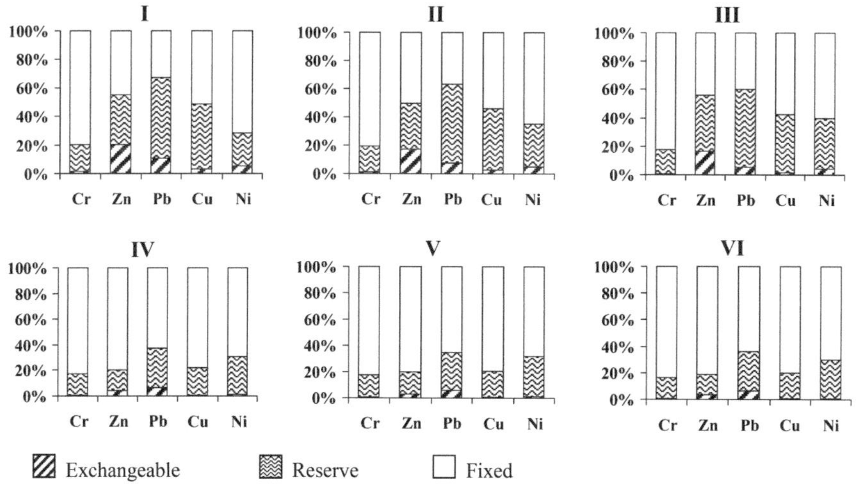

Fig. 19.1 Forms of heavy metals in soils (0–5 cm) around steelworks—**I** Mariupol, **II** Alchevsk, **III** Dniprodzerzhyns'k; and the surrounding areas—**IV** Melekine village, **V** Streltsovskaya Steppe reserve, **VI** Dnyprovsko-Orilsky reserve

content is (%): exchangeable—Pb (5.9–6.3) > Zn (3–4) > Ni (0.8–1.2) > Cr (0.7–0.8) > Cu (0.6–0.7); reserve—Pb (29–31.3) > Ni (29–31) > Cu (19–21.5) > Cr (16.4–17) > Zn (15–16.7); and the fixed form—Cr (82.8–83.4) > Zn (79.6–81.5) > Cu (77.8–80.2) > Ni (67.8–70.2) > Pb (62.5–65.1).

Using selective extractants, Samchuk et al. (1998) distinguished different forms of heavy metals in the soils around the Mariupol steelworks. The least amount of heavy metals occurred in water-soluble form (0.1–1%); in spite of the very high level of soil contamination, the concentration of heavy metals in aqueous solution is relatively low—exceeding the maximum permissible concentration only in the most contaminated spots. In the soils of background areas, most of the heavy metals are held firmly in the insoluble form.

In the town of Shostka (Fig. 19.2), the content of exchangeable forms of heavy metals decreases in the sequence %: Zn (25–62) > Pb (10–19) > Co (2–12) > Cu (5–10) > Ni (1–3) > Cr (0.04–4). The share of reserve forms decreases in the sequence %: Cu (25–75) > Co (17–51) > Zn (29 35) > Pb (28–71) > Ni (6–5) > Cr (1 7–22). The content of fixed forms decreases in the range: Cr (73–98) > Ni (73–91) > Co (48–72) > Cu (15–67) > Pb (20–53) > Zn (5–46). In the background areas, the sequence of the exchangeable form of heavy metals is (%)—Zn (10–28) > Cu (11–26) > Pb (12–19) > Co (1–3) > Ni (1–2) > Cr (0.4–2);

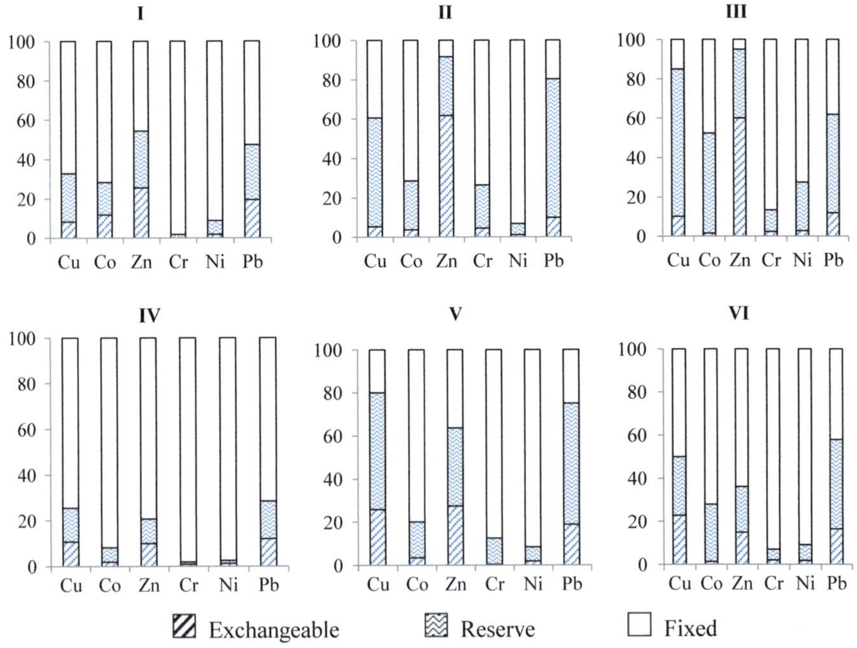

Fig. 19.2 Forms of heavy metals in soils affected by the chemical industry (0–5 cm)—**I** Svema, **II** Zirka, **III** Shostka plant chemicals; and surrounding areas—**IV** Bogdanivsky reserve, **V** Obrazhiyivka village, **VI** Lazarivka village

reserve form—Pb (17–56) > Cu (15–54) > Zn (11 36) > Co (6–27) > Cr (1–12) > Ni (1–6); and fixed form—Ni (91–98) > Cr (87–92) > Co (72–92) > Zn (36–79) > Cu (20–74) > Pb (25–71).

Soil Buffer Capacity

To investigate the buffering by soils impacted by the steel industry, we used Samchuk's method (1998) and, as a criterion for the quantitative assessment of ecological and geochemical stability of the soil solution, calculated the buffering coefficient K_b = CE/ΔpH (where CE is the capacity of the soil absorbing complex).

In both the Mariopol area and the Shostka area, technogenically contaminated areas have a lower buffer capacity than similar, uncontaminated soil landscapes (Table 19.4).

Soil Microbiology

Anthropogenic loading of heavy metals has a severe impact on ecosystems, leading to changes in the biota including, significantly, microscopic fungi. During the investigations of soil affected by the steel and chemical industries, 15 genera, 27 species, and 61 strains of microscopic fungi have been isolated and identified.

Table 19.4 Index of buffering of soils

Location of soil samples	K_b
Technogenic soils affected by the steel industry	
Mariupol area	4.2
Alchevsk area	4.8
Dniprodzerzhyns'k area	4.5
Soils of background areas	
Melekine village	55.2
Shooting Steppe nature reserve	28.8
Dnyprovsko-Orilskyy nature reserve	87.7
Technogenic soils affected by the chemical industry	
Shostka plant chemicals	3.6
Zirka	2.8
Svema	2.3
Soils of background areas	
Bogdanivsky reserve	11.8
Obrazhiyïvka village	10.5
Lazarivka village	11.5

In soil samples taken near the Azovstal Metallurgical Combine at Mariupol, the dominant fungi were *Mucor plumbeus*, *Aspergillus fumigatus*, and *Aspergillus flavus* (species not typical for background plots). The last two species are Group III pathogens able to produce mycotoxins dangerous to humans and animals. Soil samples taken near the Illich Steel and Iron Works at Mariupol were dominated by *Rhizopus stolonifer* and *Aspergillus niger; Aspergillus flavus* and *Aspergillus fumigatus* were also encountered.

Soil samples taken near the Svema plant at Shostka were dominated by *Penicillium spp.,* among them many species with high biochemical activity (*P. harzianum* Rifai, *P. Thomii* Maire, *P. Godlevsky* Zalessky, *P. decumbens* Thom) producing biologically active substances that are important for the national economy: enzymes, antibiotics, etc. Soil samples taken near Shostka Plant Chemicals were dominated by *Aspergillus sulphureus* Thom et Church, *Botrytis cinerea* Persoon ex Fries, *Chaetomium homopilatum* Omvik, dark-coloured *Cladosporium cladosporiodes*, and *Clonostacchis rosea*—all tolerant of extreme environmental conditions. We may conclude that certain sets of species of microorganisms may be sensitive indicators of soil contamination by heavy metals.

Conclusions

- Soil organic carbon, pH, and the content of adsorbed cations decrease in technogenically contaminated soil compared with background areas.
- The geochemical association of heavy metals in the topsoil of Shostka town, affected by the heavy chemicals industry, is represented by (%): Pb53 > Ni16 > Cr9 > Co5 > Ag4 > Cu2. The geochemical association of heavy metals in soils of Mariupol that are affected by the steel industry is represented by (%): Pb14 > Cu9 > Zn5 > Cr4 > Mn3.5.
- For areas affected by the steel industry, total pollution index (Z_C) is 3–581 with a mean of 38. The index for soils affected by the chemical industry at Shostka reaches 170 the industrial areas and diminishes to 2–3 at some distance from the industrial plants.
- Contamination of soil by the steel industry disturbs the natural form of heavy metals; in both *Chernozem* and *Sod podzolic soils* (*Albeluvisols*, IUSS WRB 2006; *Retisols*, WRB 2014), the incidence of mobile forms is increased. Their higher migration capacity is explained by an order of magnitude reduction in the buffering capacity of soils from emissions of anhydrides of strong acids.
- Some species of fungi are specific and may indicate concentrations of particular elements.

References

Arinushkina EV (1970) Manual of methods for soil chemical analysis. Moscow University Publishing House, Moscow (Russian)
Bondarenko GN (2011) Development of geochemistry technogenesis after the Chernobyl disaster. In: Scientific papers of the Institute of Environmental Geochemistry, vol 19. Kyiv, pp 19–40 (Russian)
Dolin VV (2011) Technogenic and ecological safety ecogeosystematics of the Bug Estuary in terms of heavy metal pollution. RAL-Poligrafiya, Kyiv-Nikolaev (Russian)
Glazovskaya MA (1990) Principles of soil classification of the danger of contamination by heavy metals. Biol Sci 9:38–52 (Russian)
Il'in VB (1997) Buffer properties of soil and the allowable level of pollution by heavy metals. Agrochemistry 11:65–70 (Russian)
IUSS WRB (2006/2014) World reference base for soil resources. World Soil Resources Reports 103/106, FAO, Rome
Kashin VK (2013) Features of accumulation of trace elements in oat grain in the Western Baikal. Agrochemistry 10:55–65 (Russian)
Kuraieva IV (2013) Forms of heavy metals in soils in Ukraine. In: Collected works UkrNDMI NANU, Kyiv, pp 331–339 (Ukrainian)
Kuraieva IV (2014) Forms of occurrence of heavy metals in technogenicaly contaminated soils affected by the steel industry (for example, the city of Alchevsk). Sci Pap Inst Environ Geochem 23:105–109 (Russian)
Kuraieva IV, Manichev VJ, Olishevskaja SV et al (2009) Biogeochemical indicators of soil in the area affected by the Kostyanetskiy lead-nickel plant. Mineral Mag 1:58–61 (Ukrainian)
Mitskevich BF (1971) Geochemical landscapes of the Ukrainian Shield. Naukova Dumka, Kiev (Russian)
Pavlotskaya FI (1974) Migration of radioactive products of global attacks on soils. Atomizdat, Moscow (Russian)
Saet YE, Revich BA, Yanin EP (1989) Environmental geochemistry. Nedra, Moscow (Russian)
Samchuk AI, Mitskevich BF, Sushchik YY, Shramarenko IF (1993) Mobile forms of heavy metals in the soils of the Kievskogo Poles'ya. Geol J 1:81–86 (Russian)
Samchuk AI, Bondarenko GN, Dolin VV et al (1998) Physico-chemical conditions of the mobile forms of toxic metals in soils. Mineral J 2:48–60 (Russian)
Vodyanitsky YN, Ladonin DV, Savichev AT (2012) Soil pollution by heavy metals. RASKhN, Moscow (Russian)
Zhovinsky EY (2002) Geochemistry of heavy metals in soils in Ukraine. Naukova Dumka, Kyiv (Russian)

Chapter 20
Dangerous Mercury Contamination Around the Former Radikal Chemicals Factory in Kyiv and Possible Ways of Rehabilitating this Area

Lyubov Maslovska, Aleksandra Lysychenko, Natal'ya Nikitina and Aleksandr Fesay

Abstract Mercury contamination of soil and buildings in and around the former Radikal chemicals factory in Kyiv, its partial ecological restoration, and the volume of waste still to be removed are described. The existing level of pollution in the Radikal JSC industrial area is a major threat to public health, human life, and the environment. The hazards of the removal of mercury-contaminated waste are assessed, and recommendations made for procedures to remove the contaminated soil and buildings to ensure compliance with regulations on public health and the environment.

Keywords Mercury · Waste · Soil · Environmental safety · Environmental protection

Introduction

Construction of the Radikal chemicals factory began in 1949 on a site then 10 km east of the city of Kyiv. Since then, the city has expanded so the factory is now within the residential and industrial area of the city. Radikal JSC (previously the Kyiv chemical plant of the Radikal State Production Association) started production in 1951, making DDT. The focus of the company was heavy chemicals but, by 1996, Radikal JSC was a monopoly producer of various synthetic materials, organic, and crop-protection chemicals. Several of the production processes were novel, at least for the Soviet Union, and many of them flawed. Environmental considerations were disregarded, and there were significant emissions and discharges of toxic substances. Production ceased on 15 July 1996. The industrial site and the soils of the

L. Maslovska (✉) · A. Lysychenko · N. Nikitina · A. Fesay
Institute of Environmental Geochemistry, National Academy of Sciences of Ukraine, 34a Palladina Av, Kyiv-142 03680, Ukraine
e-mail: elis.liubov@gmail.com

D. Dent and Y. Dmytruk (eds.), *Soil Science Working for a Living*,
DOI 10.1007/978-3-319-45417-7_20

surrounding area are polluted by highly toxic substances including mercury, which are a real threat to public health. The protection zone around the site should be 1000 m (Beletsky 2004) but, around the protection zone, there are open markets, shopping malls, and churches. The aim of this study is to provide scientific justification and recommendations for removal of the toxic waste from the site.

Site and Methods

The industrial site of Radikal JSC is located in the Desnyanskiy industrial area of Kyiv, official address: Chervonotkatska St 61, Kyiv 02100 (Fig. 20.1).

Class 1 and Class 2 hazardous substances, explosives, and flammable products were used or produced over many years. Some of these products are very stable and

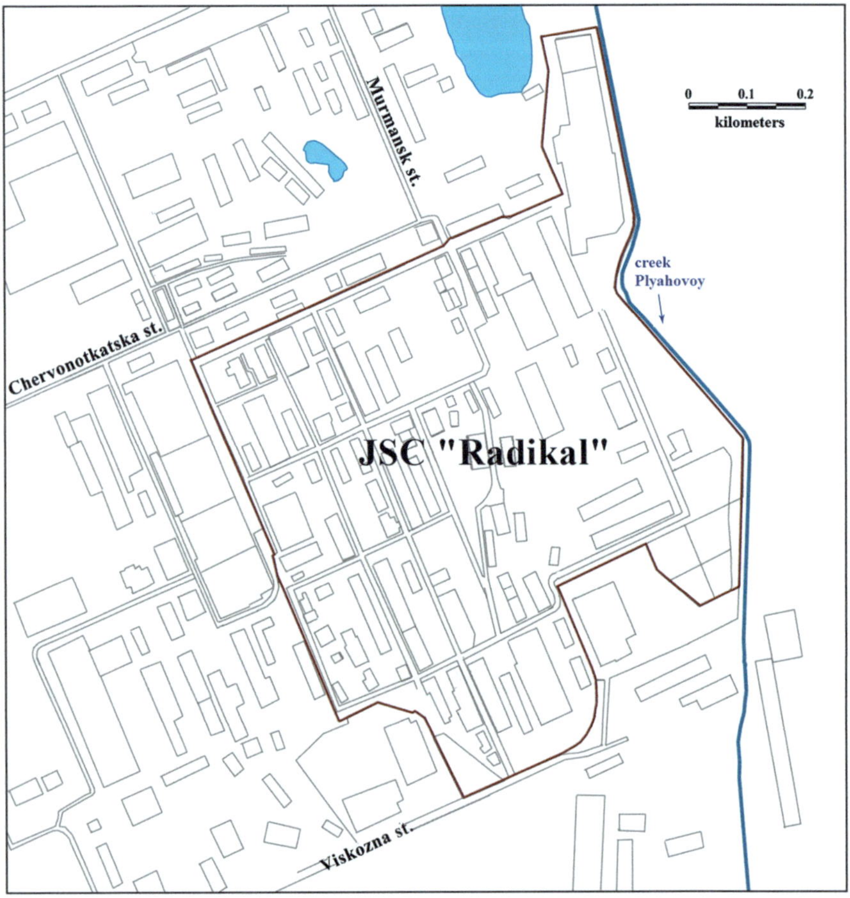

Fig. 20.1 Location of the Radikal JSC site in Kyiv (perimeter marked in brown)

accumulate in the environment with no diminution of their hazardous properties. In particular, metallic mercury was used in large quantities; mercury and its compounds are very stable and remain toxic in the environment for centuries (Opeyd and Shvayko 2008). During its 42 years of operation, the factory was responsible for *a loss* of some 700 tonne of mercury, of which about half remains on site—in the soil, buildings, and sludge pits (Panichkin et al. 2007).

Under current regulations, the maximum allowable concentration (MAC) for mercury in soil is 2.1 mg/kg; for water bodies supplying potable and community water, MAC is 0.0005 mg/l (for inorganic compounds); and average MAC in the air is 0.0003 mg/m^3 (MOH 2005). Ecological and geochemical surveys carried out periodically since 1990, found mercury significantly in excess of maximum allowable concentration (MAC) in soils, the fabric of the factory buildings, surface

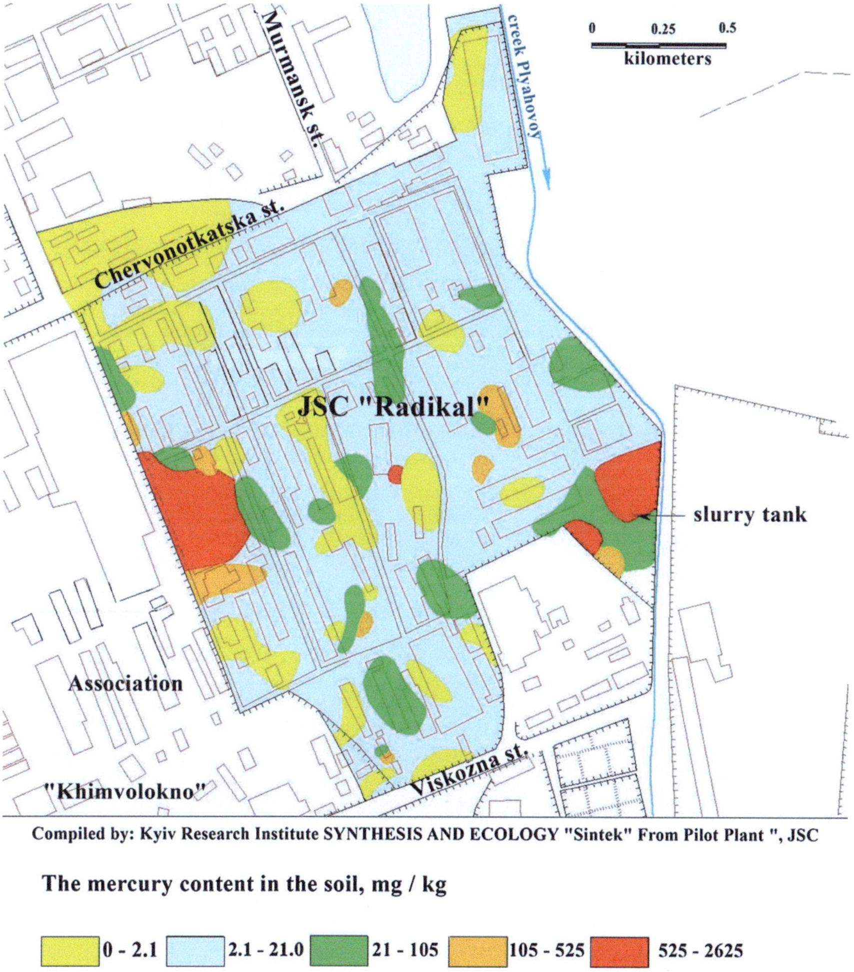

Fig. 20.2 Surface soil contamination by mercury at the Radikal JSC site, as of 2002

and groundwater, and further afield (Nikitash and Prikhodko 2002). Recognition of the anomalous mercury levels in the soil to a depth of at least 3 m across an area of 1.2 km^2 (the industrial site itself is 0.6 km^2) prompted declaration of a regional-scale emergency. Hydrogeological conditions within the Radikal site provide no protection for the groundwater against pollution (Shestopalov et al. 1978), and mercury is also transferred to the atmosphere by evaporation. Mercury contamination at the site is depicted in Fig. 20.2, drawn up by the independent organization SINTEKO KNDI in 2002.

A phased environmental protection procedure was begun in 2011. Cleaning and reorganization of the site of the electrolysis factory, the tailings pond, and the area

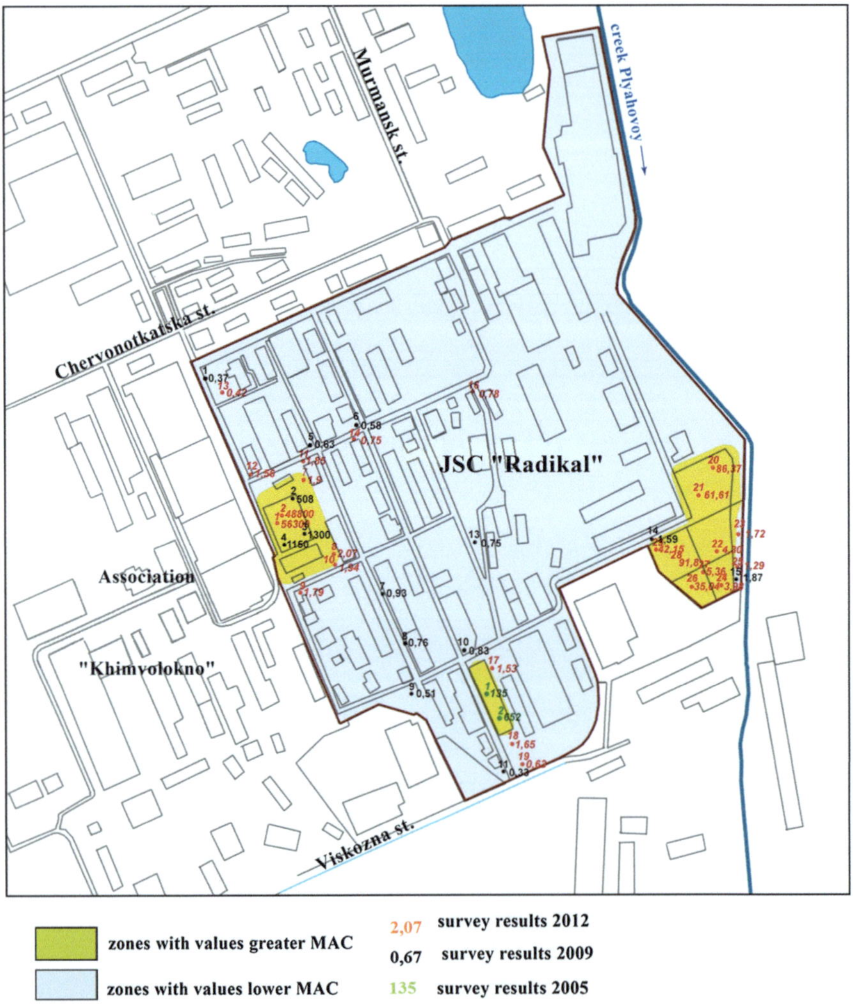

Fig. 20.3 Location of mercury pollution, 2012

Table 20.1 Volume of solid waste to be decontaminated

Description	The value of
Electrolysis factory building	
Dimensions of the building	142 × 66 m
Dimensions of the production building	114.6 × 61.5 × 16.5 m
Volume of building materials above ground	6500 m^3
Volume of construction materials comprising the floors of the building	3500 m^3
Amount of contaminated soil under the floor of the building	10,500 m^3
Amount of contaminated soil in the surrounding area	2000 m^3
Sludge storage	
Volume of enclosing banks of tailings pond	32,000 m^3
Amount of material of bottom screens and drainage systems	20,000 m^3
Volume of sludge	20,000 m^3
Amount of debris on the tailings pond	4000 m^3
Composition salt	
Amount of salt in stock	2500 m^3
Volume of sludge in the landfill	30 m^3
Soils at the industrial site	
Amount of contaminated soil to be removed (apart from material-associated industrial plant)	≅200 m^3

of salt solvent has greatly reduced contamination on site (Fig. 20.3) but this was discontinued in 2012.

The three largest concentrations, covering a total area of about 0.08 km^2, still have pollution levels above 10 MAC (21 mg/kg). The greatest intensity and depth of mercury contamination (about one g/kg) are fixed within the buildings of the chlorine production plant (the mercury content in the concrete floor of the electrolysis factory reaches more than 10 g/kg); large concentrations of mercury are also fixed within the tailings pond. The estimated amount of solid waste to be decontaminated is listed in Table 20.1.

Results and Discussion

Having established the facts, we can define the problems posed by the necessary removal of the mercury. The first is the conditions of work in the Radikal industrial site; secondly, the risks of creating further contaminated sites by moving the contaminated materials; thirdly, the lack of any Ukrainian enterprise able to fully clean-up the waste. Removal of waste to landfill in Ukraine for temporary storage is also unacceptable because of the risk of contamination of these areas. From the economic point of view, creating new business enterprises to clean up the waste is a non-starter. Any technology for removal of mercury from the waste requires the creation of landfill for the disposal of highly toxic waste. The job must be done by specialized enterprises outside Ukraine.

To prevent accidents, it is necessary to document the waste management project at the Radikal site. A plan is needed to contain accidents and deal with them on site, taking account of environmental, health and safety, fire, and territorial restrictions, and in accordance with applicable regulations of Ukraine.

Conclusions

- Mercury presents great danger to public health, human life, and the environment. The existing level of pollution in the Radikal JSC industrial area is a major threat to all these.
- Removing the contaminated waste is, in itself, a hazardous operation. Additional ecological and geochemical survey of the site is recommended before removing the waste.
- It is important to establish clear and strict requirements for organizations that conduct the decontamination. The separation of mercury from contaminated waste that is removed from the site should be carried out at specialized enterprises outside Ukraine—because there are no facilities in country that can undertake full decontamination.
- After the removal of the waste products of the Radikal site, it will be necessary to establish and monitor compliance with a strict health-protection zone around it.

References

Beletsky VS (2004) Encycolpaedic mining dictionary, vol 3. East Publishing House, Donetsk (Ukrainian)

MOH (2005) Methodical instructions: determination of mercury in production facilities, the environment and biological materials. Ministry of Health of Ukraine, No. 263 of 10 June 2005 (Ukrainian)

Nikitash AP, Prikhodko VL (2002) Assessing the impact of mercury contamination of Kyiv drinking water drawn from groundwater affected by the Radikal plant, vol 1. Pivnichheolohiya Hydrogeological Group, Kyiv (Russian)

Opeyd J, Shvayko A (2008) Glossary of chemistry. Weber, Donetsk (Ukrainian)

Panichkin VY, Ilyushchenko MA, Postolov LE (2007) Predicted mercury contamination spreading beyond the industrial area of Radikal JSC, Kyiv, and justification for a groundwater monitoring network. EuroChem JV, Kyiv (Russian)

Shestopalov V, Rybin VF, Lysychenko GV et al (1978) Report on budget theme "develop recommendations to increase the water intake of groundwater for urban water supply of Kiev, Lviv, Khmelnytsky". IGN Ukrainian Academy of Sciences, Lviv (Russian)

Chapter 21
Comparison of Groundwater and Surface Water Quality in the Area of the Bilanovo Iron Deposit

Oleksa Tyshchenko

Abstract The quality of groundwater and surface water in the country around the Bilanovo iron ore deposit has been investigated with respect to chemical and bacteriological characteristics. Comparative analysis identified various indicators that exceeded the maximum allowable concentration.

Keywords Bilanovo iron ore • River water • Bore water • Maximum permissible concentration

Introduction

Water is a vital resource. But society also needs to exploit other resources—such as the Bilanova iron ore deposit in the Poltava region of the Ukraine. Ultimately, exploitation of the Bilanovo ore deposit will require a quarry 600–700 m deep which will destroy natural habitat and intercept the water table, so we should assess its possible environmental impact. Waters that exceed the maximum permissible concentration (MPC) for potable water are subject to regulation (Cabinet of Ministers of Ukraine 1999; EU 2000; Ministry of Health 2010).

According to the project plans, the daily volume of pumped drainage water will be about 1000 m^3 (Construction Bilanovo GOK 2011). This water will be highly mineralized and may enter the aquifers, fundamentally changing the chemistry of the groundwater. This article describes the main results of hydro-geochemical research on water bodies in the country around the Bilanovo iron-mining area. Most of the chemical elements that change in concentration across the region were studied; here we compare data for the major ions and total mineralization in surface streams, wells and bore waters.

O. Tyshchenko (✉)
Institute of Environmental Geochemistry of the National Academy of Sciences,
34a Palladina Prospect, Kyiv-142 03680, Ukraine
e-mail: froi@ukr.net

D. Dent and Y. Dmytruk (eds.), *Soil Science Working for a Living*,
DOI 10.1007/978-3-319-45417-7_21

Sampling

The Bilanovo iron ore deposit is situated on the south-western edge of Dnieper-Donetsk artesian basin, impinging on four aquifers: middle–upper Quaternary alluvium, the Kharkiv aquifer, the Buchak aquifer, and the basement of fissured Pre-Cambrian crystalline rocks—the last of these was not analysed because there are no boreholes in it. Samples were selected from boreholes, wells and the rivers Psyol and Rud'ko. (Technological waters treated with pond clarifier were also sampled but we do not use these data in this article.) The observation points are shown in Fig. 21.1.

Aquifers

An unpressurized aquifer is situated in the middle–upper Quaternary alluvium comprising fine-grained, grey, quartz sand and sandy loam, overlying Kharkiv deposits. The capacity of the aquifer is 6.5–8 m; the depth of the static water level is 0.2–3.7 m. The yield of wells varies from 1.75 to 4.2 l/s, drawing down from 1.1 to 3.14 m below the surface.

The weakly pressurized Kharkiv horizon comprises layers of marl (grey-green clay and siltstones) reaching down to 30–40 m below the ground surface. Slurry from the horizon is a mixture of fine-grained quartz and glauconite clay.

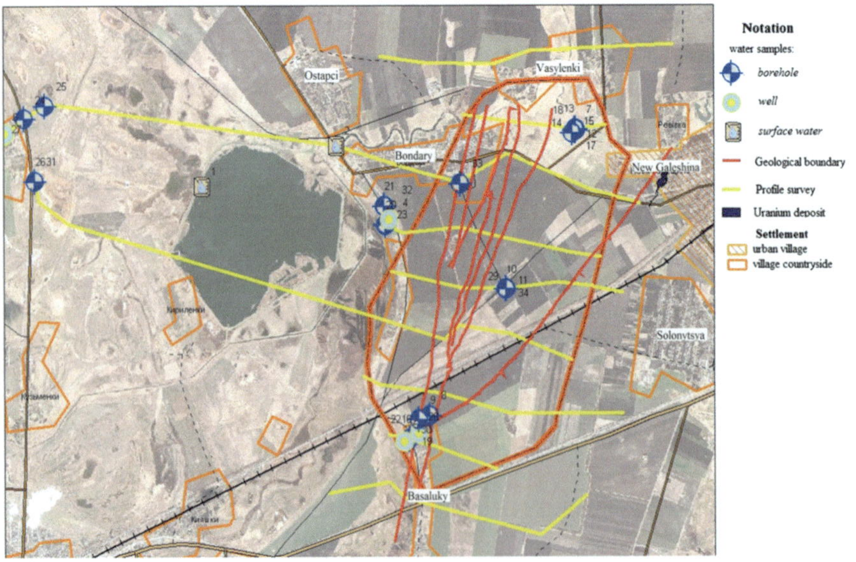

Fig. 21.1 Locations of the samples from around the Bilanovo iron ore deposit

The capacity of the aquifer is 4.5–8 m; the pressure head is about 30 m giving a static level depth of 0.7–4.5 m. Wells produce 0.5–1.23 l/s while reducing the level from 16.75 to 38.57 m.

The pressurised Buchak aquifer lies between 61.4 and 83.5 m below the surface, directly above the Pre-Cambrian basement. Its slurry is greenish grey, glauconitic quartz clay/sand. The capacity of the aquifer is 6–31.3 m; the pressure head is 73.8–81.2 m giving a static level of +1.2 to 11 m. Bores yield 0.37–2.41 l/s while reducing the level from −10 to −61.3 m.

Surface Drainage

There are two streams in study area. The Rud'ka, at Bondari village, is a tributary of the Psyol and flows through Kozelshchina and Kremenchug Districts, a distance of 34 km. The valley is broad and shallow, marshy in places; the sometimes vague channel meanders, with oxbow lakes, and is canalized downstream. The Psyol itself, by Zapsill'a village, is a left-bank tributary of the Dnieper, flowing for 717 km (692 km in Ukraine); the catchment engrosses 22,800 km^2; the valley is deep and narrow to begin with but widens to 10–20 km. Oxbows and canals divide the floodplain, which is marshy in places.

Water Quality

Water quality depends on temperature, dissolved oxygen, pH, content of suspended particles, principal ions and mineralization, biogenic and organic substances, heavy metals and possible contamination with petroleum products. While chemical composition is an integral characteristic, water quality is significantly affected by human activities—to name but a few: discharges of industrial wastes, municipal sewerage, power stations and run-off from agricultural land.

All the measured water sources are highly mineralized (3–12 g/dm^3), with a chloride–sulphate–bicarbonate–sodium composition. Large quantities of sulphates occur in mine water (from pyrite oxidation) and effluents of industrial plants that use sulphuric acid, and sulphates are added with sewage and agricultural run-off. The concentration of sulphates in surface waters shows seasonal variations related to the changing ratio between surface run-off and deep drainage, redox processes and economic activity. Increased content of nitrates is connected with leaching from fertilizers and from livestock. Samples were collected in summer, the dry season, when mineralization is highest.

The alluvial aquifer contains cold (10–11 °C) fresh water with a chloride–sulphate–hydrogen carbonate–magnesium–calcium–sodium composition, a dry residue of 0.939–1.33 g/dm^3, total hardness of 7.02 mol/m^3 and pH 7.7 (slightly alkaline). In some samples, the nitrate content reaches 210–290 mg/l compared with the maximum permissible concentration (MPC) for drinking waters of 40 mg/l. Water from the Kharkiv aquifer is cold (10–12 °C), moderately saline with sodium chloride composition, dry residue 2.42–6.46 g/dm^3, total hardness 8.99 mol/m^3 and pH 7.1 (slightly alkaline). The Buchak aquifer yields cold (10–14 °C) saline water of sodium chloride composition, with a dry residue of 6.19–9.77 g/dm^3, total hardness 16.92 mol/m^3 and pH 6.6 (slightly acid).

Overall, 34 water samples were selected: 27 from bores, 4 from wells and 3 from surface waters. Samples were collected with standard 1- and 2-l bailers. An extensive analysis of water quality was performed, including reaction (pH), colour (after filtering), dry residue (total mineralization), temperature, electrical conductivity, smell, nitrate, fluoride, adsorbed organochlorine compounds, total iron, manganese, copper, zinc, boron, beryllium, cobalt, nickel, vanadium, arsenic, cadmium, chromium, lead, tin, selenium, mercury, barium, yttrium, cyanide, sulphate, chloride, anionic surfactants, phosphate, nitrate, monoatomic phenols, non-polar extractable substances, polycyclic aromatic hydrocarbons (including phenols), pesticide substance (amount), chemical oxygen consumption, oxygen saturation, biochemical oxygen consumption, total nitrogen, ammonium ion, total organic carbon, humic substances, sorbed bacteria, faecal streptococci (enterococci), and salmonella.

The most important indicators of the groundwater quality are mineralization, cations, anions, oxygen, and organic matter. Tests were performed using an AA-8500 F, Jarrel-H atomic absorption and flame-emission dual-channel spectrophotometer.

Comparison of Integrated Indicators of Water Quality

Figures 21.2 and 21.3 present integrated analysis of the chemical parameters. Figure 21.2 shows mean values of the principle ions in rivers; wells; commercial (economic purpose, EPB) boreholes; the alluvial, Kharkiv and Buchak aquifers; and MPCs. All the main ingredients of groundwater from the aquifers around the Bilanovo deposit exceed normative permitted levels, so the groundwater cannot be dumped into the environment without purification.

Chlorine and sodium are most concentrated in the deep Buchak aquifer; the mean concentration of chlorine in Buchak water is about 2170 mg/l, 8.5 times greater than MPC. Other groundwaters also exceed MPC, so they are not fit to drink; only in river water is the concentration of chlorine less than MPC.

Figure 21.2c, d show that calcium and magnesium are most concentrated in wells, but other horizons also exceed MPC. Wells and commercial bore waters are very hard because of contamination with *ZAT Ukrtatnafta*, a water-clarifying agent that contravenes public health regulations; water hardness also arises from leaching of fertilizers.

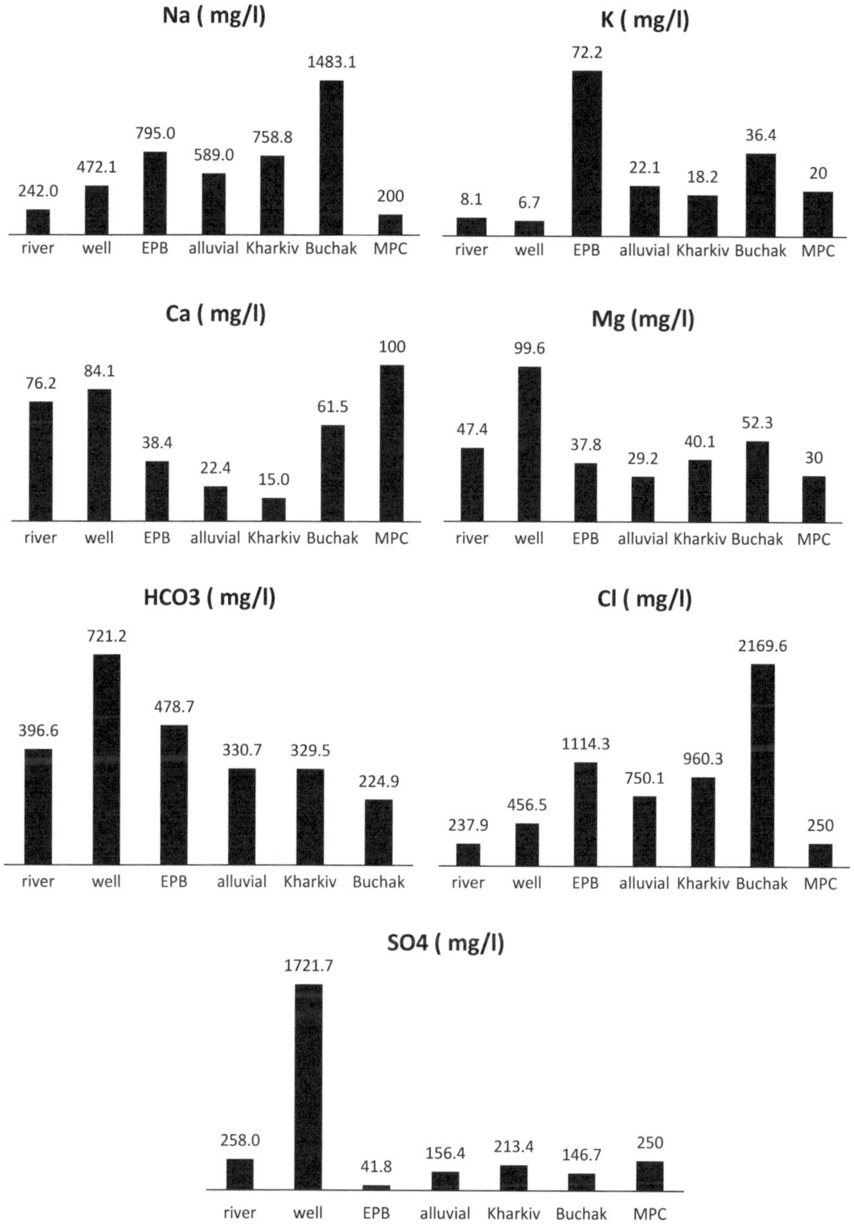

Fig. 21.2 Mean values of principal dissolved ions

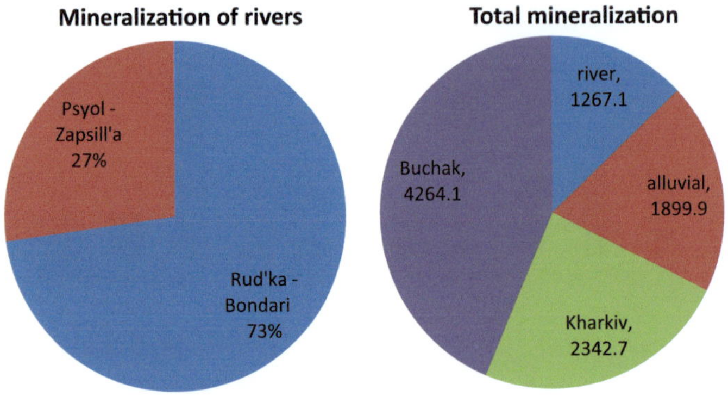

Fig. 21.3 Mineralisation of river

The content of nitrates (0.004–0.1 mg/l) did not exceed MPC; the content of ammonium ions is 2.8 mg/l, exceeding MPC 1.4 times; and the content of nitrites was determined at 0.04 mg/l.

Toxicological indicators were normal.

All the aquifers are interconnected. The relative mineralization of river water and amount of mineralization of all water sources is presented in Fig. 21.3a, b which show that mineralization (mg/l) is too high for use in drinking and industrial processes.

Figure 21.4 shows the comparative ionic composition from two sampling points on the rivers Rud'ka, at Bondari, and Psyol, at Zapsill'a. The Rud'ka at Bondari has greater concentrations of Na, K, Mg, HCO_3, Cl and SO_4 than the Psyol; this may be due to point pollution upstream, and this water is not fit to drink. The situation of the Psyol, at Zapsill'a, is better; all indicators are below MPC. There is no mains water supply, and the local water supply does not meet the needs of the inhabitants, mostly smallholders who need a significant investment in water quality. The particular situation arising from major development will be studied in further research.

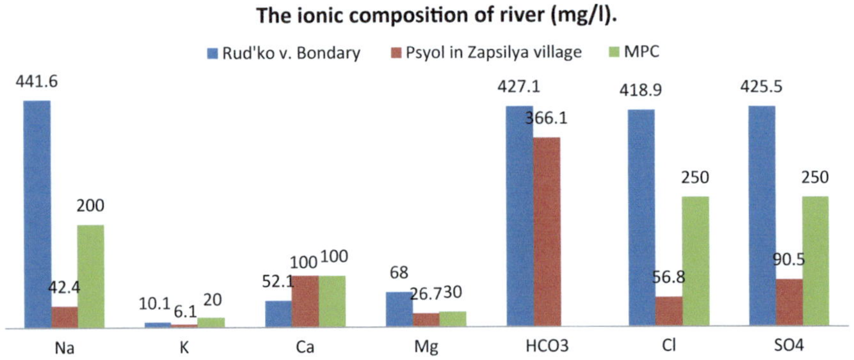

Fig. 21.4 Ionic composition of river water

Conclusions

Analysis of the chemical and bacteriological characteristics of groundwater and surface water in the area of the Bilanovo iron ore deposit shows that water from unprotected boreholes suffers from significant influence of external factors. Biochemical dilution and pollution occur simultaneously but, in general, the main chemical constituents exceed permitted levels.

References

Cabinet of Ministers of Ukraine (1999) Rules of surface water pollution from bore waters. Govt Ukraine, Kyiv (Ukrainian)

Construction Bilanovo GOK (2011) Report on the geological study and semi-technological tests on the ferruginous quartzite of the Bilanovo iron mine (Ukrainian)

Construction Bilanovo GOK (2014) The project of mining lease of the Bilanovo ferruginous quartzite. Explanatory note (Ukrainian)

EU (2000) Water framework directive 2000/60. European Commission, Brussels

Ministry of Health (2010) State sanitary norms and rules: hygienic requirements for drinking water. No. 2.2.4-171-10. Approved by Ministry of Health of Ukraine No. 400, 12 May 2010, Kyiv (Ukrainian)

GA IEG (2013) Hydrogeochemical examination of water bodies of the area of the Bilanovo Enrichment Plant, Poltava region. Institute of Environmental Geochemistry of the National Academy of Sciences of Ukraine, Kyiv (Ukrainian)

Chapter 22
Estimation of Soil Radiation in the Country Around the Bilanovo Iron and Kremenchug Uranium Deposits

Yevhen Krasnov, Valentin Verkhovtsev, Yuri Tyshchenko and Anna Studzinska

Abstract The gamma radiation and radon flux density of soils overlying the Bilanovo iron ore and Kremenchug uranium deposits have been measured in the field and laboratory. The resulting distribution maps of the main nuclides identify areas of increased radiological risk. For the most part, the parameters are not critical, but the measured levels should be considered as a basis for regular radiological monitoring.

Keywords Bilanovo deposit · Radiation monitoring · Measurements · Risk mapping

Introduction

A radiation monitoring program was initiated to assess the potential short-term and long-term radiological impact of the exploitation of the Bilanovo deposit of ferruginous quartzite. Radiation monitoring of the soils in the field and laboratory was carried out in spring–summer and autumn–winter periods. The objectives were seasonally focused collection and analysis of soil samples to determine the power of gamma radiation and radon activity in the unsaturated soil zone; and identification and mapping of areas of increased radiation risk.

Methods

Sampling points was located precisely with a global positioning system. In the field, the soils were sampled by auger to a depth of 1 m; measurements of gamma radiation were made in the auger hole over a depth of 0–1 m using an SRP-68-02

Y. Krasnov (✉) · V. Verkhovtsev · Y. Tyshchenko · A. Studzinska
Institute of Environmental Geochemistry of NAS of Ukraine,
34 a Palladina Avenue, Kyiv, Ukraine
e-mail: yevhen.krasnov@gmail.com

D. Dent and Y. Dmytruk (eds.), *Soil Science Working for a Living*,
DOI 10.1007/978-3-319-45417-7_22

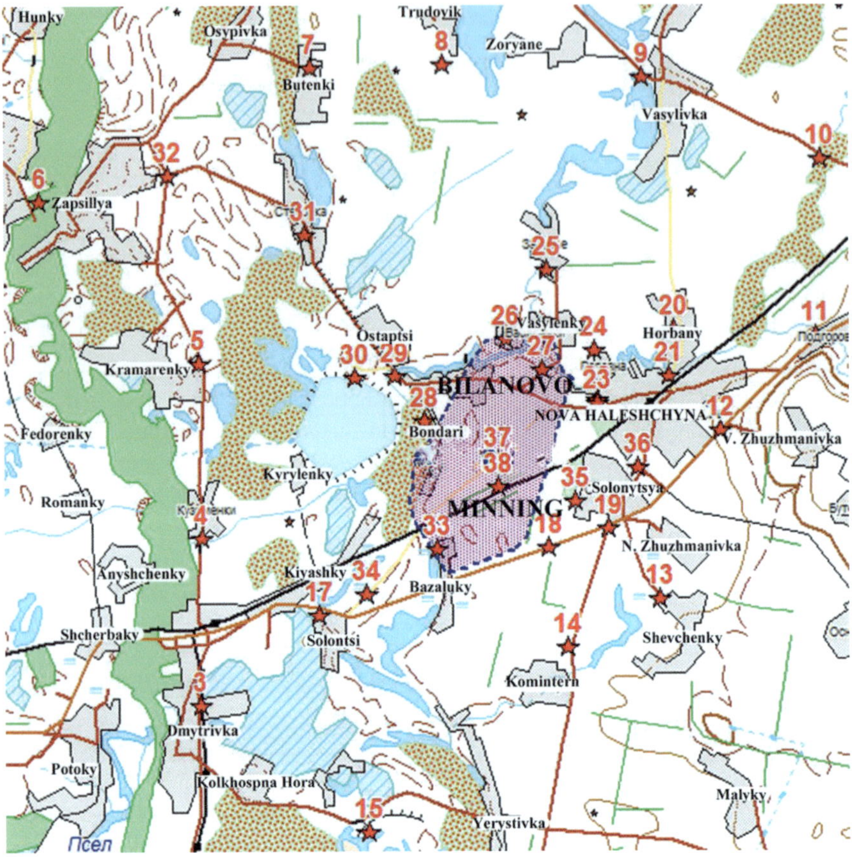

Fig. 22.1 Location of spring–summer soil sampling points (Stage 1)

instrument; and activity of radon-226 alpha radiation with an Alfarad + Radon radiometer. Laboratory research on the soil samples included: volumetric gamma activity, measured using a Food Light radiometer; gamma spectrometry using an AMA-02-F1 multichannel amplitude analyzer and a DHDK-80 detector; and alpha, beta radiometry using a Sputnik radiometer.

Results

A total of 38 points was scheduled for the spring–summer campaign (Fig. 22.1) and 60 points in the autumn–winter campaign (Fig. 22.2).

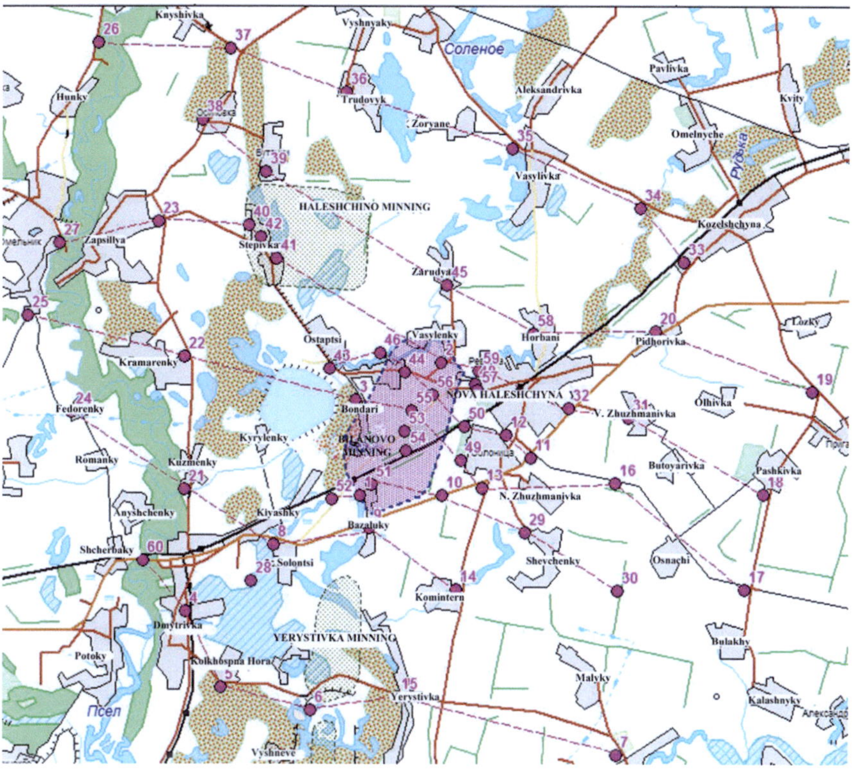

Fig. 22.2 Location of autumn–winter soil sampling points (Stage 2)

Field measurements The gamma radiation dose (GRD) of soils integrated over a depth of 0–1.0 m is presented in Tables 22.1 and 22.2 for the spring–summer period and autumn–winter period, respectively.

Laboratory measurements The total proportion of gamma radioactivity, measured by a stationary radiometer, is presented in Tables 22.3 and 22.4 for the spring–summer and autumn–winter periods, respectively.

Generalization and mapping Fieldwork and sampling during the spring–summer campaign covered 38 points, the autumn–winter campaign 60. In all, measurement and sampling were repeated at 31 points; not repeated at 7; and measurements and sampling were carried out for the first time at 29 points. The results from repeated points were averaged. Mapping of the results was carried out in the ArcMap program of ArcGIS software. Author approach was used in creating maps (Figs. 22.3, 22.4 and 22.5).

The measured radiation levels, including the dose of gamma radiation in soils, confirmed earlier results that suggested no radio-ecological threats. Despite the rather patchy distribution of soil gamma radioactivity, we observe zones of

Table 22.1 Field measurements of environmental gamma radiation dose (Stage 1)

No.	Testing point	Sample no.	Elevation, m	Latitude degrees	Longitude degrees	Date	Time	GRD in soil mkR per h	Ambient temperature, °C
2	Komsomolsk	–	91	49.00847	33.59812	27.04	8.00	38.8	18
3	Dmytrovka (well)	36-3/14	89	49.11014	33.61577	25.04	8.40	37.6	18
4	Kuzmenky (R. Rud'ka, well)	1-4/14	90	49.1432	33.61614	26.04	9.00	24.5	18
5	Kramarenko (well)	5-5/14	97	49.17783	33.61514	26.04	9.35	24.6	20
6	Zapsillya (R. Psyol, well)	4.6/14	95	49.2098	33.56547	26.04	10.35	11.8	20
7	Butenky	4-7/14	94	49.23635	33.65002	26.04	13.00	37.9	20
8	Trudovyk	8/14	91	49.2368	33.69115	26.04	14.00	21.2	22
9	Vasylivka	34-9/14	93	49.23423	33.75282	26.04	14.25	24.1	22
10	Kozelshina (R. Rud'ka, well)	35-11/14	97	49.21832	33.80788	26.04	15.30	21.2	25
11	Pidhorivka (R. Rud'ka)	27-13/14	–	49.18339	33.80642	25.04	17.45	28.1	20
12	N. Galeshina	23-14/14	107	49.16444	33.77672	25.04	18.15	32.5	20
13	Shevchenky	15/14	96	49.13126	33.75813	25.04	14.10	31.2	20
14	Comintern	8-18/14	95	49.12174	33.7293	25.04	11.20	29.6	21
15	Yerystivka	3-19/14	94	49.08505	33.66777	23.04	9.30	43.1	21
17	Solontsy (well)	2-21/14	87	49.12802	33.65279	25.04	9.22	32.5	17
18	Solonytsya	37-23/14	94	49.14132	33.72368	24.04	10.26	31.8	18
19	Solonytsya	24/14	102	49.1433	33.74187	25.04	12.00	28.4	20
20	Horbany	26/14	93	49.18451	33.76229	27.04	9.40	29.3	–
21	N. Galeshina (R. Rud'ka)	27/14	–	49.17541	33.7607	25.04	18.40	28.4	23

(continued)

Table 22.1 (continued)

No.	Testing point	Sample no.	Elevation, m	Latitude degrees	Longitude degrees	Date	Time	GRD in soil mkR per h	Ambient temperature, °C
22	N. Galeshina	29/14	100	49.1707	33.7388	27.04	12.30	21.0	–
23	N. Galeshina	21-30/14	100	49.17002	33.73837	24.04	14.10	45.3	–
24	N. Galeshina	31/14	103	49.18047	33.73748	27.04	10.35	17.0	–
25	Zaruddya	25-32/14	101	49.19631	33.72323	26.04	18.55	28.1	22
26	Vasylenky	21-33/14	94	49.18248	33.71037	26.04	19.15	23.3	–
27	Cluster 1	35/14	95	49.17672	33.7214	23.04	16.30	44.6	24
28	Cluster 4	11-39/14	96	49.16652	33.6854	24.04	9.10	47.3	20
29	Bondary (R. Rud'ka)	40/14	94	49.17522	33.67616	24.04	10.48	37.1	25
30	KPR pond	39-41/14	94	49.17505	33.66362	24.04	11.27	40.6	28
31	Stepivka	9-43/14	110	49.20333	33.64843	26.04	12.00	31.2	–
32	Zapsillya	40-44/14	92	49.215	33.60547	26.04	11.10	21.5	25
33	Cluster 3	7-46/14	89	49.14111	33.68902	23.04	14.00	17.7	29
34	Bazaluky	6-47/14	93	49.13212	33.66713	25.04	9.55	25.4	–
35	Solonytsya	13-50/14	93	49.15063	33.73187	27.04	13.55	20.3	–
36	Solonytsya	18-51/14	96	49.15733	33.75122	26.04	19.40	25.2	–
37	Cluster 2	12-54/14	94	49.15846	33.70797	24.04	16.00	52.7	24
38	Bondari	30-55/14	91	49.15346	33.70812	24.04	17.30	34.8	–

Table 22.2 Field measurements of environmental gamma radiation dose (Stage 2)

Point number	Testing point	Sample number	Latitude degrees	Longitude degrees	Date	Time	GRD in soil, mkR per h	Ambient temperature, °C
1	Cluster 3, Bazaluky	B1-46/14	49.14093	33.68742	17.09	15.50	10.8	25
2	Cluster 1	B2-35/14	49.17657	33.71973	17.09	17.15	6.8	25
3	Cluster 4, Bilany	B3-39/14	49.17656	33.71972	17.09	17.50	7.8	23
4	Dmytrivka	B4-3/14	49.10989	33.61403	18.09	10.05	14.9	18
5	Kolkhospna Hora	B5/14	49.08956	33.62882	18.09	10.30	9.3	18
6	Yerystivskyy Mining (south)	B6-19/14	49.08956	33.62882	18.09	11.00	10.1	18
7	Kobylyachok	B7/14	49.07144	33.79509	18.09	11.45	8.9	19
8	Solontsy	B8-21/14	49.12782	33.65114	18.09	13.50	11.9	20
9	Bazaluky	B9/14	49.13245	33.69131	18.09	14.15	8.4	20
10	Between villages Bazaluky and Solonytsya	10/14	49.14101	33.72212	18.09	14.30	6.8	20
11	Nyzhnya Zhuzhmanivka	B11-31/14	49.15102	33.75900	18.09	15.05	9.3	22
12	Between villages Solonytsya and N. Haleschyna	B12-31-1/14	49.15734	33.74871	18.09	15.40	8.9	22
13	Solonytsya	B13-30/14	49.14282	33.73869	18.09	16.15	8.5	20
14	Comintern	B14-38/14	49.11564	33.72782	18.09	16.45	6.2	20
15	Yerystivka	B15-44/14	49.08681	33.70312	18.09	17.25	8.4	18
16	Osnachy	B16-32/14	49.14432	33.79449	19.09	9.35	23.0	18
17	Bulakhi	B17-33/14	49.11563	33.84831	19.09	10.00	11.2	20
18	Pashkivka	B18-24/14	49.14131	33.85637	19.09	10.30	13.3	22
19	Pryharivka	B19-13/14	49.16895	33.87697	19.09	11.15	12.6	22
20	Pidharivka	B20-12/14	49.18514	33.81054	19.09	11.40	9.9	22
21	Kuzmenky	B21-35/14	49.14254	33.61367	19.09	13.35	12.2	22

(continued)

Table 22.2 (continued)

Point number	Testing point	Sample number	Latitude degrees	Longitude degrees	Date	Time	GRD in soil, mkR per h	Ambient temperature, °C
22	Kramarenky	B22-26/14	49.17818	33.61352	19.09	14.05	5.8	22
23	Zapsillya	B23-15/14	49.21460	33.60266	19.09	14.35	9.5	22
24	Fedorenky	B24-34/14	49.16315	33.56609	19.09	15.25	9.1	20
25	Omelnyk	B25-25/14	49.18935	33.54791	19.09	15.45	8.0	20
26	Gun'ky	B26-7/14	49.26232	33.57750	19.09	16.20	9.6	20
27	Omelnyk–Psel river bank	B27-14/14	49.20871	33.56092	19.09	17.05	8.5	19
28	Yerystivskyy Mining (north)	B28-43-1/14	49.11793	33.64185	20.09	9.45	13.4	16
29	Shevchenky	B29-38-2/14	49.13097	33.75652	20.09	10.20	10.4	16
30	Malyky	B30-39/14	49.11538	33.79515	20.09	11.10	13.5	18
31	Verkhnya Zhuzhmanivka	B31-23/14	49.16161	33.79996	20.09	11.40	12.2	20
32	N. Galeshina	B32-22/14	49.16432	33.77505	20.09	12.00	11.8	20
33	Kozelshina	B33-5/14	49.20358	33.82286	20.09	14.15	12.6	20
34	Kozelshina–r. Rud'ka	B34-4/14	49.21812	33.80496	20.09	15.05	10.5	20
35	Vasylivka	B35-3/14	49.23402	33.75123	20.09	15.30	5.7	20
36	Trudova	B36-2/14	49.24923	33.68155	20.09	16.14	9.2	18
37	Knyshivka	B37-1/14	49.26092	33.63279	20.09	16.40	10.0	18
38	Yosypivka	B38-8-1/14	49.24178	33.62129	21.09	9.30	13.0	16
39	Butenky	B39-8/14	49.22790	33.64757	21.09	9.55	7.3	16
40	Stepivka	B40-16-1/14	49.21351	33.64087	21.09	10.40	8.0	16
41	Stepivka	B41-16/14	49.20435	33.65247	21.09	11.05	8.4	18
42	Stepivka	B42-16-2/14	49.21029	33.64611	21.09	11.25	9.9	18
43	Ostaptsi	B43-27-1/14	49.17504	33.67450	21.09	12.00	10.3	18

(continued)

Table 22.2 (continued)

Point number	Testing point	Sample number	Latitude degrees	Longitude degrees	Date	Time	GRD in soil, mkR per h	Ambient temperature, °C
44	Bondari	B44-27-2/14	49.17420	33.70615	21.09	12.25	9.2	18
45	Zaruddya	B45-10/14	49.19729	33.72442	21.09	12.55	8.5	18
46	Vasylenko	B46-17/14	49.17926	33.69560	21.09	13.30	9.3	18
47	N. Galeshina	B47-19/14	49.16970	33.73656	21.09	14.45	9.6	18
48	N. Galeshina	B48-20/14	49.17073	33.73611	21.09	16.30	7.3	18
49	Solonytsya	B49-29/14	49.15041	33.73014	22.09	9.50	11.5	16
50	Solonytsya	B50-28-3/14	49.15962	33.73162	22.09	10.10	6.9	16
51	Bazaluky	B51-37-2/14	49.14405	33.69263	22.09	11.00	5.5	16
52	Bazaluky	B52-36/14	49.14001	33.67558	22.09	13.45	7.4	18
53	Cluster 2	B53-28/14	49.15825	33.70638	22.09	14.25	6.9	18
54	Bilanovo quarry	B54-27-3/14	49.15291	33.70689	22.09	15.40	7.0	16
55	Bilanovo quarry	B55-28-1/14	49.16365	33.70926	22.09	16.25	6.8	16
56	Bilanovo quarry	B56-28-2/14	49.16753	33.71804	22.09	16.40	7.0	16
57	N. Galeshina	B57-21-1/14	49.16915	33.73672	22.09	17.10	6.1	16
58	Horbani	B58-11/14	49.18430	33.76040	24.09	10.20	8.1	12
59	Revivka	B59-21/14	49.17446	33.73737	24.09	10.45	7.6	12
60	Shcherbaky–R. Psyol	B60-40/14	49.12362	33.59639	24.09	9.20	3.5	15

Table 22.3 Laboratory measurements of integrated specific gamma radioactivity in soils (Stage 1)

Point number	Sample number	Testing point	Specific activity, integrated, Bq/kg	
			Result	Error
3	36-3/14	Dmytrivka	36	13
4	1-4/14	Kuzmenky	55	16
5	5-5/14	Kramarenky	40	17
6	4.6/14	Zapsillya	20	13
7	4-7/14	Butenky	48	13
8	8/14	Trudovyk	57	14
9	34-9/14	Vasylivka	53	13
10	35-11/14	Kozelshina	57	17
11	27-13/14	Pidhorivka	59	18
12	23-14/14	N. Galeshina	57	14
13	15/14	Shevchenky	58	17
14	8-18/14	Comintern	53	14
15	3-19/14	Yerystivka	17	10
17	2-21/14	Solontsy	45	15
18	37-23/14	Solonytsya	63	17
19	24/14	Solonytsya	56	15
20	26/14	Horbani	52	15
21	27/14	N. Galeshina	53	13
22	29/14	N. Galeshina	63	18
23	21-30/14	N. Galeshina	59	17
24	31/14	Revivka	54	15
25	25-32/14	Zaruddya	54	19
26	21-33/14	Vasylenky	58	16
27	35/14	Cluster 1	46	14
28	11-39/14	Cluster 4	35	11
29	40/14	Bondari	35	16
30	39-41/14	KPR pond	27	18
31	9-43/14	Stepivka	46	19
32	40-44/14	Zapsillya	35	13
33	7-46/14	Cluster 3	38	17
34	6-47/14	Bazaluky	41	15
35	13-50/14	Solonytsya	62	17
36	18-51/14	Solonytsya	46	13
37	12-54/14	Cluster 2	58	15
38	30-55/14	Bondari	59	13

Table 22.4 Laboratory measurements of integrated specific gamma radioactivity in soils (Stage 2)

Point number	Testing point	Sample number	Geographic coordinates		Date and time of sampling		Specific activity, integrated Bq/kg	
			Latitude degrees	Longitude degrees	Date	Time	Result	Error
1	Cluster 3, Bazaluky	B1-46/14	49.14093	33.68742	17.09	15:50	50	12
2	Cluster 1	B2-35/14	49.17657	33.7214	17.09	17:15	66	14
3	Cluster 4, Bilany	B3-39/14	49.1665	33.6854	17.09	17:50	49	13
4	Dmytrivka	B4-3/14	49.10989	33.61403	18.09	10:05	37	14
5	Kolhospna Hora	B5/14	49.08956	33.62882	18.09	10:30	46	13
6	Yerystivskyy Mining (south)	B6-19/14	49.08956	33.62882	18.09	11:00	21	10
7	Kobylyachok	B7/14	49.07144	33.79509	18.09	11:45	62	15
8	Solontsy	B8-21/14	49.12782	33.65114	18.09	13:50	52	14
9	Bazaluky	B9/14	49.13245	33.69131	18.09	14:15	61	13
10	Between villages Bazaluky and Solonytsya	B10/14	49.14101	33.72212	18.09	14:30	73	15
11	Verkhnya Zhuzhmanivka	B11-31/14	49.15102	33.759	18.09	15:05	67	15
12	Between villages Solonytsya and N. Haleschyna	B12-31-1/14	49.15734	33.74871	18.09	15:40	54	16
13	Solonytsya	B13-30/14	49.14282	33.73869	18.09	16:15	71	17
14	Comintern	B14-38/14	49.11564	33.72782	18.09	16:45	65	18
15	Yerystivka	B15-44/14	49.08681	33.70312	18.09	17:25	45	14
16	Osnachi	B16-32/14	49.14432	33.79449	19.09	9:35	73	19
17	Bulakhy	B17-33/14	49.11563	33.84831	19.09	10:00	70	17
18	Pashkivka	B18-24/14	49.14131	33.85637	19.09	10:30	61	14
19	Pryharivka	B19-13/14	49.16895	33.87697	19.09	11:15	57	14
20	Pidharivka	B20-12/14	49.18514	33.81054	19.09	11:40	72	15
21	Kuzmenky	B21-35/14	49.14254	33.61367	19.09	13:35	56	15
22	Kramarenky	B22-26/14	49.17818	33.61352	19.09	14:05	54	14
23	Zapsillya	B23-15/14	49.2146	33.60266	19.09	14:35	48	16
24	Fedorenky	B24-34/14	49.16315	33.56609	19.09	15:25	71	14
25	Omelnyk	B25-25/14	49.18935	33.54791	19.09	15:45	60	19
26	Gun'ko	B26-7/14	49.26232	33.5775	19.09	16:20	71	16

(continued)

Table 22.4 (continued)

Point number	Testing point	Sample number	Geographic coordinates		Date and time of sampling		Specific activity, integrated Bq/kg	
			Latitude degrees	Longitude degrees	Date	Time	Result	Error
27	Omelnyk–Psyol river bank	B27-14/14	49.20871	33.56092	19.09	17:05	30	10
28	Solontsy	B28-43-1/14	49.11793	33.64185	20.09	9:45	43	12
29	Shevchenky	B29-38-2/14	49.13097	33.75652	20.09	10:20	63	12
30	Maliky	B30-39/14	49.11538	33.79515	20.09	11:10	69	14
31	Verkhnya Zhuzhmanivka	B31-23/14	49.16161	33.79996	20.09	11:40	76	16
32	N. Galeshina	B32-22/14	49.16432	33.77505	20.09	12:00	68	15
33	Kozelshina	B33-5/14	49.20358	33.82286	20.09	14:15	68	15
34	Kozelshina–R. Rud'ko	B34-4/14	49.21812	33.80496	20.09	15:05	65	19
35	Vasylivka	B35-3/14	49.23402	33.75123	20.09	15:30	65	16
36	Trudova	B36-2/14	49.24923	33.68155	20.09	16:14	58	16
37	Knyshivka	B37-1/14	49.26092	33.63279	20.09	16:40	60	14
39	Butenky	B39-8/14	49.2279	33.64757	21.09	9:55	62	13
40	Stepivka	B40-16-1/14	49.21351	33.64087	21.09	10:40	44	12
41	Stepivka	B41-16/14	49.20435	33.65247	21.09	11:05	62	12
42	Stepivka	B42-16-2/14	49.21029	33.64611	21.09	11:25	35	12
43	Ostaptsi	B43-27-1/14	49.17504	33.6745	21.09	12:00	43	13
44	Bondari	B44-27-2/14	49.1742	33.70615	21.09	12:25	49	15
45	Zaruddya	B45-10/14	49.19729	33.72442	21.09	12:55	58	12
46	Vasylenky	B46-17/14	49.17926	33.6956	21.09	13:30	37	13
47	N. Galeshina	B47-19/14	49.1697	33.73656	21.09	14:45	59	14
48	N. Galeshina	B48-20/14	49.17073	33.73611	21.09	16:30	51	13
49	Solonytsya	B49-29/14	49.15041	33.73014	22.09	9:50	53	15
50	Solonytsya	B50-28-3/14	49.15962	33.73162	22.09	10:10	59	14
51	Bazaluky	B51-37-2/14	49.14405	33.69263	22.09	11:00	63	15
52	Bazaluky	B52-36/14	49.14001	33.67558	22.09	13:45	47	13
53	Cluster 2	B53-28/14	49.15825	33.70638	22.09	14:25	63	16
54	Bilanovo quarry	B54-27-3/14	49.15291	33.70689	22.09	15:40	50	15
55	Bilanovo quarry	B55-28-1/14	49.16365	33.70926	22.09	16:25	70	16
56	Bilanovo quarry	B56-28-2/14	49.16753	33.71804	22.09	16:40	67	15
57	N. Galeshina	B57-21-1/14	49.16915	33.73672	22.09	17:10	59	14

(continued)

Table 22.4 (continued)

Point number	Testing point	Sample number	Geographic coordinates		Date and time of sampling		Specific activity, integrated Bq/kg	
			Latitude degrees	Longitude degrees	Date	Time	Result	Error
58	Horbani	B58-11/14	49.1843	33.7604	24.09	10:20	64	14
59	Revivka	B59-21/14	49.17446	33.73737	24.09	10:45	63	17
60	Shcherbaky–R. Psyol	B60-40/14	49.12362	33.59639	24.09	9:20	15	10

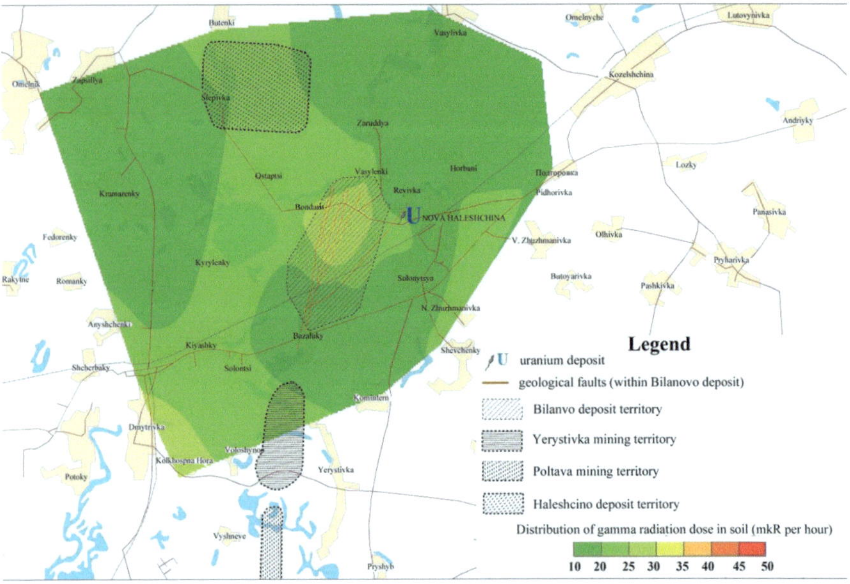

Fig. 22.3 Distribution of gamma radiation dose in soil (2014)

increased activity in the north part of Bilanovo deposit. Therefore, we propose a scheme of integrated environmental monitoring of the county around the Bilanovo mining operations, which may be based on the already completed research.

Proposed Dosimetry Monitoring of Radionuclides in the Soil Layer

Monitoring should include periodic control of the quality of groundwater and surface water, soil cover and environment; dose control of radon-226 activity in air and soils; and uranium and thorium radionuclide activity in the soils. The objective

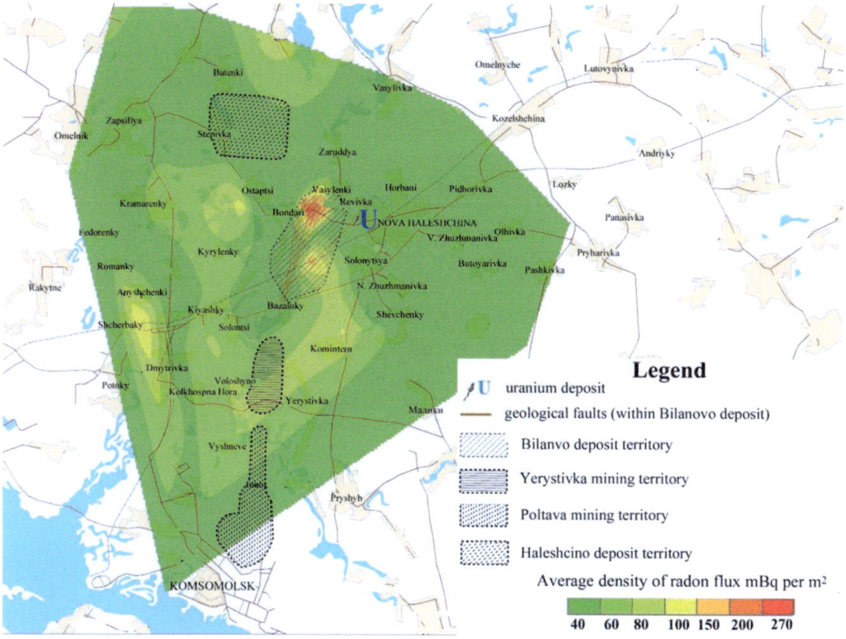

Fig. 22.4 Average density of radon flux from soil (2014)

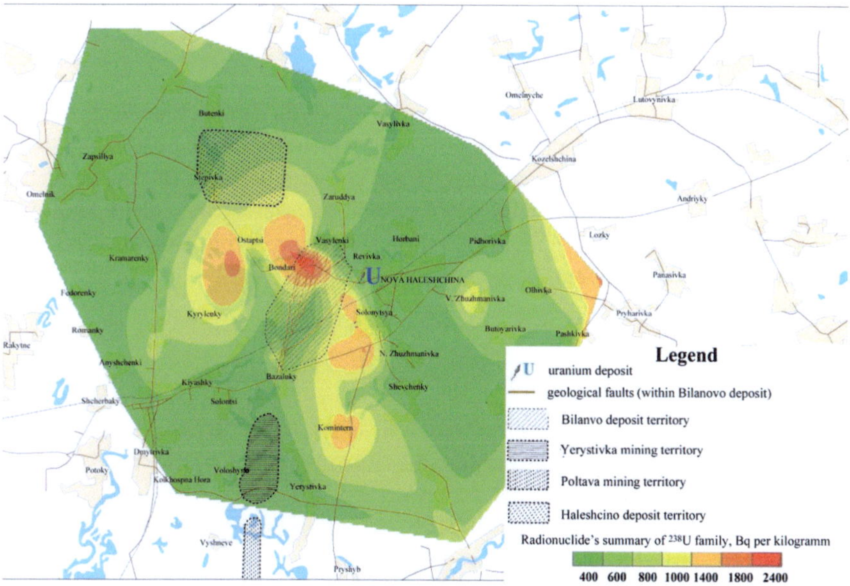

Fig. 22.5 Total specific radioactivity of gamma-emitting ^{238}U radionuclide in soil samples (2014)

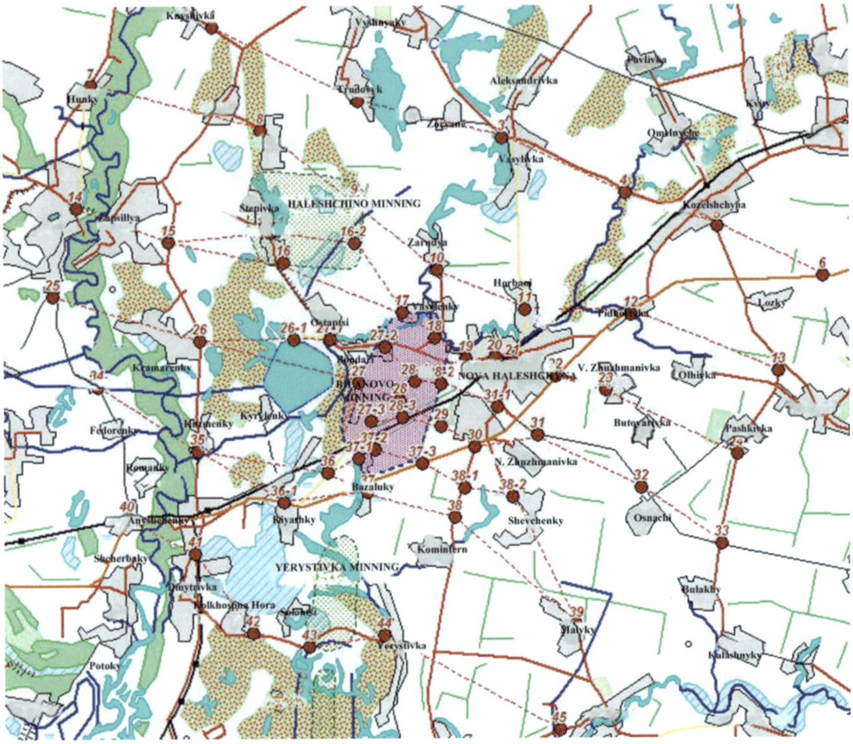

Fig. 22.6 Testing surface and soils location points

is to check activity in the soil profiles at the 59 points depicted in Fig. 22.6. The analyses should comprise field measurement of the environmental gamma radiation dose at 1 m from the surface and, also, in the soil at a depth of 0–1 m, as well as sampling and laboratory gamma-spectrometric measurements, alpha and beta radiometry. The frequency should be twice a year: in spring (March–April) and autumn (September–October).

Conclusions

- Measurements of radiation around the Bilanovo mining site in 2014 were consistent with previous years' surveys.
- Geo-positioning analysis of dosimetry, gamma-spectrometric and radiometric measurements revealed zones of increased activity of alpha- and gamma-emitting radionuclides.

- Dose levels from external radiation sources are regulated by domestic and international radiation safety standards. The levels measured in the country around the Bilanovo deposits do not exceed permitted levels.
- In some cases, radon levels in the soils exceed the regulation levels for housing. Continuing radiation monitoring is needed to ensure a safe environment for the inhabitants of the area.
- The specific activity of radionuclides of uranium and thorium in the soils does not exceed the weighted average planetary indicators.

References

IEG of NAS of Ukraine (2012) Natural radio-geo-ecological research and evaluation of risks associated with radiation danger of the development of Bilanovo iron-ore deposit, Poltava region (final). IEG research report 63/1, Kyiv (Ukrainian)

IEG of NAS of Ukraine (2013) Hydrogeochemical survey of groundwater of the territory of Bilanovo mining plant, Poltava region. IEG research report 05/2013, Kyiv (Ukrainian)

IEG of NAS of Ukraine (2013) Study of the possible impact of uranium mineralization in the Kremenchug ore formation on the natural background of radiation in the Bilanovo mining and processing enterprise area (final). IEG research report 7/2013, Kyiv (Ukrainian)

Kovalenko GD (2008) Radioecology of Ukraine. Monograph. Inzhek, Kharkiv (Ukrainian)

Tyshchenko O (2013) Actual problems of development of Bilanovo iron ore deposits. Technog Environ Saf Civ Def (journal of IEG of NAS of Ukraine, Kyiv) 6, 100–108 (Ukrainian)

Verkhovtsev V, Krasnov Y, Studzinska A et al (2013a) Main results of instrumental radio geo-ecological research of Bilanivske iron ore and Kremenchug uranium deposits. Technog Environ Saf Civ Def 6, 89–99 (Ukrainian)

Verkhovtsev V, Lysychenko G, Tyshchenko Y et al (2013b) Comprehensive radiogeoecological research of the territory of Bilanovo iron ore deposits. Technog Environ Saf Civ Def 5, 28–40 (Ukrainian)

Chapter 23
Estimation of Soil Radiation in the Country Around the Dibrova Uranium–Thorium–Rare Earth Deposit

Yuliia Yuskiv, Valentin Verkhovtsev, Vasily Kulibaba and Oleksandr Nozhenko

Abstract Radiological measurements were made on the soils of the country around the Dibrova uranium–thorium–rare earth deposit. The dose of gamma radiation and radon flux density were measured in the field; gamma spectrometric analysis and measurements of alpha and beta activity were undertaken in the laboratory; and maps were drawn of the distribution of the main nuclides, identifying areas of radiation risk. Apart from locally high radon concentration, most parameters are not critical, but the measured levels should be considered as a baseline for regular radiological monitoring.

Keywords Field · Well · Radiation monitoring · Radiation · Soil

Introduction

The radiation characteristics of the soil, atmospheric and soil air around the Dibrova uranium–thorium–rare earth deposit have been investigated in order to develop search criteria and evaluate the current and potential short- and long-term radiation exposure of the environment. Details of the research are presented by (Verkhovtsev et al. 2014).

Materials and Methods

The study area is located in the south of Pidhavrylivka village, Pokrovsky District, Dnipropetrovsk Region. The highest density of point sampling was undertaken across the area of the uranium–thorium–rare earth deposit itself (Figs. 23.1 and 23.2).

Y. Yuskiv (✉) · V. Verkhovtsev · V. Kulibaba · O. Nozhenko
Institute of Environmental Geochemistry of the National Academy of Sciences of Ukraine, 34a Palladina Avenue, Kyiv 03680, Ukraine
e-mail: yuskiv_yu@ukr.net

D. Dent and Y. Dmytruk (eds.), *Soil Science Working for a Living*,
DOI 10.1007/978-3-319-45417-7_23

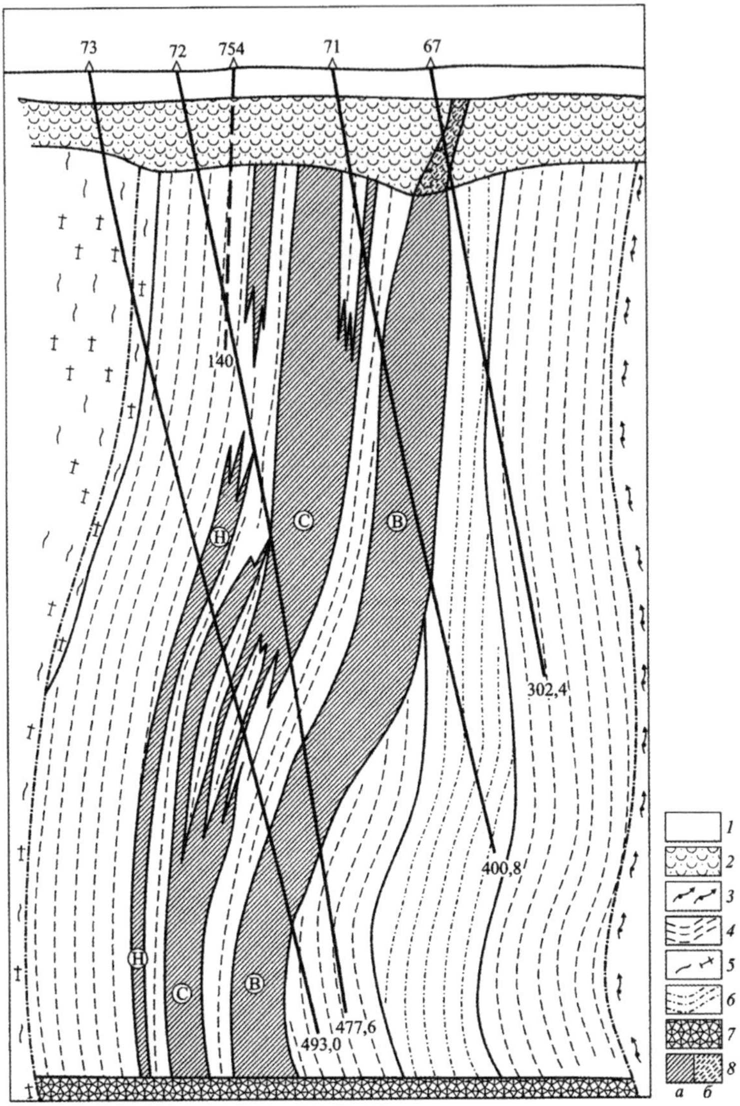

Fig. 23.1 Geological structure of the Dibrova area (after Kulish et al. 2010). *1* Sedimentary cover rocks; *2* weathering crust of crystalline rocks; *3* amphibole–biotite schists; *4* apoquartzite conglomerate, apogritstone and sandy sillimanite–biotite, sillimanite–muscovite and fuchsite with felspar and pyrite; *5* biotite migmatites; *6* ore rocks, garnet–quartz–amphibole–magnetite; *7* Dibrova ore-tectonic branch of the Devladivskiy fault zone and tectonic foliation: boudinage, blastomylonites and mylonites, breccia and breccia blasts (mainly monomikt, cataclasites, blastoma cataclasites) and intensely fractured rocks; and *8* uranium ore zone (a) and its continuation in the weathering crust (b); H—lower crust, C—average, B—upper

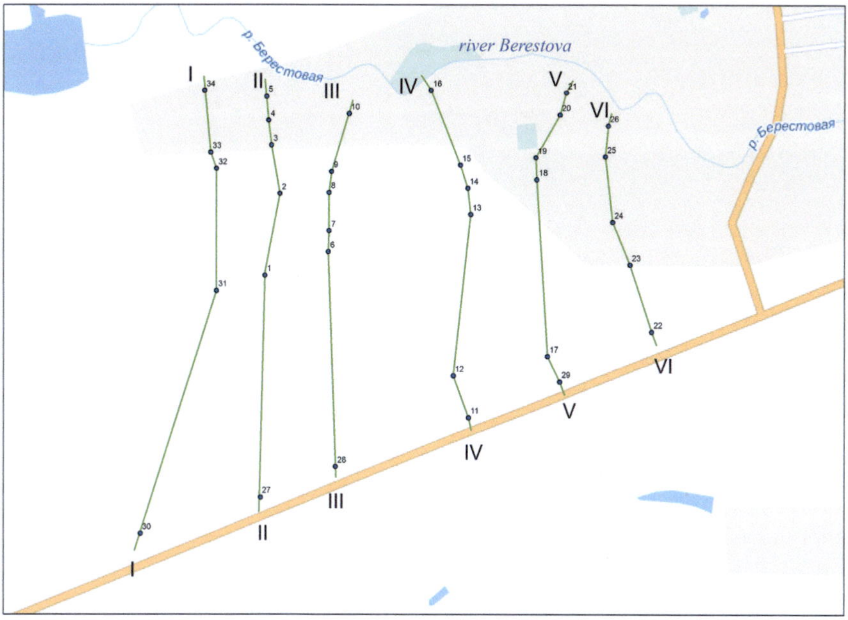

Fig. 23.2 Radiation research area (profiles and complex research points)

In the field, gamma radiation was measured using a mobile, rapid radioactivity research environment:

- *Gamma ray dose* (gamma background) was measured by a *Vector* 68-01 scintillation radiometer according to methodological requirements at a height of 1 m from the soil surface for at least 5 minutes in all sampling and testing locations
- *Gamma ray dose rate in soil* was measured by the *Vector* instrument at hand-drilled boreholes integrally at a depth of 0–1 m
- *Equivalent dose rate* was measured by the *Vector* instrument according to methodological requirements in the air and in the ground for at least 5 minutes in all sampling and testing locations
- *Volume activity of radon-222* was measured at the ground surface using an *Alfarad* + instrument. A normalized rate of flow in excess of 80 Bq/m^3 is not allowed in civil construction (Kovalenko and Rudha 2008).

Measurement and sampling were conducted as a part of a complex profile study in which profiles were selected according to the results of previous studies. Soil samples down profile were collected by a hand auger and integrated over the upper metre of the soil profile; total of about 0.5 kg soil was collected per profile. Sampling points are shown in Figs. 23.2 and 23.3. In total, 34 samples were taken.

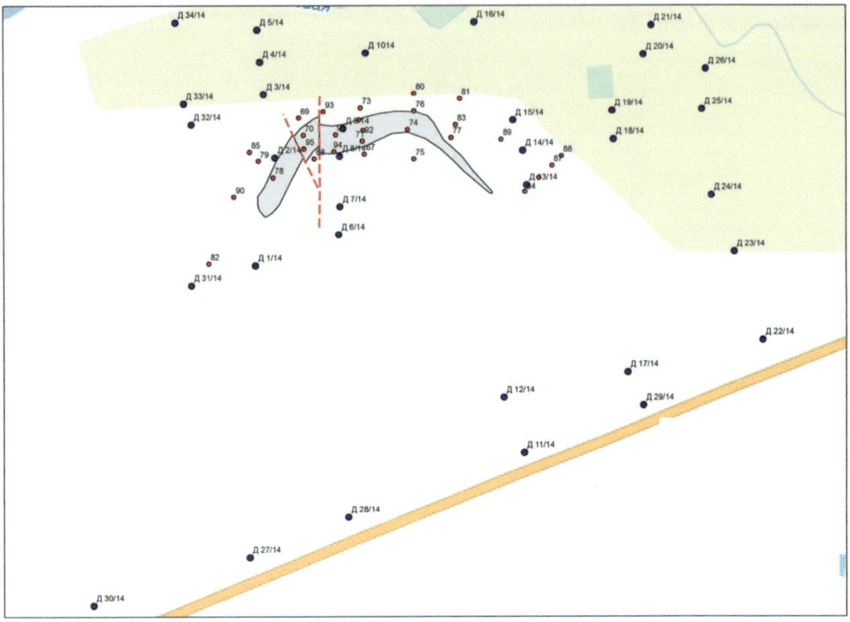

Fig. 23.3 Location of research profiles, soil sampling points, boreholes and outline of the Dibrova field

Laboratory measurements were conducted during July–August of 2014 at the laboratory of the Institute of Environmental Geochemistry of the National Academy of Sciences of Ukraine, in Kyiv:

- Measurement of soil alpha and beta activity was taken on the *Sputnik* radiometer (in imp/min)
- Gamma spectrometric analysis was executed on an *AMA-02-F1* multichannel amplitude analyser with a *DHDK-80* detector. Here we report only data on specific distribution of the total radioactivity of gamma-emitting radionuclides, family ^{235}U in soil samples.
- Total gamma soil radioactivity was measured using a stationary *Food Light* radiometer.

Results

The results of field measurements are shown on the maps: measurements of soil alpha and beta activity are shown in Figs. 23.4, 23.5 and 23.6, and the distribution of radon activity in the soil air is shown in Fig. 23.7.

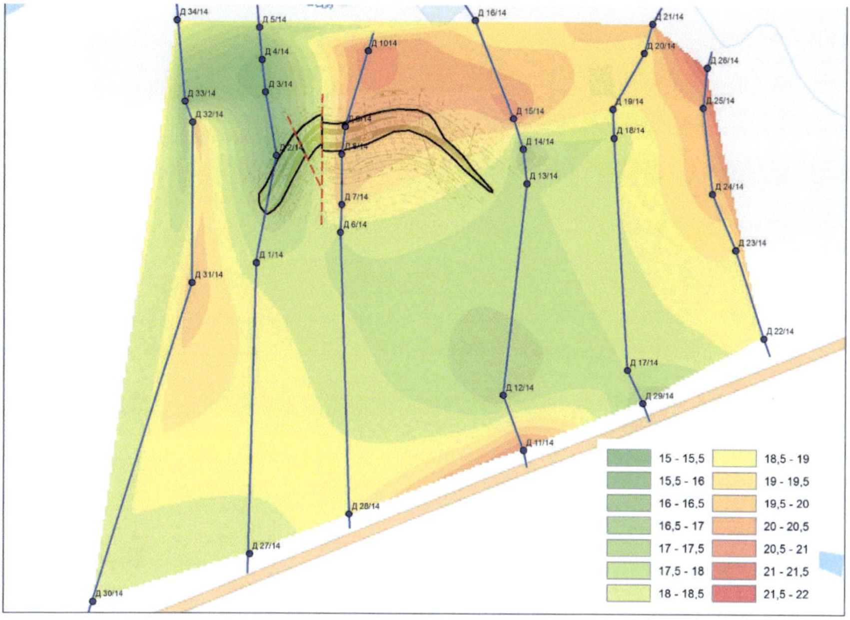

Fig. 23.4 Distribution of gamma radiation dose in soil (imp./min)

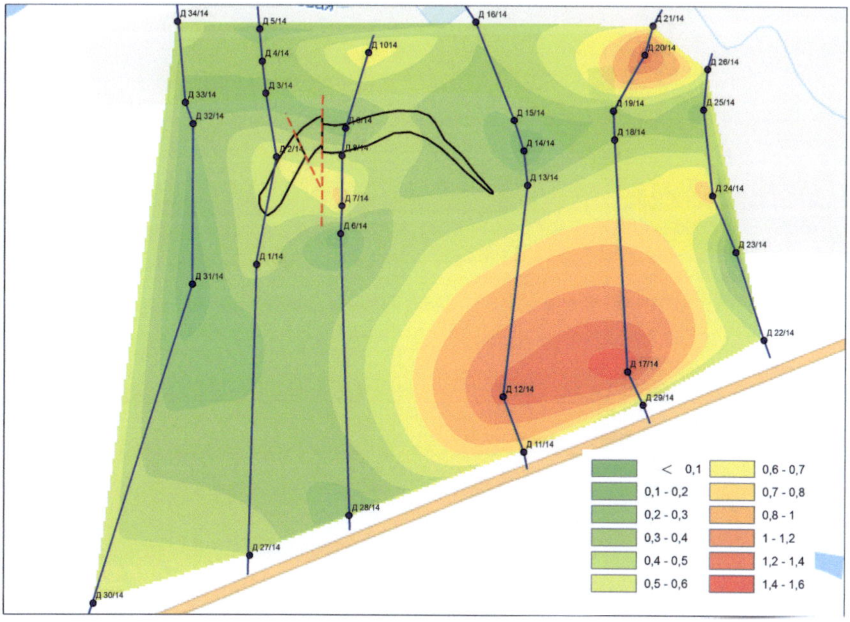

Fig. 23.5 Distribution of alpha radiation dose in the soil (imp./min)

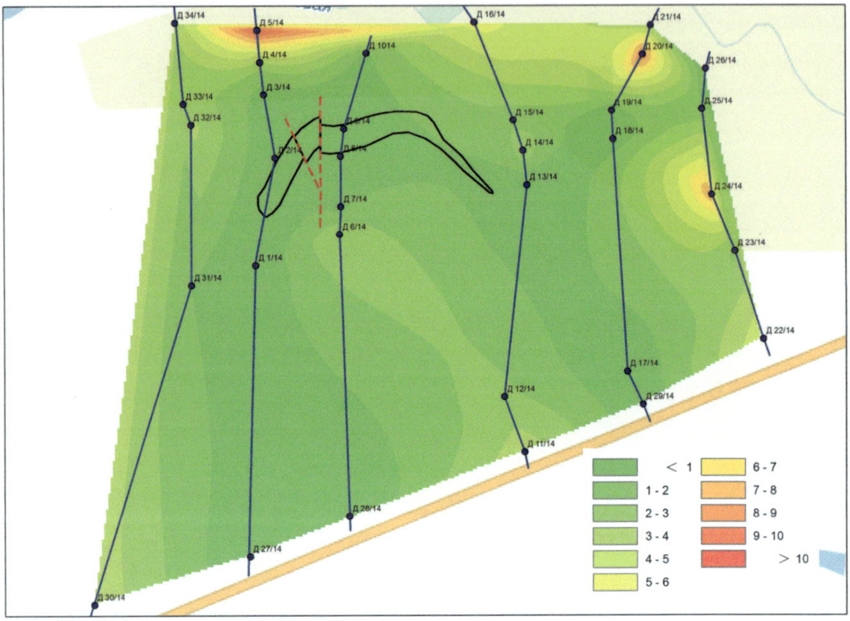

Fig. 23.6 Distribution of beta decays in the soil (imp./min)

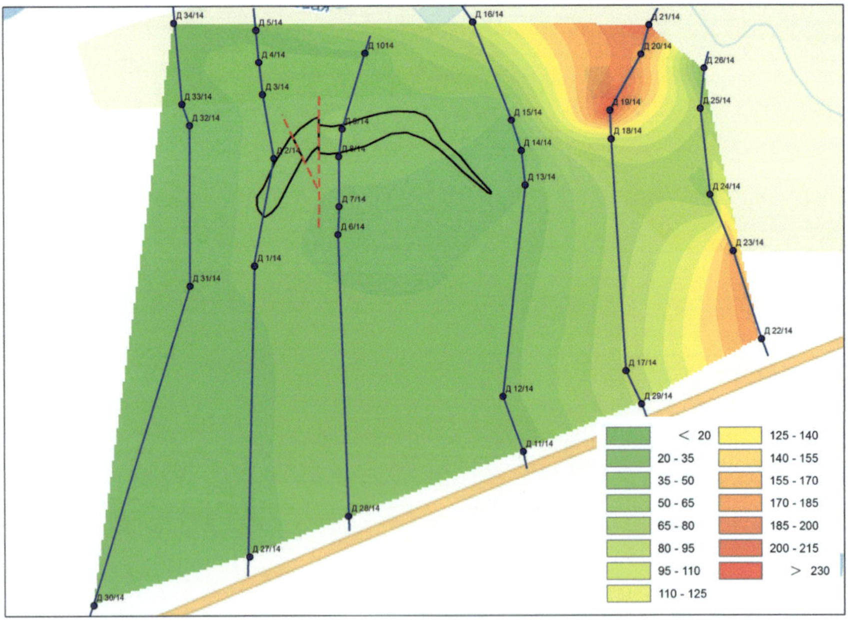

Fig. 23.7 Volume radon activity in the soil air (Bq/m^3, use multiplier ×21.05)

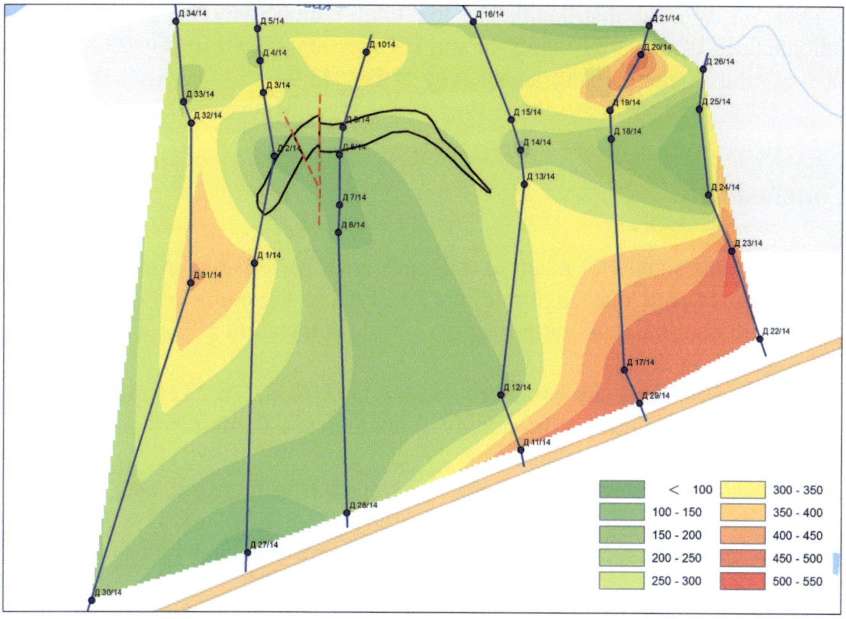

Fig. 23.8 Total specific radioactivity of gamma-emitting radionuclides family ^{235}U in soil samples (Bq/kg)

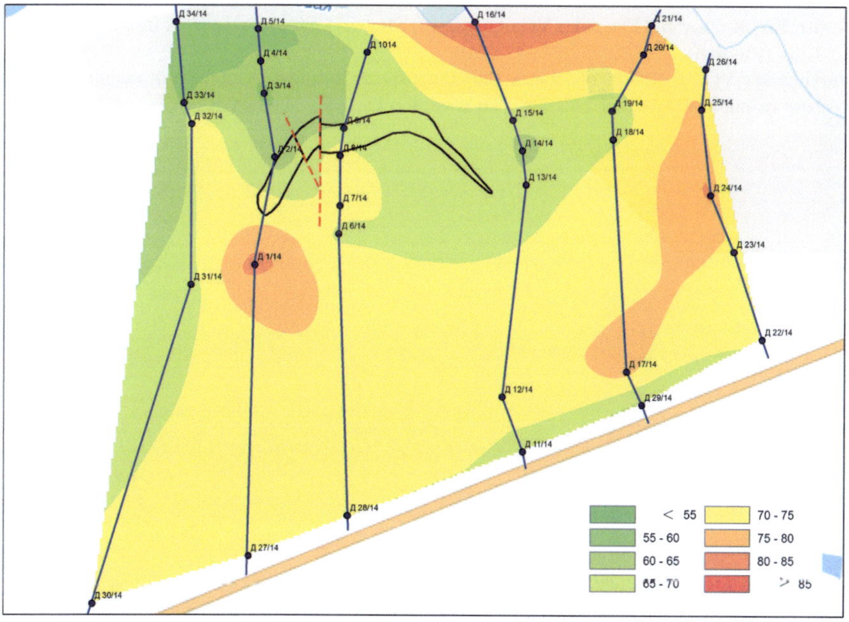

Fig. 23.9 Integrated specific gamma radioactivity in the soils (Bq/kg)

Data on specific distribution of the total radioactivity of gamma-emitting radionuclides of the ^{235}U family in soil samples are shown in Fig. 23.8, and the total specific of gamma soil radioactivity in Fig. 23.9.

Conclusions

- The volume activity of radon in soil air exceeds permissible levels for regulated housing. In particular, high rates are observed to the north-east of the Dibrova uranium–thorium–rare earth deposit. We recommend that this territory should be studied in detail.
- Radiation displays are produced mostly by U-238, Th-232 and their decay products. The soil's specific activity of uranium and thorium radionuclide series does not exceed the global norms.
- Geopositional analysis of dosimetry, gamma spectrometric and radiometric measurements leads us to allocate conditional zones of increased activity of alpha- and gamma-emitting radionuclides related to natural factors.

References

Kovalenko GD, Rudha K (2008) Radioecology of Ukraine. INSC of Ukraine, Kyiv (Russian)
Kulish EA, Komov IL, Yatsenko VG et al (2010) Strategic mineral resources of Ukraine. Logos, Kyiv (Russian)
Verkhovtsev VG, Lysychenko GV et al (2014) Prospects for the development of uranium resource base of nuclear power of Ukraine. Naykova dumka, Kyiv (Ukrainian)

Chapter 24
Lead and Zinc Speciation in Soils and Their Transfer in Vegetation

Oksana Vysotenko, Liudmyla Kononenko and German Bondarenko

Abstract Studies of the dynamics of mobile lead and zinc species in *Sod-podzolic* and *Chernozem* soils demonstrate vertical migration of heavy metals and their transfer to plants. The degree of pollution significantly affects the rate of immobilization of heavy metals and their vertical migration.

Keywords Heavy metals · Mobile species · Dynamic · Vertical migration · Immobilization

Introduction

Lead (Pb) and zinc (Zn) are among the foremost heavy metals (HMs) that pollute the environment. The sources of pollution are mines and quarries extracting polymetallic ores, smelters, coal-burning power plants, and vehicular emissions. The most concentrated streams of heavy metals occur around smelters and metal-working enterprises.

Zn and Pb differ in their biological absorption coefficient K_b, defined by BB Polynov as the ratio of the element's content in vegetation-ash to its content in rocks. This is a quantitative measure of the uptake of elements by plants; the K_b values for Zn and Pb are 11.8 and 1.5, respectively. The high value of K_b for Zn testifies to its physiological role in plants and its selective absorption from the soil; it participates in the synthesis of ribonucleic acid and chlorophyll, and Zn-containing enzymes are involved in carbohydrate and phosphate metabolism (Dobrovolsky 1998). Zn deficiency inhibits plant growth and is commonly manifest in leaf curl. In contrast, Pb has no physiological role. It is absorbed

O. Vysotenko (✉) · L. Kononenko · G. Bondarenko
Institute of Environmental Geochemistry, National Academy of Science, 34a Palladina Av., Kyiv 03680, Ukraine
e-mail: vysotenkooksana@gmail.com

D. Dent and Y. Dmytruk (eds.), *Soil Science Working for a Living*,
DOI 10.1007/978-3-319-45417-7_24

passively along with other trace elements—but it is toxic, damaging the root system, causing brittle stems and leaves, and changes in the water balance; in animals, it interferes with many metabolic processes (Panin and Sapakova 2004; Dobrovolsky 1987).

Absorption of heavy metals by plants through the root system and their upward flow through the soil–plant system is determined by the ratio of fixed to mobile HM species in the soil. HMs released into the environment by industrial processes are commonly in unnatural forms that may subsequently be transformed in the soil; mobile species of technogenic origin may greatly exceed natural species of these metals in the soil (Zhovinsky and Kuraeva 2002).

Experimental Site and Methods

Field investigations of the transformation of Zn and Pb were performed on *Sod-podzolic*[1] and *Chernozem* soils under natural vegetation near to Plesetskoe village in Vasilkovsky District, Kyiv Region. The clay content of the topsoil of the *Sod-podzolic* soil is 39.8%, its organic carbon content 1.1%, and pH 5.2; the clay content of the *Chernozem* is 69.5%, organic carbon 4%, and pH 6.6. The geochemical characteristics of the soils suggested that they were relatively free of pollutants (Vysotenko and others 2004).

Plots of 1 m^2 were contaminated by spraying a solution of Zn and Pb nitrates uniformly on the surface of the soil in single doses equivalent to 11, 22 and 45 g/m^2 of each metal. Control plots remained untreated. The soil cover and natural vegetation on the test plots were not disturbed. Sampling of the 0–5 cm layer of soil at each site was performed annually for 4 years after introduction of HMs. The total contents of Pb and Zn were determined, as well as their water-soluble and ion-exchangeable species. Extraction mobile Pb and Zn species from the soil was carried out using molar ammonium acetate solution. Elements not-extractable with this reagent were considered non-exchangeable absorbed (fixed) species.

Above-ground vegetation was sampled at the same time as soil sampling. On each occasion, the whole green mass was cut, dried and chopped. A portion of the dry matter was ashed in a muffle furnace at 450 °C, dissolved in water, and the concentrations of the elements determined. The content of Pb and Zn in soil and vegetation was determined by atomic absorption spectrometry using a Nippon Jarrell-Ash AA-8500 atomic absorption spectrophotometer.

---

[1] *Albic luvisols* in the World Reference Base for Soil Resources 2006, *Retisols* in WRB 2014 (IUSS WRB 2006/14).

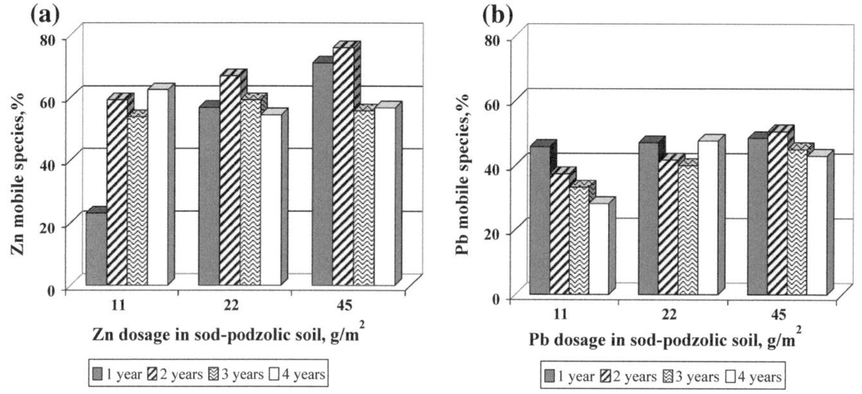

Fig. 24.1 Dynamics of mobile Zn (**a**) and Pb (**b**) introduced into *Sod-podzolic* soil (0–5 cm)

Results and Discussion

Dynamics of the Speciation of Zinc and Lead in Soils

The dynamics of mobile species of Zn and Pb introduced into the *Sod-podzolic* soil are shown in Fig. 24.1. The data on mobile species taken from the 0–5 cm layer are presented as a percentage of the total HM concentration in the sample, adjusted for the content of HM in unadulterated soil. The maximum content of mobile Zn is reached 2 years after introduction. A slight maximum of mobile Pb was observed only for the highest dose (45 g/m^2); in all other cases, there was a general, gradual decrease of mobile HM over time and, as a corollary, a gradual increase in the proportion of non-exchangeable sorbed species.

Figure 24.2 shows the dynamics of mobile Zn and Pb introduced into *Chernozem,* using the same conventions as in Fig. 24.1. In *Chernozem,* there is a decrease in mobile species over time through immobilization in the soil. In the case of Zn, we observe a smooth decrease in the proportion of mobile species; 2 years after introduction, their content is almost halved (except for the lowest dose of 11 g/m^2, where decrease was slower). Immobilization of lead occurred faster; a year after introduction, the content of mobile Pb was one-half to one-third that of mobile Zn and did not exceed 20% for all doses. Thereafter, there was a general slight decrease in the proportion of mobile Pb through continued immobilization.

Comparison of the behaviour of Zn and Pb shows that rate of immobilization of Zn depends on the degree of contamination; at a dose of 11 mg/m^2, the percentage of its mobile species decreased from 100 to 34% in the first year after introduction. With the highest degree of contamination (45 mg/m^2), the proportion of mobile zinc species decreases not so much (to 46%). In contrast, the dynamics of mobile species of Pb is independent of the initial dose.

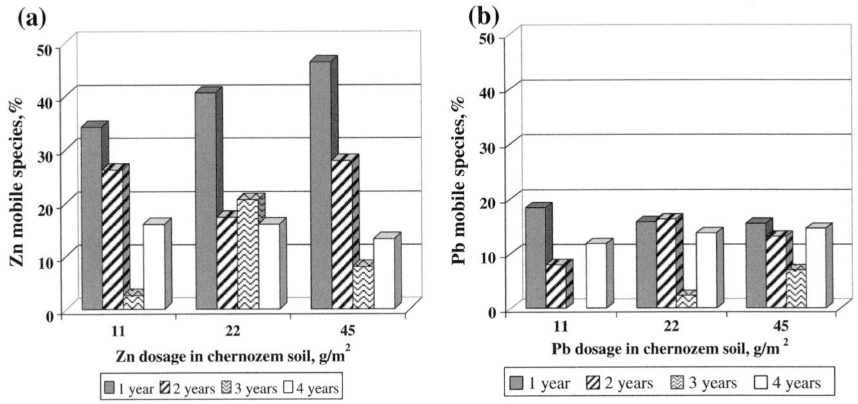

Fig. 24.2 Dynamics of mobile species of Zn (**a**) and Pb (**b**) artificially introduced into *Chernozem* (0–5 cm layer)

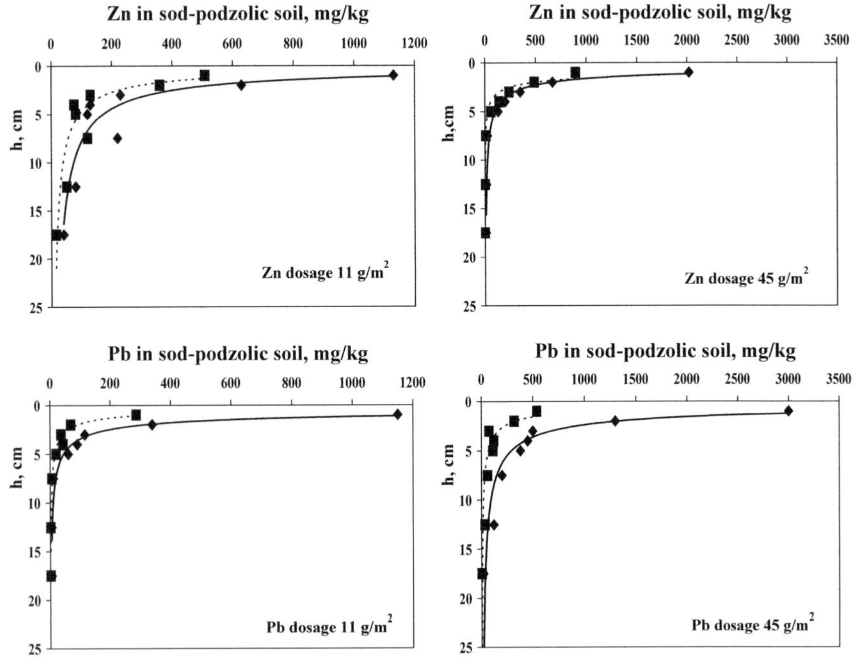

Fig. 24.3 Vertical distribution of Zn and Pb and their mobile species 3 years after introduction to *Sod-podzolic* soil *square*—mobile species, *diamond*—total concentration

Vertical Migration of Zinc and Lead in Soils

To determine the vertical distribution of the introduced Zn and Pb, vertical soil columns were sampled 3 years after the introduction of solutions of their salts to the surface of the experimental sites. Layered samples (0–5, 5–10, 10–15 and 15–20 cm) were taken from each column, and total HM and concentrations their mobile species were determined. Figures 24.3 and 24.4 illustrate the vertical distribution of Zn and Pb in *Sod-podzolic soil* and *Chernozem*, respectively, 3 years after artificial introduction.

Comparing these data with those of mobile species (Fig. 24.1), it appears that in *Sod-podzolic* soil the degree of contamination greatly affects both the rate of immobilization of HM and their vertical migration. At low doses of Pb, there was a significant decrease in the proportion of its mobile species over time, that is to say immobilization occurred which resulted in a significant HM concentration in the 0–5 cm layer of soil. With higher doses of Zn and Pb, where the higher contents of their mobile species were observed, more intense migration occurred and vertical migration of mobile species was in advance of migration of total Zn and Pb. The trends of vertical distribution of total Zn and its mobile species at a high degree of contamination are practically identical because of the very high percentage of mobile Zn.

In the case of Chernozem (Fig. 24.4), introduction of contaminants in solution resulted in rapid vertical migration of HM and further immobilization in underlying soil layers. Therefore, the trends of the vertical distribution of total content of heavy metals and their mobile species are almost identical. The vertical migration of Zn was faster than that of Pb; and the greater the degree of contamination of soil with Zn, the higher is its rate of vertical migration. Any influence of the degree of contamination by Pb on its vertical migration is less noticeable.

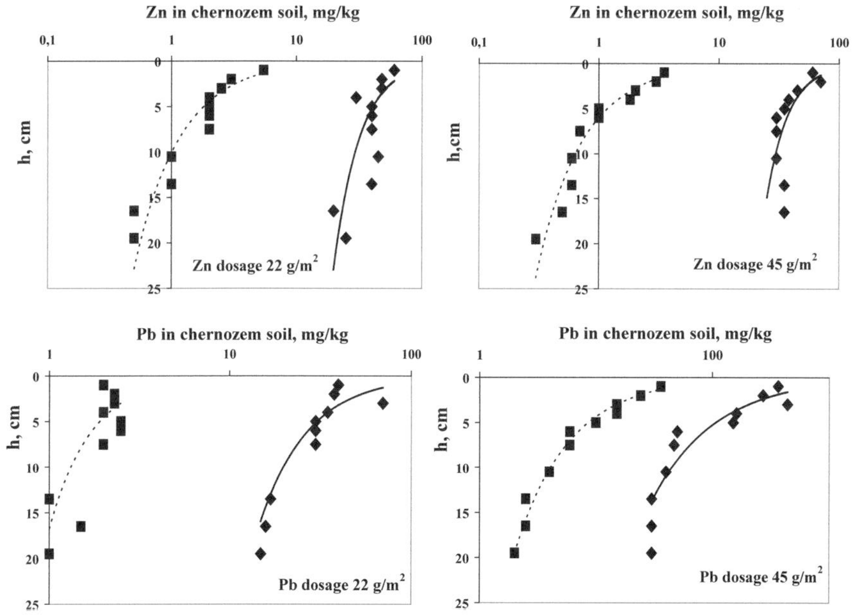

Fig. 24.4 Vertical distribution of Zn and Pb and their mobile species 3 years after introducing their soluble salts into *Chernozem square*—mobile species, *diamond*—total concentration

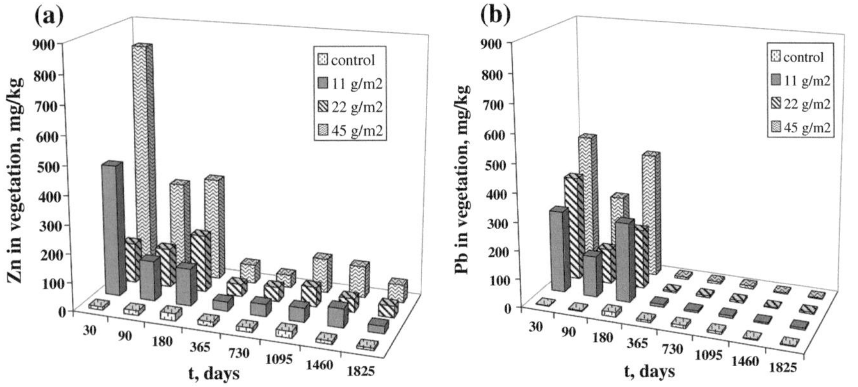

Fig. 24.5 Dynamics of contamination of vegetation on *Sod-podzolic* soil with zinc (**a**) and lead (**b**)

HM Contamination of Vegetation

Lead and zinc contents in the vegetation of the experimental sites are shown in Fig. 24.5. Vegetation sampled on *Sod-podzolic* soil after 30 days was significantly enriched in Zn. After a dose of 11 g/m^2, a 40-fold excess of Zn was observed in vegetation compared to the control. Thereafter, there was a gradual reduction of pollution in the vegetation throughout the year, corresponding with the trend of mobile Zn in the upper 5-cm soil layer where the bulk of the metal was concentrated. Subsequently, there was a slight enrichment of plants with Zn, probably due to metal penetration into the deeper layers of the soil. Thirty days after dosage of 45 g/m^2 in the *Sod-podzolic* soil, the vegetation was strongly enriched in Zn—up to 815 mg/kg calculated as ash, which corresponds to more than 70-fold excess relative to the control.

The degree of Pb contamination of vegetation was lower than that observed for Zn, corresponding to a lesser content of mobile Pb compared with Zn. In the first months after the introduction of HM solutions into the soil, the Pb content in vegetation was much higher than in the control but, a year later, its content did not exceed the control (Fig. 24.5).

Figure 24.6 depicts the contamination of vegetation of the experimental sites on *Chernozem*. In the first days after the introduction of dissolved HM, vegetation abundantly absorbed the pollutants. In the vegetation samples taken 30 days after a dose of 11 g/m^2, the Pb content was two orders of magnitude greater than the control. Thereafter, there was a gradual decrease of Pb in the vegetation so that, after a year, the Pb content in vegetation was no greater than the background. On the site receiving 45 gPb/m^2, extremely high enrichment of vegetation with Pb (up to 1450 mg/kg calculated as ash) was observed after 30 days; over the year, the Pb content of the vegetation decreased to 20 mg/kg and gradually decreased further so that, after 3 years, it was the same as the control. The dynamics of decreasing Pb pollution in

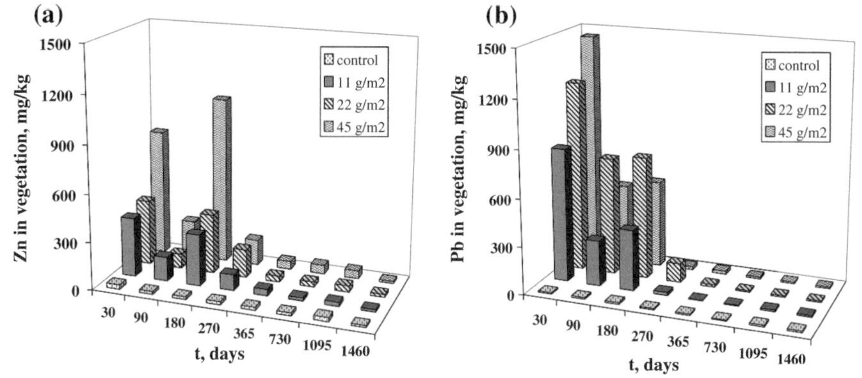

Fig. 24.6 Dynamics of contamination of vegetation on *Chernozem* with zinc (**a**) and lead (**b**)

the vegetation is consistent with the dynamics of its immobilization in *Chernozem*—a significant decrease in the percentage of mobile species in the first year after its introduction into the soil and subsequent slow further decrease (Fig. 24.2).

Decrease of the Zn content of vegetation occurred more slowly than in the case of Pb, in accord with the trends of their mobile species. A year after introduction of 11 gZn/m^2, the Zn content in vegetation was double the background value but, after 2 years, its content decreased to the background value. Following introduction of 45 gZn/m^2, the Zn content in vegetation reached 2 or 3 times the background value and heightened values were maintained for 3 years after the introduction of the contaminant, decreasing to the background value only after 4 years.

In *Chernozem*, after an intense absorption of HM by vegetation, there is a gradual self-cleaning process due to immobilization of HM in the soil. The rate of self-cleaning decreased with an increase in the degree of HM contamination.

Conclusions

1. In *Sod-podzolic* soil, a high content of mobile Zn persists for a long time after initial contamination. That is to say, immobilization of Zn in this soil is not very effective.
2. Immobilization of Pb and Zn is more effective in *Chernozem*, probably due to higher contents of clay and organic carbon. In *Chernozem*, a rapid decrease in the percentage of mobile Pb occurred in the first year after its introduction and there was a gradual further decrease over several years; the rate of decrease hardly depended on the degree of contamination.
3. Vertical migration of Pb and Zn is determined by the dynamics of their immobilization. Vertical migration of Zn is faster than Pb and depends on the degree of soil contamination.

4. Decrease of HM pollution of the vegetation with time, or self-cleaning of the vegetation, is in accord with the dynamics of mobile species in soil. Pb contamination of vegetation decreased almost to background value over one year, whereas reduction of Zn contamination in vegetation is much slower and the rate of self-cleaning decreased with the increasing degree of HM contamination of the soil.

References

Dobrovolsky VV (1987) Lead in the environment. Nauka, Moscow (Russian)

Dobrovolsky VV (1998) Basic biogeochemistry. Vysshaia Shkola, Moscow (Russian)

IUSS WRB 2006/14 World reference base for soil resources. FAO World Soil Resources Report 103/106, Rome

Panin MS, Sapakova AK (2004) The effect of different dosage of lead species in the soil—barley sprouts system. In Modern problems of soil pollution. Moscow State University, Moscow (Russian), pp 148–150

Vysotenko OO, Manichev VY, Kononenko LV et al (2004) Lead and zinc immobilization by sod-podzolic and chernozem soils: nature and laboratory studies. Collected papers of the Institute of Environmental Geochemistry, vol 9, pp 87–94 (Ukrainian)

Zhovinsky EY, Kuraeva IV (2002) Geochemistry of heavy metals in soils in Ukraine. Naukova dumka, Kyiv (Ukrainian)

Chapter 25
Copper and Zinc in the Soils of the Olviya Archaeological Site

Tatyana Bastrygina, Larysa Demchenko, Nataliia Mitsyuk and Olga Marinich

Abstract The behaviour of heavy metals in soil around the copper and bronze artifacts at archaeological sites provides analogues of the modern contamination of the environment. Being components of copper and bronze artifacts, Cu and Zn are the principal contaminants of soils of archaeological sites. Migration of Cu and Zn as a result of corrosion of copper and bronze artefacts is caused by soil–medium factors that affect electro-chemical processes at the metal surface. Due to adsorption of the dissolved elements on clay minerals and deposition of basic salts on the surface of the source metal, Cu and Zn migration from the solid source does not exceed 5–8 cm from the medium–metal interface.

Keywords Archaeological site · Anthropogenesis · Cultural layer soils · Heavy metals

Introduction

Man's contamination of the natural habitat in early times marked the beginning of human influence on the soil. A cultural layer (CL) is anthropogenically transformed soil of an archaeological site; and the CL bronze may provide useful information on anthropogenic activity and soil evolution.

In soil, metal reacts with soil components which retain not only the solid but, also, water-soluble metal compounds in the form of organo-mineral, mineral, sorption–chemisorption, and chemi-depositional compounds (Ostroverkhov 1989). Physical and chemical processes occurring at the metal–environment boundary are characterized by the metal corrosion rate and products of corrosion released into the

---

T. Bastrygina (✉) · L. Demchenko · N. Mitsyuk · O. Marinich (✉)
Institute of Environmental Geochemistry, National Academy of Sciences of Ukraine, 32a Palladina Prospect, Kyiv 03142, Ukraine
e-mail: tatyana.bastrygina@gmail.com

O. Marinich
e-mail: marinich2010@ukr.net

D. Dent and Y. Dmytruk (eds.), *Soil Science Working for a Living*,
DOI 10.1007/978-3-319-45417-7_25

environment. The corrosion rate of antique copper and bronze products is about 0.01 mm/year but, in certain conditions, it may be as high as 0.4 mm/year (Hallberg et al. 1988; Bresle et al. 1983). Soil–metal assays help to determine site boundaries and to identify the impact of human activity on the soils and landscapes (Zlobenko et al. 1999); furthermore, the distribution of exchangeable and sorbed forms of the heavy metals in the environment close to an artefact may throw light on the future behaviour of modern contamination (Sycheva 1994).

Archaeological excavation of the Olviya settlement in the South of Ukraine has revealed extensive handicraft production including workshops manufacturing copper and bronze objects, as well as small ore smelters with the remains of furnaces. The archaeological soils of Olviya were formed from the same parent material as the background undisturbed soil but they experienced intensive human disturbance and inherited a considerable amount of biologically incorporated and trace elements (Lakimenko and Bogachenko 2003). The aim of this research was to study the behaviour of copper and zinc in soil close to the copper and bronze artifacts—as analogues of modern anthropogenic contamination of the environment.

Site and Methods

Olviya, one of the antique historical sites excavated in the South of Ukraine, is located in the steppe zone in North Prychornomorya (Fig. 25.1).

Olviya is situated in Mykolaiv Oblast, on the right bank of the Yuzhnyi Bug river close to its confluence with the Dnieper (Fig. 25.2).

The climate is continental with a mean annual precipitation of 320–370 mm; the soils are *Southern chernozem, Dark-chestnut* and *Chestnut* soils combined with *Solonetz* and *Soloth*. But there have been significant changes in groundwater regime and the intensity of soil-forming processes—the extent of climate change has been enough to restructure the soil profile.

The greater part of the Upper Town of Olviya is located on the loess plateau and most of the excavated sites are seated on loess. The Quaternary layer is 20 m thick; the thickness of the humus soil horizon is 25–50 cm. Pollen analysis reveals that the climate in Northern Prychornomorya changed from cold and arid to warm and wet in the fourth and fifth centuries BC (the period of town construction). From the middle of the third to the end of the first century BC, the climate was very hot and dry; then, over the turn of the millennium, temperatures dropped and mean annual precipitation increased giving a temperate climate with a positive moisture balance; in the fourth century AD, an arid period set in, similar to the present day (Demchenko et al. 2005).

Soil samples were taken from the archaeological excavations of the Fourth-and-Fifth-century BC settlement. We examined samples from the cultural layer of the sites documented by archaeological findings as dwelling houses, castings yard (site of bronze manufactures), and suburbs–both anthropogenically

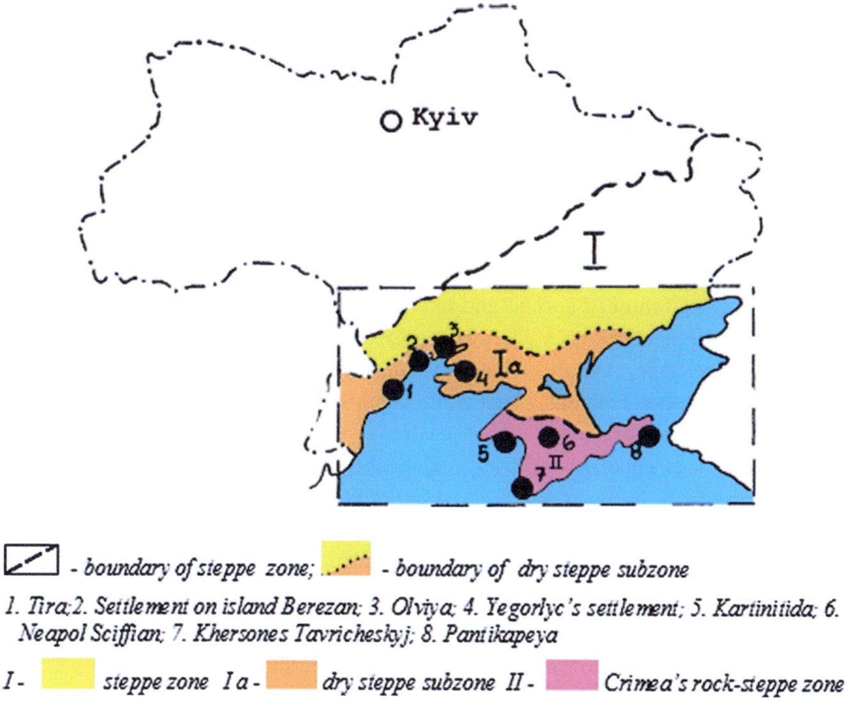

Fig. 25.1 Schematic map of the location of the Olviya archeological site in the North Prychornomorya

Fig. 25.2 Schematic plan of the Olviya's archeological dig

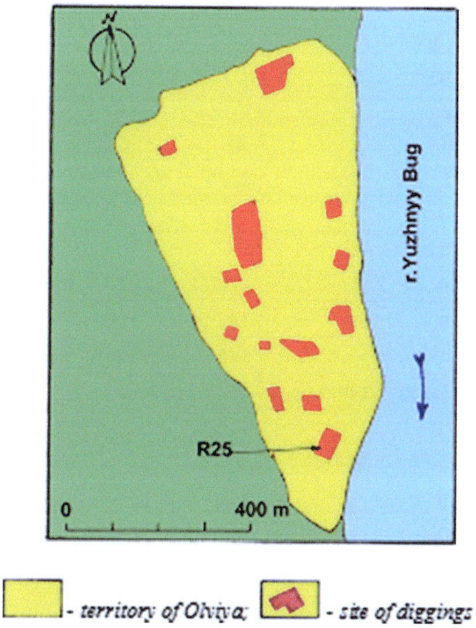

transformed CL soil and undisturbed background soil. Analyses included extraction by 1 M HCl (sorbed forms) and 1 M NH_4—acetate (exchangeable forms); the Cu and Zn concentration in each fraction was determined by atomic absorption spectrometry.

Results and Discussion

The spatial distribution of sorbed and exchangeable Cu and Zn in background soil and CL is presented in Table 25.1. Background soil, situated in the accumulative part of the landscape, contains minor amounts of contaminants, whereas the CL is enriched with these elements, especially samples from anthropogenically transformed soil located under the ancient castings yard and dwelling house.

Ecological and geochemical sampling of the soil from the cultural layer at the excavation sites (a, b, c) show that the contamination is local—and connected with workshops for smelting non-ferrous metals, remains of furnaces, and slag heaps. Close to these sources, soils contain increased concentrations of copper, zinc and lead. In the CL close to the furnace, concentrations of copper and lead of 2290 and 300 mg/kg, respectively, were detected; beyond the workshop, the content of lead in the soil was below 60 mg/kg and that of zinc from 97 to 186 mg/kg, and the maximum value for copper was 360 mg/kg.

Table 25.1 Radiocarbon dates and heavy metal concentrations (ppm) in collected soil samples

Sites in CL	Sampling points					
(a)						
Workshops	1	2	3	4	5	6	7
Cu	270	340	360	340	270	175	110
Zn	97	115	168	126	115	110	100
Pb	30	40	60	60	40	30	30
CL age, years	2025 ± 60						
(b)							
Casting yard	8A	10	10A	10Б	11A	13A	
Cu	2290	57	54	43	700	480	
Zn	122	118	95	87	97	112	
Pb	125	100	20	20	300	50	
CL age, years	1870 ± 65						
(c)							
Dwelling house	1	2	3	4	5		
Cu	110	80	78	50	50		
Zn	95	84	90	90	90		
Pb	30	30	30	30	30		
CL age, years	1930 ± 60						

Heavy metals introduced into the soil react with its components. Heavy metals speciation in the CL is represented by water-soluble organo-mineral, mineral, and other compounds which were analysed in soil samples from different sites (Table 25.2). The amount of exchangeable and sorbed copper and zinc in CL soils was found to be 10% at the sites of workshops, and 6% in the yards of dwellings. The central part of the city was more heavily polluted compared with the residential suburbs.

Contamination of the soil surrounding a copper or bronze object is due to electrochemical corrosion on the metal surface, effected by the pore water and controlled by redox reactions in the copper. The distribution of copper in the surrounding soil layer is controlled by mass transfer under chemical equilibrium conditions when reactants move to the metal surface and the dissolved copper is removed from the surface. We have studied distribution of sorbed Cu and Zn in the soil profile beneath the local sources of contamination (Fig. 25.3); their concentration is greatest in the upper and middle part of the section beneath artifacts or smelted slag.

Table 25.2 Distribution of exchangeable and sorbed Cu and Zn (ppm) in soils of Olviya archaeological complex

Forms	Exchangeable				Sorbed			
Casting yard								
Cu	7.0	3.2	5.3	3.8	15.5	4.2	13.5	5.4
Zn	0.8	0.3	1.2	1.2	2.0	2.4	8.0	10.0
Dwelling house								
Cu	1.5	1.3	2.9	2.6	0.6	1.0	2.4	3.2
Zn	0.3	0.4	0.9	0.7	2.6	8.3	4.2	6.3

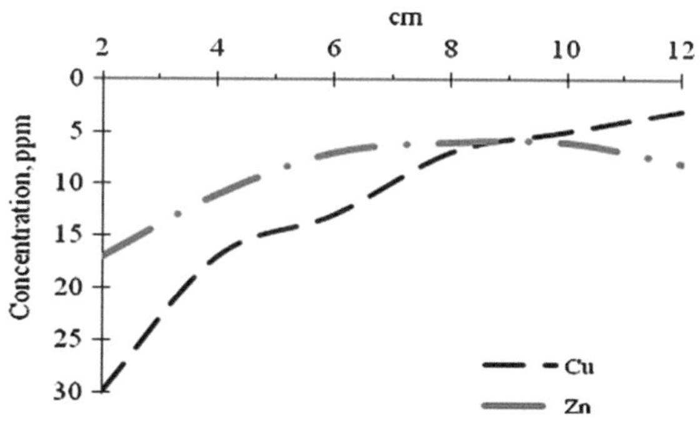

Fig. 25.3 Distribution of sorbed forms of Cu and Zn in CL soil profile beneath the source

The migration depth of Cu and Zn in loess-like loams reaches 7–9 cm during the whole historical period. These data tally with those published by King (1995) and King and Kolar (1995) who measured Cu concentration profiles in marine sediments near a bronze gun affected by saltwater at 7 ± 5 °C for 310 years. Interfacial copper in concentrations of about 10 (ppm) had diffused into the sediments to the depth not exceeding 5 cm. In the Olviya CL, low migration ability of the elements is caused by the high carbonate content (6.8% CaO), increasing Na_2O content (0.6–1.4%) with depth, and an alkaline soil reaction (pH 9.1–9.2). The absorption capacity is 19–23 me/100 g soil, and Ca comprises 67–69%, Mg 22–26%, and Na 4% of the exchangeable bases.

Conclusions

The highest Cu and Zn concentrations were found in CL soils in the castings yard while, on the dwelling house site, the concentrations are comparable with those in the undisturbed environment.

Migration of Cu and Zn in the soil as a result of corrosion of copper and bronze goods in the CL is caused by soil factors that affect electro-chemical processes at the metal surface. The depth of Cu and Zn migration from the macroscopic solid-phase source does not exceed 5–8 cm from the '*medium*–metal' interface.

Movement of heavy metal species is constrained by adsorption of the dissolved elements on clay minerals and deposition of the basic salts of these elements on the surface of the source metal.

References

Bresle A, Saers J, Arrhenius B (1983) Studies on pitting corrosion on archaeological bronzes. 215 in SKB Technical Report 83–05

Demchenko LV, Zlobenko B, Manichev V, Kadoshnikov V, Spasova L (2005) Anthropogenic analogues for geological disposal of high-level and long-lived waste. In: Final report of coordinated research project 1999–2004. IAEA-TECDOC-1481, Austria, pp 55–71

Hallberg R, Östlund P, Wadsten T (1988) Inferences from a corrosion study of a bronze cannon, applied to high-level nuclear waste disposal. Appl Geochem 3:273–280

King F (1995) A natural analogue for the long-term corrosion of copper nuclear waste containers —reanalysis of a study of a bronze cannon. Appl Geochem 10:477–487

King F, Kolar M (1995) Prediction of the lifetimes of copper nuclear waste containers under restrictive mass transport and evolving redox conditions. In: CORROSIO-95 NACE international, Houston, p 425

Lakimenko O, Bogachenko Y (2003) Trace elements as tracers of human impact in soils of ancient settlements. In: Proceedings of 7th international conference on the biogeochemistry of trace elements, Uppsala, pp 62–63

Ostroverkhov A (1989) Production and use of window glass in antique towns of the Northern Prichernomorie. Essays Nat Sci Tech Hist 37:68–75 (Russian)

Sycheva S (1994) Soil-geomorphologic aspects of the ancient settlements cultural layer formation. Pochvovedenie 3(8):28–33 (Russian)

Zlobenko B, Kadoshnikov V, Manichev V, Demchenko L, Golovko T, Krapivina V (1999) Study of materials stability surrounding with loess- clay-loam rocks on an example of "OLVIYA" monument of Ukranian Northern Prichernomorya. In: Proceeding international youth nuclear congress, Bratislava, Slovakia p 285

Chapter 26
Environmental Assessment of Soil Based on Fractional–Group Composition of Heavy Metals

Salgara Mandzhieva, Tatiana Minkina, Galina Motuzova, Mykola Miroshnichenko and Anatoly Fateev

Abstract In a pot experiment, *Ordinary chernozem*, artificially contaminated with Zn and Pb salts, was ameliorated by addition of chalk and glauconite. The fractional and group composition of metal compounds in soil extracts were determined by combined fractionation to reveal the increase in the environmental hazard when the soils were contaminated—and a decrease when ameliorants were applied. Both strongly and loosely fixed metal fractions are involved in the mobility of heavy metals in soils. From these data, mobility coefficients (MCs) of the heavy metals in the soils and the stability coefficients (SCs) of the soils in respect of heavy metals were calculated. MC characterizes the environmental vulnerability of soils to the impact of heavy metals: SC characterizes the environmental sustainability of soils in respect of heavy metal contamination.

Keywords Contamination · Fractionation · Soil · Sorbents · Trace element

Introduction

The elaboration of indices to assess the degree of soil contamination is a topical issue (Kar and Berenjian 2013). Information about the compounds of metallic elements, their presence in different soil components, and a selective account of all the metal forms contributes to: (1) better understanding of geochemical cycles in technogenic landscapes (Saito 2013); (2) identification of diagnostic groups of

S. Mandzhieva (✉) · T. Minkina
Southern Federal University, Rostov-on-Don, Russia 344090
e-mail: msaglara@mail.ru

G. Motuzova
Soil Science Faculty, Lomonosov Moscow State University, Vorobjevy Gory, Moscow, Russia 119991

M. Miroshnichenko · A. Fateev
Sokolovskyi Institute of Soil Science and Agrochemistry, 4 Tchaikovsky St, Kharkiv 61024, Ukraine

D. Dent and Y. Dmytruk (eds.), *Soil Science Working for a Living*,
DOI 10.1007/978-3-319-45417-7_26

metal compounds to indicate the severity of adverse effects on the environment (Al-Sharafat et al. 2012; Babakhouya et al. 2010); and (3) assessment of soil stability in respect of contamination by heavy metals. Study of the transformations of contaminant metal compounds when they are inactivated by ameliorants should reveal the mechanisms responsible for the effects of sorbent application (Minkina et al. 2008a; Mousavi et al. 2013).

Materials and Methods

Zn and Pb are the main contaminants of the soils of the country around Rostov (Minkina et al. 2008a), so an experiment was carried out to study the transformation of Zn and Pb compounds in soils with high levels of contamination when amendments are applied to fix these metals in the soil (Motuzova and Hong Van 1999). The experiment was conducted with the topsoil of arable *Ordinary chernozem* (calcareous, low-humus, clay loam) from the steppe zone of Oktyar'skii region, Rostov Oblast. The 0–20-cm layer was characterized by pH_{water} 7.2; 59% of particles <0.01 mm; 1% $CaCO_3$; 2.3% organic carbon; 290 mmol/kg exchangeable Ca^{2+}, 30 mmol/kg exchangeable Mg^{2+}, and 1 mmol/kg exchangeable Na^+.

In a series of one-litre polythene vessels, the bottoms were covered by 30 cm of washed glass, for drainage, overlain by 1 kg of 5-mm sieved soil mixed with dry metal salts. As pollutants, Pb and Zn acetates (10,000 mg/kg of each element) were applied separately and together and the soil was kept moist at 60 per cent of field capacity. Two months later, 50 g/kg of chalk or glauconite was added as an amendment, accounting for 5% of the soil mass. Then, the soil was incubated for a year at the same level of moisture. The scheme of the experiment was: (1) without heavy metals; (2) with Zn or Pb; (3) with metal + $CaCO_3$; (4) with metal + glauconite; (5) with Zn + Pb. The experiment was performed in triplicate.

Samples were taken from the whole soil mass for determination of the total content of heavy metals (HM) and their extracted forms. The total HM content was determined by X-ray fluorescence. Their mobile forms were determined by the atomic absorption spectrophotometry. To detect the group and fractional composition of HM compounds, a combined scheme of fractionation was applied (Minkina et al. 2008b). Special attention was paid to the ratio of firmly and loosely bound HM. The loosely bound (LB) metal includes exchangeable, complex, and specifically absorbed compounds. The group of firmly bound (FB) Pb mainly includes compounds fixed in a crystal lattice of primary and secondary minerals, as well as hardly soluble salts and stable organo-mineral complexes.

The methods of sequential and parallel fractionation of metal compounds are widely applied separately but, when used in combination, it is advisable to use the Tessier scheme of sequential fractionation (Tessier et al. 1979; Minkina et al. 2008b) in combination with parallel extraction using individual reagents. According to this scheme, the loosely bound forms were represented by HM compounds extracted by 1 M ammonium acetate buffer solution (AAB) at pH 4.8; 1% EDTA in

1 M AAB; and 1 M HCl. These extracting agents are not selective. AAB at pH 4.8, as proposed by Krupskyi and Aleksandrova (1957), presumably extracted exchangeable HM. One per cent EDTA can extract metals present in soil as organic complexes; HM in organic–mineral complexes was estimated as the difference between the metal content in 1% EDTA in 1 M AAB and that in AAB (Minkina et al. 2008b). HM extracted in 1 M HCl was taken to be the potential amount of labile metal compounds in the soil; acid-soluble HM compounds were, presumably, represented by exchangeable and specifically absorbed HM compounds, the latter being determined by means of the difference between their amounts extracted by 1 M HCl and AAB. Firmly bound HM compounds were identified by their constituents, represented by compounds connected with organic matter, Fe and Mn (hydr)oxides, and silicates.

Results

In the soil under study, the total content of Zn averaged 67 mg/kg and Pb 25 mg/kg (Table 26.1), which corresponds to their initial level in *Ordinary chernozem* (Minkina et al. 2008a; Nikityuk 1998; Samokhin 2003). The major part of Zn and Pb in the soil was firmly bound. The share of loosely bound forms was 13 and 18% of their total content, respectively (Table 26.2). The loosely bound heavy metals were mostly specifically absorbed forms (69 and 85% of the total loosely bound compounds, respectively); in the case of Zn, the share of loosely bound compounds was particularly high. The content of exchangeable and complex compounds of these metals was insignificant.

Ordinary chernozem artificially contaminated with Zn and Pb showed an increase in the total content reaching almost 10,000 mg/kg. There was, also, an absolute increase of all three mobile fractions of metal compounds: the amount of

Table 26.1 Total and loosely bound Zn and Pb compounds (mg/kg) in *Ordinary chernozem* contaminated by these metals and after application of sorbents

Experimental treatments	Exchangeable		Complex		Specifically sorbed		Total content	
	Zn	Pb	Zn	Pb	Zn	Pb	Zn	Pb
Without metal addition	0.7	0.9	0.6	0.5	7.1	3.0	67	25
Metal (Me)[a]	2478	2962	1628	2443	4386	2568	9890	9851
Me + $CaCO_3$	782	1062	660	1267	2578	1986	9791	9826
Me + glauconite	1034	1462	916	1385	2537	2047	9633	9776
Zn + Pb	2844	2658	1545	2434	4183	2334	9776	9940
$LSD_{0.95}$ for numerator	235	244	158	232	323	205	317	459

[a]Separate application of metals (Zn and Pb) at a rate of 10,000 mg/kg

Table 26.2 Indices for the group composition of Zn and Pb compounds in mono- and polymetal-contaminated *Ordinary chernozem*

Experimental treatments	Loosely bound compounds/firmly bound compounds[b]		Exchangeable/complex/specifically sorbed[c]	
	Zn	Pb	Zn	Pb
Without metal addition	13/87	18/82	8/7/85	20/11/69
Metal (Me)[a]	86/14	81/19	29/19/52	37/31/32
Zn + Pb	88/12	75/25	33/18/49	36/33/31

[a]Separate application of Zn and Pb at a rate of 10,000 mg/kg
[b]% of the total content
[c]% of the content of loosely bound compounds

exchangeable forms increased 3500-fold for Zn and 3000-fold for Pb; the content of complex compounds 3000- and 5000-fold; and specifically absorbed compounds 600- and 800-fold, respectively (Table 26.1).

Discussion

In the course of soil contamination, the relative proportions of firmly bound and loosely bound HM compounds changed—loosely bound compounds increased up to 86–81% of the total content. This change was accompanied by an increase in the share of mobile compounds among the loosely bound forms. The process appears to be accelerated for Zn in the case where it was applied together with Pb (Table 26.2).

The ratio of loosely bound compounds in the group was also changed. The mobility of the metals increased—mostly through such mobile compounds as exchangeable Zn and Pb complex forms. The share of exchangeable compounds proved to be higher by 3.6–4.1 for Zn and 1.8 times for Pb. The complex forms showed an increase by 2.7 for Zn and 2.8–3.0 for Pb, but the specifically absorbed forms of these metals were almost halved (Table 26.2).

Contamination of the soil by these heavy metals is dangerous because not only is their content increased but, also, the mobility of pollutants is greatly increased (Glazovskaya 1994). Zn mobility increased more than the mobility of Pb, which indicates a relatively high mobility of technogenic zinc; this increased Zn mobility is connected with an increase in the (mobile) exchangeable fraction of this element. It is clear that qualitative changes take place in the mobility of metals upon contamination and amelioration of the soil; this is confirmed by changes in the content of different fractions in the composition of loosely bound compounds.

Trends in metal absorption by soil can be considerably changed if cations of other elements are present. This is because of competition between ions interacting with reactive surfaces. The intensity of such interaction between heavy metals is determined by the size of reactive surfaces of soil components; the metal ions are absorbed when they are close to the sorbent surface. In the model experiment with high doses of pollutant, practically all the reactive surface seemed to be filled up. According to Ladonin and Plyaskina (2009), Zn is highly affected by the other elements—testified by its lower absorption when Zn is present in the solution together with Cu and Pb (Pinskii et al. 2010).

When amendments are applied to the contaminated soils, the total HM content is unchanged. However, the amount of loosely bound Zn and Pb compounds and their mobility are decreased (Tables 26.1 and 26.2). Chalk application fixes metals more effectively than glauconite. Thanks to the chalk amendment, the content of loosely bound Zn and Pb decreased by 45 and 37%, respectively, but with the glauconite amendment, only by 39 and 31%, respectively (Table 26.2). The ratio of loosely bound compounds was changed as well. Both amendments (chalk and glauconite) stimulated a decline in the share of exchangeable compounds and an increase in specifically absorbed compounds (Table 26.2). The share of complex compounds remained almost unchanged.

In a field experiment on chernozem artificially contaminated with Zn and Pb (Minkina et al. 2008a), application of the same amendments at rates of 25 and 50 t/ha, respectively, created an identical situation. In contrast to glauconite application, the content of exchangeable Zn and Pb was lower in the first year of the chalk application—the share of loosely bound compounds in contaminated soils reached the same level as in uncontaminated soils, and because of the decreased metal mobility, the yield and quality of barley were improved for 3 years after the chalk application.

An overview of changes in the fractional and group composition of HM compounds in contaminated soils throws light on the mechanism of amendment effects on metal immobilization, the redistribution of compounds, and the direction of their transformation processes. The amendments have a significant influence on transformation of Zn and Pb compounds in soil; a decreased content of loosely bound compounds is accompanied by the formation of firmly bound compounds (Table 26.3).

The chalk application decreased the share of exchangeable Zn and Pb compounds in the group of loosely bound compounds, at the same time increasing the share of their compounds bound with carbonates. Differences in the action of amendments were well seen in the group composition of firmly bound compounds. Absolute and relative amounts of these metals bound with silicates became higher in the group composition; there was a marked increase in this fraction in case of glauconite application. The role of Fe hydroxides became more significant in Zn fixation; the share of these compounds increased twofold thanks to chalk application and by 1.7 times with glauconite (Table 26.3). This indicates differences between the mechanisms responsible for fixing these metals by the different amendments. Any metal taking part in ion exchange can be bound within glauconite by inter-structural fixation. The input of molecules to an adsorbing cave is observed in when their size is lower than the entrance window, and it is known that zeolites adsorb a higher amount of HM salts because of their cation exchange capacity accounting for 100–300 mg/kg (Pinskii 1997).

Table 26.3 Fractional and group composition of Zn and Pb compounds in contaminated *chernozem* after the application of amendments, mg/kg

Experimental treatments	Loosely bound compounds				Firmly bound compounds			Sum of fractions
	Exchangeable	Complex	Specifically sorbed		With organic matter	With Fe and Mn (hydr)oxides	With silicates	
			On carbonates	On Fe and Mn (hydr)oxides				
Zn								
Without metal addition	0.3	0.6	6.3	0.8	1.1	6.1	55.5	71
Metal (Me)[a]	2226	1628	2714	1672	127	867	278	9512
Me + $CaCO_3$	656	660	1988	590	809	1797	3113	9613
Me + glauconite	923	916	1826	711	635	1521	3241	9773
Pb								
Without metal addition	0.4	0.5	1.6	1.4	6.4	1.4	14.7	26
Metal (Me)[a]	2443	2443	1476	1092	2055	523	250	10,282
Me + $CaCO_3$	948	1267	1758	228	2781	1204	2229	10,415
Me + glauconite	1379	1385	1666	381	1992	732	2461	9996

[a]Separate application of metals (Zn and Pb) in the rate of 10,000 mg/kg

The glauconite-induced increase in the absolute content of heavy metals in fractions enriched with non-silicate forms of Fe and Mn and organic substances is probably conditioned by different mechanism fixations of chemical elements. The results of the long-term field experiment showed that the efficiency of glauconite in metal fixation increases in the second year after its application (Minkina et al. 2008a); metal fixing in the crystal lattice continues over a long period of time which, probably, accounts for the weak effect of glauconite as compared with chalk.

The changes in the group composition of HM compounds in contaminated soils caused by carbonate application, and chalk in particular, occur thanks to mechanisms increasing the soil's absorbing capacity (Minkina et al. 2010). Thus, the analysis of fractional and group composition of HM compounds in contaminated and uncontaminated soils enabled identification of the mechanisms of amendment effects on the HM mobility in soils.

Conclusion

Application of high doses of Zn and Pb (10,000 mg/kg) changes the group composition of their compounds. An increase of exchangeable forms is most hazardous.

Because the content of heavy metals bound with silicates is more stable, even at a high level of soil contamination, the relative HM content in soil can serve as an indicator of technogenic effects on the environment.

From a comprehensive analysis of the group composition of HM compounds, the mechanisms of amendments' effects on the HM mobility in soil may be identified, and their efficiency assessed. Application of chalk and glauconite to contaminated soils decreases the content of loosely bound compounds and increases the content of firmly bound compounds—which, in turn, decreases Pb and Zn mobility. The effect depends on the amendment: chalk proved to be the better sorbent.

Testing of a combined fractionation method for determining the group composition of metal compounds in soils under different conditions (original soil, polluted soil, and the soil after amendments) confirmed the validity of the combined scheme of HM fractionation and the information value of the determined parameters as an index of soil ecological state. The method of parallel extractions can be used for the rapid assessment of heavy metal mobility to monitor soil condition. Combined fractionation is particularly useful for studying the transformation of metals in contaminated soils because it separates the metal compounds with different mobility under specific conditions and, also, it can predict the behaviour of pollutants.

Acknowledgments This work was supported by projects of the Russian Foundation for Basic Research, No. 13-05-00583; the Southern Federal University, No. 213-01/2015-05; the Leading Scientific School, No. 9072.2016.11. Analytical work was carried out on the equipment of Centres for collective use of Southern Federal University "High Technology" and "Biotechnology, biomedical, and environmental monitoring".

References

Al-Sharafat A, Altarawneh M, Altahat E (2012) Effectiveness of agricultural extension activities. Am J Agric Biol Sci 7(2):194–200

Babakhouya N, Boughrara S, Abad F (2010) Kinetics and thermodynamics of Cd (II) ions sorption on mixed sorbents prepared from olive stone and date pit from aqueous solution. Am J Environ Sci 6(5):470–476

Glazovskaya MA (1994) Criteria for classification of soils according to lead-pollution risk. Eurasian Soil Sci 26(1):58–74

Kar SZ, Berenjian A (2013) Soil formation by ecological factors: critical review. Am J Agric Biol Sci 8(2):114–116

Krupskyi NK, Aleksandrova AM (1957) On definition of mobile forms of microcells. Microcells in life of animals, plants and the person. Naukova Dumka, Kiev (Russian)

Ladonin DV, Plyaskina OV (2009) Heavy metal pollution of urban soils. Eurasian Soil Sci 42 (7):816–823

Minkina TM, Motuzova GV, Nazarenko OG et al (2008a) Forms of heavy metal compounds in soils of the steppe zone. Eurasian Soil Sci 41(7):708–716

Minkina TM, Motuzova GV, Nazarenko OG et al (2008b) Combined approach for fractioning metal compounds in soils. Eurasian Soil Sci 41(11):1171–1179

Motuzova GV, Van Hong NT (1999) The geochemistry of major and trace elements in the agricultural terrain of South Viet Nam. J Geochem Explor 66(1–2):407–411

Mousavi SM, Bahmanyar MA, Pirdashti H (2013) Lead and cadmium availability and uptake by rice in response to different biosolids and inorganic fertilizers. Am J Agric Biol Sci 5(1):25–31

Nikityuk NV (1998) Mobility of heavy metals in chernozem carbonate soils and ways of its assessment: PhD review, KGAU, Krasnodar (Russian)

Pinskii DL (1997) Ion-exchange processes in soils. Publishing house of ONTI PNT Russian Academy of Sciences, Pushchino (Russian)

Pinskii DL, Minkina TM, Gaponova YuI (2010) Comparative analysis of mono- and polyelement adsorption of copper, lead, and zinc by an Ordinary chernozem from nitrate and acetate solutions. Eurasian Soil Sci 43(7):748–756

Saito CH (2013) Environmental education and biodiversity concern: Beyond ecological literacy. Am J Agric Biol Sci 8(1):12–27

Samokhin AP 2003 Transformation of compounds of heavy metals in soils of Nizhny-on-Don. PhD review, Rostov-on-Don (Russian)

Tessier A, Campbell PGC, Bisson M (1979) Sequential extraction procedure for the speciation of particulate trace metals. Anal Chem 51(7):844–850

Chapter 27
Regularity of Transformations of Oil-Contaminated Microbial Ecosystems by Super-Oxidation Technology and Subsequent Bio-remediation

Elena Maklyuk, Ganna Tsygichko, Ruslan Vilnyy and Alex Mojon

Abstract Super-oxidation technology uses chemical reagents to break down heavy residues of oil hydrocarbons. Its effects on the soil microbial flora were investigated for potential recovery of soil biotic activity. The dynamics of microbial ecosystems in oil-contaminated soils under application of the patented bio-remediation technology of the Man Oil Group (Switzerland) were investigated in the field and laboratory in turf and sand with different degrees of contamination by oil products over 5 years. Neither oil contamination nor super-oxidation technology sterilized the soil, but the structure of the microbial populations of peat and sandy soil was much altered, depending on soil type, characteristics of pollutant, period and pollution intensity.

Keywords Soil contamination · Bio-remediation · Microbial flora

Introduction

Nowadays, there are many of different bio-remediation technologies. Their major advantage over physico-chemical treatments of oil-contaminated soil is that they are environment-friendly; they depend on bio-preparations or activation of the native soil microbes (Rogozina et al. 2010; Golodyaev et al. 2009). But rapid and effective removal of oil and oil products from soil is difficult using only bio-remediation because of physiological constraints on the main object of bio-remediation—the

E. Maklyuk (✉) · G. Tsygichko · R. Vilnyy
Sokolovskyi Institute for Soil Science and Agrochemistry Research,
4 Tchaikovsky St, Kharkiv 61024, Ukraine
e-mail: maklyukelena@gmail.com

A. Mojon
Man Oil Group, 8 Bleichistrasse, 6300 Zug, Switzerland

D. Dent and Y. Dmytruk (eds.), *Soil Science Working for a Living*,
DOI 10.1007/978-3-319-45417-7_27

microbial flora itself—under certain soil and climatic conditions (Kochergin et al. 2009; Filler 2008). Therefore, improving the effectiveness of bio-remediation is our goal (Karpenko et al. 2009; Belkov 2000).

Oil in the soil has a selective effect on the native microbial flora, inhibiting sensitive groups but promoting oil-pollution-resistant microorganisms that can utilize *n*-alkanes and benzene hydrocarbons as an energy substrate. However, thanks to bio-degradation of the hydrocarbons, the effect passes over the years. Therefore, we tested a combined method of the remediation in which, first, the patented super-oxidation technology developed by Man Oil Company (Switzerland) acts upon the long-chain hydrocarbons, tearing the chain into smaller lengths, followed by a further bio-remediation stage. Choice of chemical reagents is important because they must be non-toxic to the environment and should not create harmful residues that may accumulate in the soil. The rate of application is also important to allow the recovery of the native soil microbial flora.

Materials and Methods

We investigated oil-contaminated samples of raised-bog peat and sand (contaminated for 5 years), from the Western Siberian oil slurry pits (Yuganskneftegaz) remediated by the Man Oil Group. The level of contamination in the samples was different: a mean of 12% in the peat but 68% (1%—10,000 ppm) in the sand. The little-decomposed *Sphagnum* peat had an ash content of 6.6% and elementary composition of: N 2.58%, C 55.8%, H 5.99%, S 0.35% and O 35.3%. The composition of turf characterized by the low consists of bitumens 3.1%, the total composition of water- soluble and easy hydrolysable substances is 47%. According to Arkhipov (1998), the typical peat of the region is low in bitumens (3%); the total content of water-soluble and easily hydrolysable substances is 47%; the content of humic and fulvic acids is—17.0% and 16.4%, respectively; and combined cellulose and lignin 15.5%. The total composition of cellulose and lignin is 15.5%. The sandy soil of the area is characterized by available nitrogen (nitrate and ammoniacal) 30–40 mg/kg; available phosphorus and potassium (Kirsanov) 120–170 and 150–200 mg/kg, respectively; humus 2.0–2.2%; cation exchange capacity 4–8 meq/100 g of soil; and pH_{water} 5.7 and pH_{CaCl_2}—4.2 (Tseh and Hintermayer-Erhard 2009).

It is generally accepted that pollution by oil leads to sterility of the soil; it is for this reason that remediation is required. Employment of chemical reagents has two effects: the first, negative—even greater damage to the soil biome; the second, positive—break down of petroleum into lighter fractions of hydrocarbon that are readily available for nutrition of microorganisms. Therefore, the task is to choose technological conditions that do not reduce the numbers of the resistant microbial flora that can thrive within an environment of active hydrocarbon degradation. The MOG technology was carried out with different concentrations of reagents (Fig. 27.1).

	Concentration of Reagent A, 1 kg/2 l H_2O				
Concentration of reagent B, %		**0**	**0.1**	**0.3**	**1.0**
	0	V1	V2	V3	V4
	1	V5	V8		
	10	V6		V9	
	30	V7			V10

Fig. 27.1 Experimental design

Laboratory analyses were carried out in the Soil Microbiology Laboratory of Sokolovskyi Institute for Soil Science and Agrochemistry Research. The structure of the microbial ecosystem before and after application of the MOG super-oxidation technology was assessed according to the numbers of microorganisms of ecological and trophic groups that actively participate in the destruction of the oil hydrocarbons and reconstituting the biological activity of soil in these specific conditions. Data were obtained by seeding on selective culture media: for fungi (th/g) on the

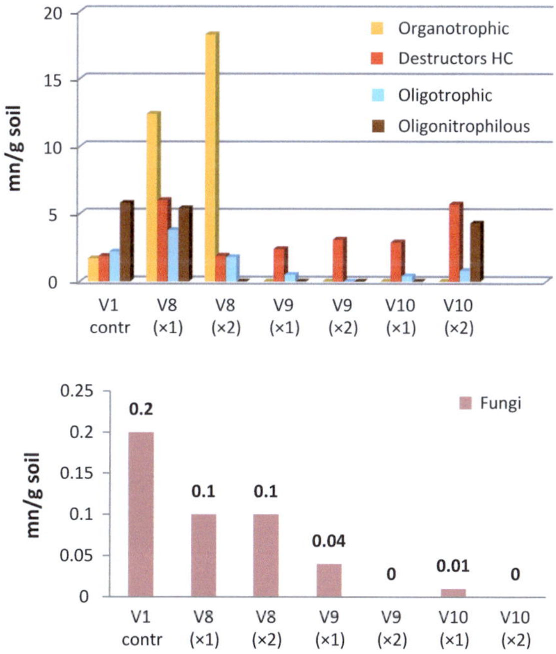

Fig. 27.2 Impact of the reagents and their concentration on the main ecological—trophic groups of microorganisms, including bio-destructors, in the peaty soil

Richter medium; for m/o, which assimilate organic forms of the nitrogen (colony mn/g) on agar-meat infusion; destructors HC on the medium for hydrocarbon destructors (patent № RU2390555); oligotrophic microbes on starvation agar; and oligonitrophilous microbes on the Ashby medium (Iutyns'ka 2006).

Results and Discussion

Microbiological analysis revealed colonies of all microorganism groups in the contaminated peat before application of the technology—variant control. In the soil, over a long time, stable microbial ecosystems formed where the number of groups of microorganisms is almost the same—but the dominant bacteria are oligonitrophilous (adapted for the nitrogen content) and fungi.

After using the MOG oxidation technology at low concentrations of reactants (V8) and with only a single treatment, the number of microorganisms in the soil was increased, but fungi lost their ecological niche. We observed significantly

Fig. 27.3 Influence of strengthening concentration of oxidizing reagents on the turf microflora

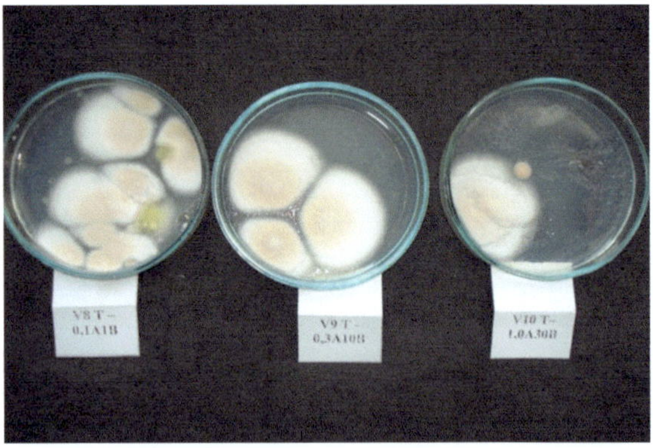

(a) Fungal colonies

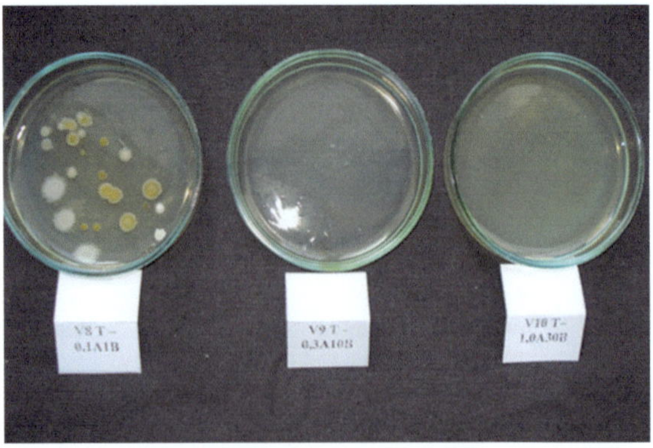

(b) Colonies of m/o which assimilate organic forms of the nitrogen

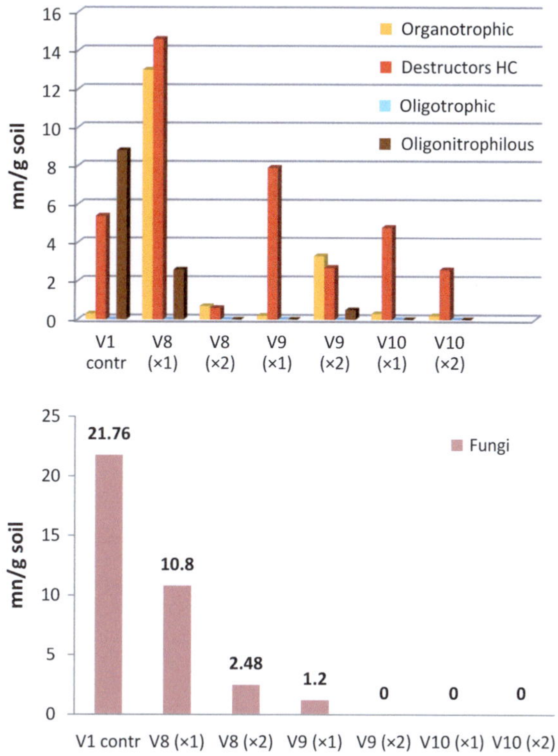

Fig. 27.4 Impact of the reagents and their concentration on the main ecological—trophic groups of microorganisms, including bio-destructors, in the sandy soil

greater numbers of organotrophs—destructors of oil hydrocarbons—and high increase in organotrophs was promoted by the double treatment of the turf (V8 × 2). With strengthening of the concentration of reagents (V9, V10), soil life was significantly decreased (Fig. 27.2).

Even so, the turf, as an organic soil, maintains conditions suitable for bio-destructors, even with increased concentration of reagents. In short, use of the technology can keep the resistant microflora of turf, but high reagent concentrations destroy microflora (Fig. 27.3). That is why we need to choose such norms of reagents that will be most beneficial for subsequent bio-remediation—where the treatment will activate oil-destructors within the native soil microbial flora.

At the same time, we need to consider that each type of soil has its specific soil microbiota—soil characteristics have great impact on the effectiveness of the bio-remediation. Sand is characterized by poor microflora, and oil contamination contributes to the selection of hydrocarbon destructors and oligonitrophilous microbes in the microbial ecology (Fig. 27.4). With strengthening of the concentration of reagents, we observed a decrease in microbial and fungal numbers, but the number of bio-destructors was at the high level, suggesting a significant opportunity for the use of this group in bio-remediation.

Conclusions

Our data suggest a possible solution for intensification of bio-remediation by combining technologies—where scientifically grounded norms of chemical components in the oxidation of the heavy hydrocarbon components increase the amount of light hydrocarbons available for bio-remediation.

At the same time, we must consider soil characteristics, period of pollution and pollution intensity to arrive at the scientifically based norm for using chemical reagents. Further research is needed on other impact factors such as characteristics of pollutant and application methods.

References

Arkhipov VS (1998) Composition and properties of the turf typical in the central part of Western Siberia. In Arkhipov VS, Maslov SG (eds) Chemistry of plant raw materials, vol 4, pp 9–16. Tomsk (Russian)

Belkov VM (2000) Technology methods and concept of recycling carbonaceous industrial and solid waste. http://www.promeco.h1.ru/stati/38.shtml. Accessed 3 Nov 2002 (Russian)

Filler M (2008) Bioremediation of oil hydrocarbons in cold regions. Cambridge University Press, Cambridge

Golodyaev GP, Kostenkov NM, Oznobihin VI (2009) Bioremediation of oil-contaminated soil by composting. Pochvovedenie 8:996–1005 (Russian)

Iutyns'ka HO (2006) Soil microbiology. Aristey, Kyiv (Ukrainian)

Rogozina EA, Andreeva OA, Zharkova SI, Martinova DA (2010) Comparative characteristics of native bio-preparations proposed for cleanup of soil and grounds from pollution. O&G geology. Theory and practice 5, 3. http://www.ngtp/rub/7/37_2010.pdf (Russian)

Karpenko OY, Shchehlova NS, Novikov VP (2009) The biological evaluation of soil from oil hydrocarbons using BioPAR and calcium peroxide. http://vlp.com.ua/files/17_51.pdf (Russian)

Kochergin IE, Oznobihin VI, Savelev AV, Kereev VO (2009) Experience of bioremediation of oil-contaminated soil at the northern Sakhalin experimentation. Collector of articles REA1, pp 84–96 (Russian)

Tseh V, Hintermayer-Erhard G (2009) Soils of the world: Atlas. Study guide: nanotechnology (German)

Chapter 28
Transformation of Non-polar Hydrocarbons in Soils

Viktoriia Shkapenko, Vadim Kadoshnikov and Irayida Pysanskaia

Abstract IR spectroscopy was used to monitor biodegradation of non-polar hydrocarbons in the presence of clay minerals. It reveals IR absorption bands (1170, 1270, 1610 cm^{-1}) not present in the transformer fluid and identified as oscillations of aromatic hydrocarbon rings; C=O groups formed by oxidation of aliphatic chains also appeared. In aerobic soil, the main mechanism of oxidation of hydrocarbons is incorporation of oxygen into the hydrocarbon molecule, replacing bonds with low rupture energy (C–C, C–H) by bonds with higher rupture energy (C–O, H–O). At low temperatures and in the presence of ammonium, these intermediate products may be transformed to humic substances. This offers a mechanism of biodegradation akin to natural humification. The process may be accelerated by introducing specially formulated clay–bio-decomposer inoculum.

Keywords Soil · Clay · Microorganisms · Hydrocarbons · Transformation · Oil pollution

Introduction

Contamination of soil, water and bottom sediments may occur in the course of oil exploration, production, transport and processing. Under natural conditions, decontamination takes a long time. The process may be accelerated by field bio-remediation technologies but, in a European climate, 3–4 growing seasons are needed. Over large areas where the industry operates, it is not possible to carry out bio-remediation throughout the year so we seek technologies that can shorten the time needed and reduce the area of contamination. Bio-reactors use a community of microorganisms compatible with the indigenous microorganisms (IMO) and able to degrade oil. They achieve rapid remediation by maintaining optimal conditions for

V. Shkapenko (✉) · V. Kadoshnikov · I. Pysanskaia
Institute of Environmental Geochemistry, National Academy of Sciences, 34a Palladina Prospect, Kyiv-142 03680, Ukraine
e-mail: vika.shk@yandex.ru

D. Dent and Y. Dmytruk (eds.), *Soil Science Working for a Living*,
DOI 10.1007/978-3-319-45417-7_28

the microorganisms in the reactors but, at present, there is only a limited amount of published data on the design characteristics of bio-reactors used for remediation of oil-polluted environments (Glazovskaya1988; Marfenina 1991; Zvolynsky 2005).

Analysis of the transformation of oil spills is needed to understand the mechanisms of self-remediation of soils. Such understanding will help to identify the time of contamination and terms of remediation and increase the efficiency of monitoring contaminated environments. The main oil-degradation mechanism is microbiological oxidation (mineralization), the end products of which are carbon dioxide, water and, in some cases, complex resinous substances resistant to degradation. The process takes place in several stages, and complete destruction of oil products in soil takes more than 5–6 years.

The first stage, up to 1.5 years, is characterized by suppression of the microflora. Physico-chemical processes include penetration of the oil into soil, evaporation of light fractions, leaching, oxidation with atmospheric oxygen and photochemical decomposition of petroleum hydrocarbons. During the first few months, depending on soil and climatic conditions, the oil concentration decreases by 40–50%. In the second stage, over 3–4 years under oxidizing conditions, the amount of residual oil decreases to 10 per cent and most petroleum hydrocarbons are transformed into low-toxic substances. Destruction of methane-naphthenic factions (the most toxic components) is accomplished by indigenous hydrocarbon-oxidizing soil microorganisms, which multiply 25-fold. In this period, transformation of *n*-alkanes (C 17–30) takes place in the residual oil and the oxidation of simple aromatic hydrocarbons begins. Various metabolites are produced: esters, ketones, aldehydes and acids. Under natural conditions, if no additional measures are taken, comes the third stage—which remains poorly understood. Four to five years after the spill, the most stable high molecular weight compounds, resistant to microbiological attack, remain in the contaminant. This stage is characterized by microbiological degradation of the less-toxic hydrocarbons and resin–asphaltene components that form on the polluted surface, the so-called *kirs*—hard scales that prevent penetration of air into the root zone, leading to lack of oxygen for plants and soil fauna (Glazovskaya 1988).

Recovery of ecosystems proceeds in line with these stages of biodegradation but at different rates in different compartments of ecosystems; a saprotrophic complex of animals develops much more slowly than the microflora and vegetation. As a rule, complete reversibility of the process is not observed. The most dramatic increase in microbial activity occurs in the second stage of biodegradation but, with further reduction of the numbers of all groups of microorganisms compared with reference values, the number of hydrocarbon-oxidizing microorganisms still remains abnormally high for many years.

To accelerate the first and second stages of oil degradation and avoid formation of *kirs*, introduction of a special group of oil-oxidizing microorganisms is recommended. For this purpose, Ukraine produces biological preparations compatible with the IMO. The aim of our work is to study oil biodegradation by biological decomposers produced in Ukraine.

Materials and Methods

For our investigation, we used the following bio-preparations:

Econadin manufactured by the EKONAD scientific production association, Odessa. Designed to clean up oil from soil and water surfaces, it contains bacterial decomposers and milled peat. The preparation is a brown powder dispersed with fibre, floating, hydrophobic.
Rodex manufactured by BTU-Centre, Ladyzhyn, Vinnytsia Region. Designed for biodegradation of crude oil and oil products contaminating soil, water and interior surfaces of tanks, it contains a group of oil-oxidizing microorganisms. To isolate the bacterial phase in *Rodex,* the samples of microorganisms were inoculated on a common growth medium consisting of fish flour hydrolysate (15 g/l), casein hydrolysate (10 g/l), yeast extract (2 g/l), NaCl (3.5 g/l), glucose (1 g/l) and agar-agar (10 g/l). Then a growth medium of peptone (10 g/l), yeast extract (5 g/l) and NaCl (10 g/l) was used. Microbiological analysis showed that the microorganisms belong to *Pseudomonas* spp.: gram-negative, rod-shaped, aerobic, flagellated, non-spore-forming bacteria that are used for various purposes.

For the mineral phase, to simulate soil clays, we used bentonite from the Dashbent deposit. As described by Kadoshnikov et al. (2013), its clay fraction contains over 75% alkaline montmorillonite as well as palygorskite and fine-grained illite; the sand fraction is mainly quartz and calcite with small amounts of accessory minerals and amorphous iron and manganese (hydr)oxides. The fine clay contains 100-μm amorphous spherulites of amorphous silica, iron and titanium. The total amount of non-clay minerals does not exceed 10%.

Oil from the Hlynsko-Rozbysheskiy field (Kachanovo, Poltava Region, Ukraine) was used in the experiment. This is a light oil, density 0.85 g/cm^3, containing 30% of light fractions. Infrared spectroscopy revealed that the IR spectra (Fig. 28.1a) present bands belonging only to C–H vibrations (Table 28.1). The absence of absorption bands in the ring-oscillation range in the IR spectra is because the content of aromatic components in these oils is less than 0.8% (Marfenina 1991), which is below sensitivity level of IR spectroscopy.

To analyse oil bio-destruction by microorganisms in vitro, the following mixtures were prepared:

- Check sample EC1—*Econadin* + 2.5 ml oil
- Sample ECB5—*Econadin* + 5 g bentonite and 2.5 ml oil
- Sample ECB10—*Econadin* + 10 g bentonite and 2.5 ml oil
- Check sample PC1—*Rodex* + 2.5 ml oil
- Sample RB5—*Rodex* + 5 g bentonite and 2.5 ml oil
- Sample RB10—*Rodex* + 10 g bentonite and 2.5 ml oil.

In all mixtures, the oil content was about 5%. To activate the microorganisms, the preparation was diluted with water, covered with semi-permeable film and placed in an incubator at 37 °C for 14 days. Then the samples were dried to the

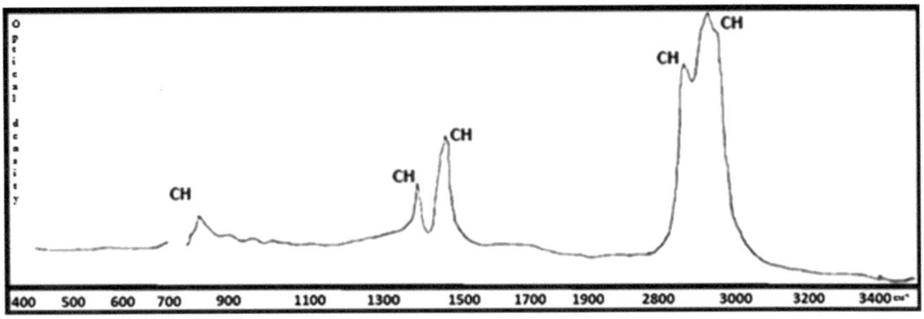

(a) IR-spectrum of Rozbyshevska oil sample

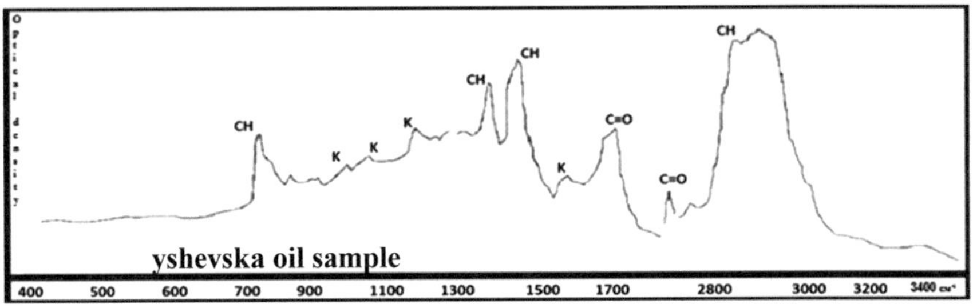

(b) IR-spectrum of sample EC1

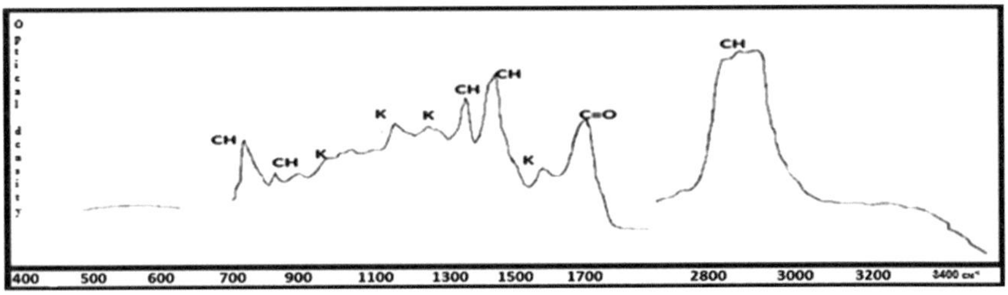

(c) IR-spectrum of sample ECB5

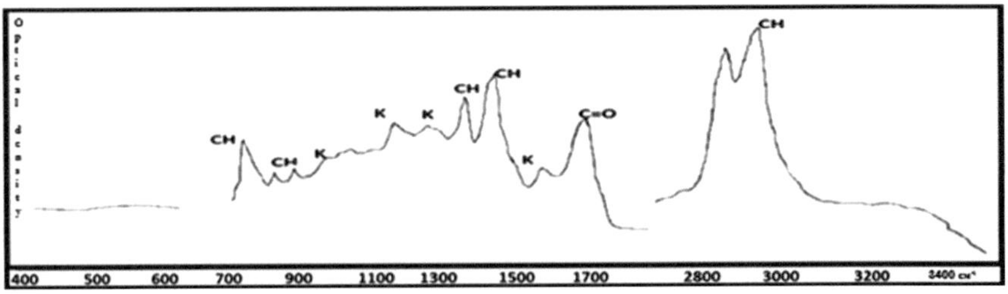

(d) IR-spectrum of sample ECB10

Fig. 28.1 The results of IR-spectroscopic studies of samples

Table 28.1 IR absorption bands of dry residue after chloroform extraction

No.	Functional group	Oil	Absorption region cm^{-1}					
			Econadin			*Rodex*		
			EC1	ECB5	ECB10	RC1	RB5	RB10
1	CH deformation vibrations	730	730	730	730	730	730	730
2	CH deformation vibrations	820	820	820	820	815	818	819
3	CH deformation vibrations	–	890	880	880	875	880	880
4	CH and ring deformation vibrations	–	980	980	980	980	980	980
	C–O–C and C–OH stretching vibrations	–	–	–	–	1040	1040	1040
5	Ring vibrations	–	1170	1173	1175	1170	1170	1170
6	C–O–C ring stretching vibrations	–	1270	1270	1270	–	–	–
7	CH deformation vibrations	1380	1380	1380	1380	1380	1380	1380
8	CH deformation vibrations	1470	1470	1470	1470	1470	1470	1470
9	C=C, C=O, OH ring deformation vibrations	–	1610	1613	1620	1650	1650	1650
10	C=O stretching vibrations	–	1710	1710	1720	1720	1720	1720
11	C=O stretching vibrations	–	1740	1740	1740	–	–	–
12	CH stretching vibrations	2870	2870	2870	2870	2800	2800	2800
13	CH stretching vibrations	2935	2935	2935	2935	3000	3000	3000

constant weight. To estimate the amount of oil remaining, the sample was milled to 100 μm, wrapped in a *blue ribbon* filter paper cartridge and placed in a Soxhlet extractor. Chloroform extraction was carried out at 30 ± 5 °C for 5 h; the chloroform-extracted residue was placed in a flask for later analysis. As the control sample, dry residue of a chloroform solution of crude oil was used. IR spectroscopy was used to measure the destruction of non-polar hydrocarbons by the bio-decomposers.

Results and Discussion

Figure 28.1a presents the IR spectrum of clean oil. Figure 28.1b–d shows the IR spectra of oil samples after processing with *Econadin*. Figure 28.2a–c shows the IR spectra of oil samples after processing with *Rodex-T*. In the figures, "K" indicates oscillations of the aromatic hydrocarbon ring.

Our earlier investigation (Lysychenko et al. 2011) showed that *Rodex* is more effective than *Econadin*. The above spectra of the oil products residue suggest that influence of the both bio-decomposers on the destruction mechanism of

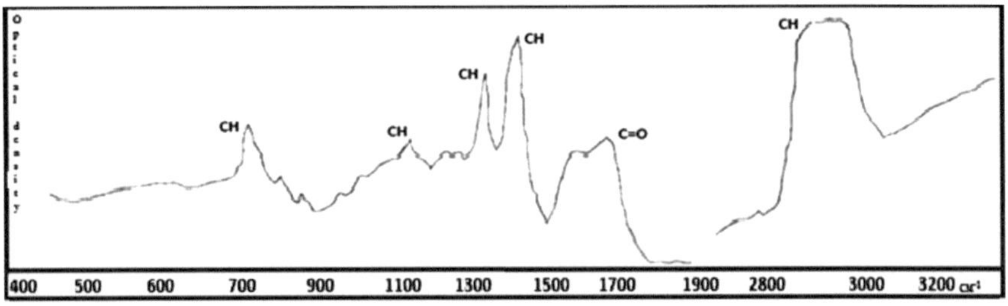

(a) IR-spectrum of sample PC1

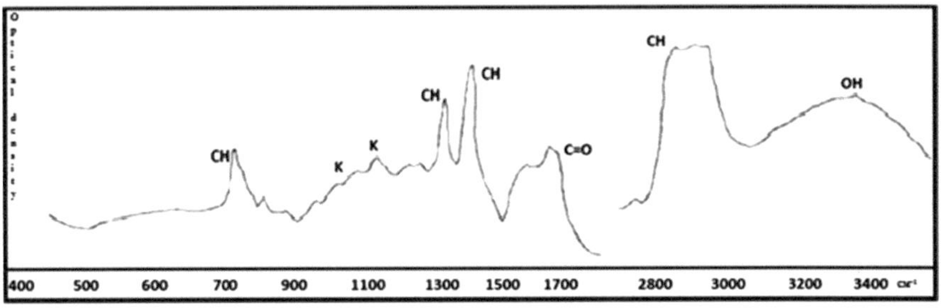

(b) IR-spectrum of sample RB5

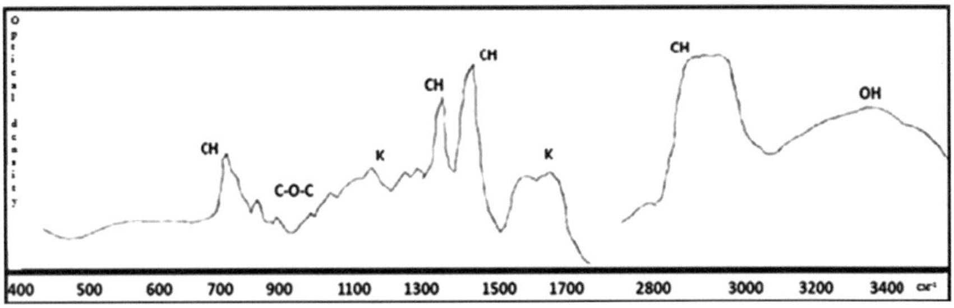

(c) IR-spectrum of sample RB10

Fig. 28.2 The results of IR-spectroscopic studies of samples

hydrocarbons is similar, but its intensity is somewhat different for each preparation. After chloroform extraction, new bands, which were not found in crude oil spectra, were observed in the spectra of the extracts and identified as aromatic ring vibrations (1170, 1270, 1610 cm^{-1}). We also observed bands for C=O groups formed as a result of oxidation of aliphatic chains.

Under aerobic conditions similar to those at the Earth's surface, the main mechanism of oxidation of hydrocarbons is incorporation of oxygen into the hydrocarbon molecule—substitution of bonds with low rupture energy (C–C, C–H) for bonds with high rupture energy (C–O, N–O) (Glazovskaya 1988). Biogenic hydrocarbon oxidation processes are complex and, as yet, there is no clear understanding of their behaviour. The problem is so challenging because the direction of these processes is affected by many factors: medium acidity, redox conditions, temperature, light, osmotic pressure, etc. We also need to recognize physiological patterns of the microorganisms exhibited during oxidation of separate hydrocarbons and their mixtures (Khotimsky and Akopyan 1970). Microorganisms are selective towards different hydrocarbons, depending on not only the difference in the structure of the substance but, even, the number of carbon atoms. For example, *Bacterium aliphaticuman* and *B. aliphaticum liquefaciens* (as isolated and described by I Tauch and V Petrov) oxidize *n*-hexane, *n*-octane, decane, hexadecane, triakontan, whereas *B. paraffinicum* (isolated by the same workers) oxidizes only higher homologues of this series, starting with hexadecane (Feist and Hegeman 1969; Cerniglia 1992; Kosheleva et al. 2000).

Oxidation of hydrocarbons by most known microorganisms involves adaptive enzymes. In the course of oxidative degradation, formation of intermediate products such as alcohols, aldehydes, ketones, carboxylic acids, esters, and bifunctional compounds—aldehyde acids, keto acids, etc.,—is possible. At first sight, the best outcome of biodegradation of spilled oil might be complete oxidation to CO_2. However, if CO_2 is evolved in an aqueous environment, it practically completely dissolves in water, causing build-up of mineral carbon (CO_3^{2-}, HCO^{3-}), and, with partial oxidation, humic polymers or insoluble resinous substances *(kirs)* may be formed. However, under natural conditions, microorganisms oxidize substances using both aerobic and anaerobic respiration (Lysak 2007); hydrocarbon humification at low temperatures and lack of oxygen is of interest because the end products are natural soil components. Humification of hydrocarbon intermediate oxidation products is also an efficient method of biochemical transformation of oil hydrocarbons (Illarionov 2006). In northern latitudes, humification accounts for 80% of the transformation of hydrocarbons and mineralization 20%; in southern latitudes, the situation is reversed with mineralization accounting for 70% and humification 30% of transformation (Panicheva et al. 2012).

Here, we propose a concept of remediation of oil-contaminated soils based on the transformation of oil compounds into non-toxic humic compounds that can accelerate remediation of the disturbed ecosystem. The possibility of humification is based on the fact that, when oxygen is deficient, anaerobic oxidation of hydrocarbons is accomplished by nitrates and nitrites, yielding ammonia. Reductive amination of intermediate hydrocarbon oxidation products with formation of nitrogen-containing organic compounds, including amino acids, predetermines humification. We may mention that discovery of the bacteria *Corynebacterium glutamicum* which secrete L-glutamic acid was the beginning of a new era in industrial use of the processes of partial oxidation of carbohydrates and hydrocarbons for amino acid synthesis (URL: http://micro.moy.su/publ.).

The mechanism promoting synthesis of amino acids from α-keto acids and ammonium was named trans-reamination by AE Braunstein. Specifically, it is reductive amination of α-ketoglutaric acid with the formation of glutamic acid (the reaction is catalysed by glutamate dehydrogenase) and the subsequent transamination of glutamate with α-keto acid. As a result, L-amino acid, which is consistent with the original keto acid, is formed, and α-ketoglutaric acid (able to accept a new ammonia molecule) is released. In many bacteria and plants, the system of ammonium nitrogen incorporation into organic compounds, involving transfer of amino groups, takes place when the concentration of ammonium ions in the medium is very low (less than 1 mmol/l), likewise for the immobilization of N_2. The efficiency of amination of intermediate oxidation products of organic compounds (and, consequently, humification) depends on the content of ammonium ions in the system formed in the process of dissimilative reduction of nitrates and anaerobic oxidation of hydrocarbons and organic residues. In northern environments, soils contain a significant amount of ammonia which accounts for dominance of humification over mineralization of organic matter. In southern environments, humification is limited by lack of ammonium. So, to improve soil fertility and reduce the toxic influence of oil products, biodegradation of oil products in soils should be carried out under conditions of in high ammonia-nitrogen content. This may be achieved by introduction of ammonium fertilizers into the oil-contaminated soil.

Hydrocarbon destruction by microorganisms proceeds at the mineral–solution interphase boundary, so increase in the interphase boundary area by introduction of natural dispersed clay minerals (vermiculite, zeolite and glauconite) accelerates the microbial degradation of oil hydrocarbons in aqueous media. The use of oil-oxidizing germ culture of *P. putida* on the surface of zeolite increases its quantity in water and prolongs the activity of bacteria (Pisarchuk et al. 2011).

Our IR data show that, under laboratory conditions, oil bio-decomposers fortified with montmorillonite promote the synthesis of aromatic, polycyclic compounds (Figs. 28.1b, c and 28.2c, d). Formation of chloroform-soluble cyclic structures from naphthenic hydrocarbons is activated by microorganisms in the presence of clay minerals. However, based on these results, it is difficult to explicitly estimate the influence of clay minerals on these processes. Figure 28.3 shows the influence of montmorillonite on the growth of micro-fungi *Thielavia terrestris*—an efficient bio-decomposer of oil pollution; a low content of montmorillonite in the culture medium accelerates the growth of micro-fungi dramatically, while with further increase in the clay component, the growth of the fungi sharply reduces.

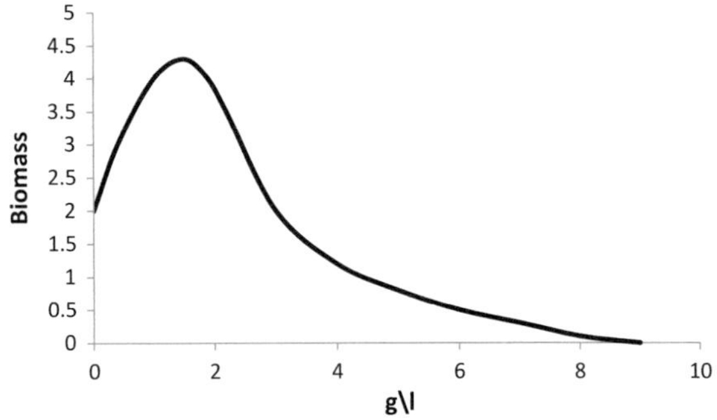

Fig. 28.3 Influence of montmorillonite on the growth of micro-fungi *Thielavia terrestris*

The observed increase in the activity of the oil-decomposing microorganisms in the presence of montmorillonite allowed us to develop, on the basis of the commercial *Rodex* preparation, a special non-phosphate bio-mineral mixture for clean-up of oil pollution from solid mineral surfaces. Laboratory analyses have shown that the new mixture removes 90% of oil pollution from a gravel surface and 95% from a tiled surface.

Conclusions

IR spectroscopic studies suggest that cyclic aromatic structures are formed by the action of oil-oxidizing microorganisms on the aliphatic chains of oil hydrocarbons. At low temperatures and oxygen deficiency, these cyclic aromatic compounds may later transform into humic substances.

Dispersed clay minerals catalyse oil destruction in soils and water solutions. The catalytic activity depends on the nature of the microorganisms and concentration of clay minerals. The nature of clay minerals influence on the growth and activity of microorganisms requires further research.

A highly efficient mixture for oil pollution clean-up from solid mineral surfaces has been developed using *Rodex* bio-decomposer and bentonite.

References

Cerniglia CE (1992) Biodegradation of polycyclic aromatic hydrocarbons. Biodegradation 3:351–368

Feist CF, Hegeman GD (1969) Phenol and benzoate metabolism by *Pseudomonas putida* of tangential pathways. J Bacteriol 100:869–877

Glazovskaya MA (ed) (1988) Remediation of oil contaminated ecosystems. USSR Academy of Sciences, Science Council on Biosphere, Nauka (Russian)

Illarionov SA (2006) Transformation of oil hydrocarbons in the soil damp zone. Thesis, Dr of Biological Sciences, Syktyvkar (Russian)

Kadoshnikov VM, Shekhunova SB, Zadverniuk GP, Manichev VI (2013) Authigenic bentonite clay minerals from the Cherkasy deposit. Mineral J 35(3):54–60

Khotimsky BG, Akopyan AP (eds) (1970) Transformation of oil by microorganisms. In: Proceedings of All-Union Oil Research Institute of Geological Exploration, AUORIGE, Leningrad (Russian)

Kosheleva IA, Balashova NV, Izmalkova TY et al (2000) Degradation of phenanthrene by mutant strains—naphthalene decomposers. Microbiology 6:783–789 (Russian)

Lysak VV (2007) Microbiology. Tutorial. BSU Minsk 426 p.(Russian)

Lysychenko GV, VM Kadoshnikov, VV Shkapenko et al (2011) Destruction of oil products on silicate-containing materials by bio-mineral decomposers. In: VII International research conference environmental safety: problems and solutions, vol 2. UkrSRIPI. Rider, Kharkiv, pp 32–37 (Ukrainian)

Marfenina OE (1991) Microbiological aspects of soil protection. MSU Publishing House, Moscow (Russian)

Panicheva LP, Moiseenko TI, Kremleva TI, Volkova SS (2012) S.S. Biochemical transformation of petroleum hydrocarbons in the waters of Western Siberia. Bulletin of Tyumen State University, vol 12, pp 38–48 (Russian)
Pisarchuk AD, Tereshchenko NN, Lushnikov SV (2011) Efficiency in use of hydrocarbon oxidizing bacteria *Pseudomonas putida* and modified vermicompost-based sorbent for decontamination of oil-polluted soil. Bull Tomsk State Univ Biol 3(15):180–182 (Russian)
Zvolynsky V (2005) If oil spills. Oil of Russia 5:49–51. http://micro.moy.su/publ (Russian)

MIX
Papier aus verantwortungsvollen Quellen
Paper from responsible sources
FSC® C105338

FSC
www.fsc.org

If you have any concerns about our products,
you can contact us on
ProductSafety@springernature.com

In case Publisher is established outside the EU,
the EU authorized representative is:
Springer Nature Customer Service Center GmbH
Europaplatz 3, 69115 Heidelberg, Germany

Printed by Libri Plureos GmbH
in Hamburg, Germany